P9-CBI-331

Pitman Research Notes in Mathematics Series

Submission of proposals for consideration
Suggestions for publication, in the form of outlines and representative samples, are invited by the Editorial Board for assessment. Intending authors should approach one of the main editors or another member of the Editorial Board, citing the relevant AMS subject classifications. Alternatively, outlines may be sent directly to the publisher's offices. Refereeing is by members of the board and other mathematical authorities in the topic concerned, throughout the world.

Preparation of accepted manuscripts
On acceptance of a proposal, the publisher will supply full instructions for the preparation of manuscripts in a form suitable for direct photo-lithographic reproduction. Specially printed grid sheets are provided and a contribution is offered by the publisher towards the cost of typing. Word processor output, subject to the publisher's approval, is also acceptable.

Illustrations should be prepared by the authors, ready for direct reproduction without further improvement. The use of hand-drawn symbols should be avoided wherever possible, in order to maintain maximum clarity of the text.

The publisher will be pleased to give any guidance necessary during the preparation of a typescript, and will be happy to answer any queries.

Important note
In order to avoid later retyping, intending authors are strongly urged not to begin final preparation of a typescript before receiving the publisher's guidelines and special paper. In this way it is hoped to preserve the uniform appearance of the series.

Longman Scientific & Technical
Longman House
Burnt Mill
Harlow, Essex, UK
(tel (0279) 26721)

Titles in this series

1 Improperly posed boundary value problems
A Carasso and A P Stone

2 Lie algebras generated by finite dimensional ideals
I N Stewart

3 Bifurcation problems in nonlinear elasticity
R W Dickey

4 Partial differential equations in the complex domain
D L Colton

5 Quasilinear hyperbolic systems and waves
A Jeffrey

6 Solution of boundary value problems by the method of integral operators
D L Colton

7 Taylor expansions and catastrophes
T Poston and I N Stewart

8 Function theoretic methods in differential equations
R P Gilbert and R J Weinacht

9 Differential topology with a view to applications
D R J Chillingworth

10 Characteristic classes of foliations
H V Pittie

11 Stochastic integration and generalized martingales
A U Kussmaul

12 Zeta-functions: An introduction to algebraic geometry
A D Thomas

13 Explicit *a priori* inequalities with applications to boundary value problems
V G Sigillito

14 Nonlinear diffusion
W E Fitzgibbon III and H F Walker

15 Unsolved problems concerning lattice points
J Hammer

16 Edge-colourings of graphs
S Fiorini and R J Wilson

17 Nonlinear analysis and mechanics: Heriot-Watt Symposium Volume I
R J Knops

18 Actions of fine abelian groups
C Kosniowski

19 Closed graph theorems and webbed spaces
M De Wilde

20 Singular perturbation techniques applied to integro-differential equations
H Grabmüller

21 Retarded functional differential equations: A global point of view
S E A Mohammed

22 Multiparameter spectral theory in Hilbert space
B D Sleeman

24 Mathematical modelling techniques
R Aris

25 Singular points of smooth mappings
C G Gibson

26 Nonlinear evolution equations solvable by the spectral transform
F Calogero

27 Nonlinear analysis and mechanics: Heriot-Watt Symposium Volume II
R J Knops

28 Constructive functional analysis
D S Bridges

29 Elongational flows: Aspects of the behaviour of model elasticoviscous fluids
C J S Petrie

30 Nonlinear analysis and mechanics: Heriot-Watt Symposium Volume III
R J Knops

31 Fractional calculus and integral transforms of generalized functions
A C McBride

32 Complex manifold techniques in theoretical physics
D E Lerner and P D Sommers

33 Hilbert's third problem: scissors congruence
C-H Sah

34 Graph theory and combinatorics
R J Wilson

35 The Tricomi equation with applications to the theory of plane transonic flow
A R Manwell

36 Abstract differential equations
S D Zaidman

37 Advances in twistor theory
L P Hughston and R S Ward

38 Operator theory and functional analysis
I Erdelyi

39 Nonlinear analysis and mechanics: Heriot-Watt Symposium Volume IV
R J Knops

40 Singular systems of differential equations
S L Campbell

41 N-dimensional crystallography
R L E Schwarzenberger

42 Nonlinear partial differential equations in physical problems
D Graffi

43 Shifts and periodicity for right invertible operators
D Przeworska-Rolewicz

44 Rings with chain conditions
A W Chatters and C R Hajarnavis

45 Moduli, deformations and classifications of compact complex manifolds
D Sundararaman

46 Nonlinear problems of analysis in geometry and mechanics
M Atteia, D Bancel and I Gumowski

47 Algorithmic methods in optimal control
W A Gruver and E Sachs

48 Abstract Cauchy problems and functional differential equations
F Kappel and W Schappacher

49 Sequence spaces
W H Ruckle

50 Recent contributions to nonlinear partial differential equations
H Berestycki and H Brezis

51 Subnormal operators
J B Conway

52 Wave propagation in viscoelastic media
F Mainardi

53 Nonlinear partial differential equations and their applications: Collège de France Seminar. Volume I
H Brezis and J L Lions

54 Geometry of Coxeter groups
H Hiller

55 Cusps of Gauss mappings
T Banchoff, T Gaffney and C McCrory

56 An approach to algebraic K-theory
A J Berrick

57 Convex analysis and optimization
J-P Aubin and R B Vintner

58 Convex analysis with applications in the differentiation of convex functions
J R Giles

59 Weak and variational methods for moving boundary problems
C M Elliott and J R Ockendon

60 Nonlinear partial differential equations and their applications: Collège de France Seminar. Volume II
H Brezis and J L Lions

61 Singular systems of differential equations II
S L Campbell

62 Rates of convergence in the central limit theorem
Peter Hall

63 Solution of differential equations by means of one-parameter groups
J M Hill

64 Hankel operators on Hilbert space
S C Power

65 Schrödinger-type operators with continuous spectra
M S P Eastham and H Kalf

66 Recent applications of generalized inverses
S L Campbell

67 Riesz and Fredholm theory in Banach algebra
B A Barnes, G J Murphy, M R F Smyth and T T West

68 Evolution equations and their applications
F Kappel and W Schappacher

69 Generalized solutions of Hamilton-Jacobi equations
P L Lions

70 Nonlinear partial differential equations and their applications: Collège de France Seminar. Volume III
H Brezis and J L Lions

71 Spectral theory and wave operators for the Schrödinger equation
A M Berthier

72 Approximation of Hilbert space operators I
D A Herrero

73 Vector valued Nevanlinna Theory
H J W Ziegler

74 Instability, nonexistence and weighted energy methods in fluid dynamics and related theories
B Straughan

75 Local bifurcation and symmetry
A Vanderbauwhede

76 Clifford analysis
F Brackx, R Delanghe and F Sommen

77 Nonlinear equivalence, reduction of PDEs to ODEs and fast convergent numerical methods
E E Rosinger

78 Free boundary problems, theory and applications. Volume I
A Fasano and M Primicerio

79 Free boundary problems, theory and applications. Volume II
A Fasano and M Primicerio

80 Symplectic geometry
A Crumeyrolle and J Grifone

81 An algorithmic analysis of a communication model with retransmission of flawed messages
D M Lucantoni

82 Geometric games and their applications
W H Ruckle

83 Additive groups of rings
S Feigelstock

84 Nonlinear partial differential equations and their applications: Collège de France Seminar. Volume IV
H Brezis and J L Lions

85 Multiplicative functionals on topological algebras
T Husain

86 Hamilton-Jacobi equations in Hilbert spaces
V Barbu and G Da Prato

87 Harmonic maps with symmetry, harmonic morphisms and deformations of metrics
P Baird

88 Similarity solutions of nonlinear partial differential equations
L Dresner

89 Contributions to nonlinear partial differential equations
C Bardos, A Damlamian, J I Díaz and J Hernández

90 Banach and Hilbert spaces of vector-valued functions
J Burbea and P Masani

91 Control and observation of neutral systems
D Salamon

92 Banach bundles, Banach modules and automorphisms of C*-algebras
M J Dupré and R M Gillette

93 Nonlinear partial differential equations and their applications: Collège de France Seminar. Volume V
H Brezis and J L Lions

94 Computer algebra in applied mathematics: an introduction to MACSYMA
R H Rand

95 Advances in nonlinear waves. Volume I
L Debnath

96 FC-groups
M J Tomkinson

97 Topics in relaxation and ellipsoidal methods
M Akgül

98 Analogue of the group algebra for topological semigroups
H Dzinotyiweyi

99 Stochastic functional differential equations
S E A Mohammed

100 Optimal control of variational inequalities
V Barbu
101 Partial differential equations and dynamical systems
W E Fitzgibbon III
102 Approximation of Hilbert space operators. Volume II
C Apostol, L A Fialkow, D A Herrero and D Voiculescu
103 Nondiscrete induction and iterative processes
V Ptak and F-A Potra
104 Analytic functions – growth aspects
O P Juneja and G P Kapoor
105 Theory of Tikhonov regularization for Fredholm equations of the first kind
C W Groetsch
106 Nonlinear partial differential equations and free boundaries. Volume I
J I Díaz
107 Tight and taut immersions of manifolds
T E Cecil and P J Ryan
108 A layering method for viscous, incompressible L_p flows occupying R^n
A Douglis and E B Fabes
109 Nonlinear partial differential equations and their applications: Collège de France Seminar. Volume VI
H Brezis and J L Lions
110 Finite generalized quadrangles
S E Payne and J A Thas
111 Advances in nonlinear waves. Volume II
L Debnath
112 Topics in several complex variables
E Ramírez de Arellano and D Sundararaman
113 Differential equations, flow invariance and applications
N H Pavel
114 Geometrical combinatorics
F C Holroyd and R J Wilson
115 Generators of strongly continuous semigroups
J A van Casteren
116 Growth of algebras and Gelfand–Kirillov dimension
G R Krause and T H Lenagan
117 Theory of bases and cones
P K Kamthan and M Gupta
118 Linear groups and permutations
A R Camina and E A Whelan
119 General Wiener–Hopf factorization methods
F-O Speck
120 Free boundary problems: applications and theory, Volume III
A Bossavit, A Damlamian and M Fremond
121 Free boundary problems: applications and theory, Volume IV
A Bossavit, A Damlamian and M Fremond
122 Nonlinear partial differential equations and their applications: Collège de France Seminar. Volume VII
H Brezis and J L Lions
123 Geometric methods in operator algebras
H Araki and E G Effros
124 Infinite dimensional analysis–stochastic processes
S Albeverio
125 Ennio de Giorgi Colloquium
P Krée
126 Almost-periodic functions in abstract spaces
S Zaidman
127 Nonlinear variational problems
A Marino, L Modica, S Spagnolo and M Degiovanni
128 Second-order systems of partial differential equations in the plane
L K Hua, W Lin and C-Q Wu
129 Asymptotics of high-order ordinary differential equations
R B Paris and A D Wood
130 Stochastic differential equations
R Wu
131 Differential geometry
L A Cordero
132 Nonlinear differential equations
J K Hale and P Martinez-Amores
133 Approximation theory and applications
S P Singh
134 Near-rings and their links with groups
J D P Meldrum
135 Estimating eigenvalues with *a posteriori/a priori* inequalities
J R Kuttler and V G Sigillito
136 Regular semigroups as extensions
F J Pastijn and M Petrich
137 Representations of rank one Lie groups
D H Collingwood
138 Fractional calculus
G F Roach and A C McBride
139 Hamilton's principle in continuum mechanics
A Bedford
140 Numerical analysis
D F Griffiths and G A Watson
141 Semigroups, theory and applications. Volume I
H Brezis, M G Crandall and F Kappel
142 Distribution theorems of L-functions
D Joyner
143 Recent developments in structured continua
D De Kee and P Kaloni
144 Functional analysis and two-point differential operators
J Locker
145 Numerical methods for partial differential equations
S I Hariharan and T H Moulden
146 Completely bounded maps and dilations
V I Paulsen
147 Harmonic analysis on the Heisenberg nilpotent Lie group
W Schempp
148 Contributions to modern calculus of variations
L Cesari
149 Nonlinear parabolic equations: qualitative properties of solutions
L Boccardo and A Tesei
150 From local times to global geometry, control and physics
K D Elworthy

151 A stochastic maximum principle for optimal control of diffusions
U G Haussmann
152 Semigroups, theory and applications. Volume II
H Brezis, M G Crandall and F Kappel
153 A general theory of integration in function spaces
P Muldowney
154 Oakland Conference on partial differential equations and applied mathematics
L R Bragg and J W Dettman
155 Contributions to nonlinear partial differential equations. Volume II
J I Díaz and P L Lions
156 Semigroups of linear operators: an introduction
A C McBride
157 Ordinary and partial differential equations
B D Sleeman and R J Jarvis
158 Hyperbolic equations
F Colombini and M K V Murthy
159 Linear topologies on a ring: an overview
J S Golan
160 Dynamical systems and bifurcation theory
M I Camacho, M J Pacifico and F Takens
161 Branched coverings and algebraic functions
M Namba
162 Perturbation bounds for matrix eigenvalues
R Bhatia
163 Defect minimization in operator equations: theory and applications
R Reemtsen
164 Multidimensional Brownian excursions and potential theory
K Burdzy
165 Viscosity solutions and optimal control
R J Elliott
166 Nonlinear partial differential equations and their applications. Collège de France Seminar. Volume VIII
H Brezis and J L Lions
167 Theory and applications of inverse problems
H Haario
168 Energy stability and convection
G P Galdi and B Straughan
169 Additive groups of rings. Volume II
S Feigelstock
170 Numerical analysis 1987
D F Griffiths and G A Watson
171 Surveys of some recent results in operator theory. Volume I
J B Conway and B B Morrel
172 Amenable Banach algebras
J-P Pier
173 Pseudo-orbits of contact forms
A Bahri
174 Poisson algebras and Poisson manifolds
K H Bhaskara and K Viswanath
175 Maximum principles and eigenvalue problems in partial differential equations
P W Schaefer
176 Mathematical analysis of nonlinear, dynamic processes
K U Grusa
177 Cordes' two-parameter spectral representation theory
D F McGhee and R H Picard
178 Equivariant K-theory for proper actions
N C Phillips
179 Elliptic operators, topology and asymptotic methods
J Roe
180 Nonlinear evolution equations
J K Engelbrecht, V E Fridman and E N Pelinovski
181 Nonlinear partial differential equations and their applications. Collège de France Seminar. Volume IX
H Brezis and J L Lions
182 Critical points at infinity in some variational problems
A Bahri
183 Recent developments in hyperbolic equations
L Cattabriga, F Colombini, M K V Murthy and S Spagnolo
184 Optimization and identification of systems governed by evolution equations on Banach space
N U Ahmed
185 Free boundary problems: theory and applications. Volume I
K H Hoffmann and J Sprekels
186 Free boundary problems: theory and applications. Volume II
K H Hoffmann and J Sprekels
187 An introduction to intersection homology theory
F Kirwan
188 Derivatives, nuclei and dimensions on the frame of torsion theories
J S Golan and H Simmons
189 Theory of reproducing kernels and its applications
S Saitoh
190 Volterra integrodifferential equations in Banach spaces and applications
G Da Prato and M Iannelli
191 Nest algebras
K R Davidson
192 Surveys of some recent results in operator theory. Volume II
J B Conway and B B Morrel

Recent developments in hyperbolic equations

L Cattabriga, F Colombini,
M K V Murthy & S Spagnolo (Editors)

Recent developments in hyperbolic equations

Proceedings of the conference on Hyperbolic Equations, University of Pisa, 1987

Longman Scientific & Technical

Copublished in the United States with
John Wiley & Sons, Inc., New York

Longman Scientific & Technical
Longman Group UK Limited
Longman House, Burnt Mill, Harlow
Essex CM20 2JE, England
and Associated Companies throughout the world.

Copublished in the United States with
John Wiley & Sons, Inc., 605 Third Avenue, New York, NY 10158

First published 1988

AMS Subject Classifications: (main) 35L, 35S, 58G
(subsidiary) 35P, 70H

ISSN 0269-3674

British Library Cataloguing in Publication Data
Conference on Hyperbolic Equations (1987:
University of Pisa, Italy)
Recent developments in hyperbolic
equations.
1. Hyperbolic differential equations
I. Title II. Cattabriga, L.
515.3′53

ISBN 0-582-03491-4

Library of Congress Cataloging-in-Publication Data
Recent developments in hyperbolic equations.
(Pitman research notes in mathematics series; 183)
Bibliography: p.
1. Differential equations, Hyperbolic--Congresses.
2. Differential equations, Partial--Congresses.
I. Cattabriga, Lamberto. II. Series.
QA377.R42 1988 515.3′53 88-9025
ISBN 0-470-21169-5

Printed and bound in Great Britain by
Biddles Ltd, Guildford and King's Lynn

Contents

List of contributors

E. Bernardi	Department of Mathematics, University of Bologna, Piazza di Porta S. Donato, 5, 40127 Bologna, Italy.
A. Bove	Department of Mathematics, University of Bologna, Piazza di Porta S. Donato, 5, 40127 Bologna, Italy.
J-Y. Chemin	Centre de Mathématiques, Ecole Polytechnique, 91128 Palaiseau, France.
Y. Choquet-Bruhat	Institut de Mécanique, Université Paris VI, 16 Ave. d'Alembert, 92160 Antony, France.
M. Cicognani	Dipartimento di Matematica, Università di Bologna, Piazza di Porta San Donato 5, 40127 Bologna, Italy.
A. Corli	Università di Ferrara, Dipartimento di Matematica, Via Machiavelli, 35, I-44100 Ferrara, Italy.
P. Gérard	Département de Mathématiques, Ecole Normale Supérieure, 45, rue d'Ulm, F-75230 Paris Cedex 05, France.
D. Gourdin	Université des Sciences et Techniques de Lille Flandres Artois, U.F.R. de Mathematiques Pure et Appliquées, 59655 Villeneuve d'Ascq Cedex, France.
Y. Hamada	Département de Mathématiques, Université Technologique de Kyoto, Matsugasaki, Sakyo-Ku, Kyoto, 606, Japan.
B. Helffer	Department of Mathematics, Université de Nantes, 2, Chemin de la Houssinière, F-44072 Nantes, France.
E. Horst	Abteilung Mathematik im FB IV, Postfach 3825, 5500 Trier, West Germany.

J-L. Joly — Department of Mathematics, University of Bordeaux I, 351 Cours de la Liberation, 33405, Talence, France.

K. Kajitani — Institute of Mathematics, University of Tsukuba, Ibaraki 305, Japan.

H. Komatsu — University of Tokyo, Faculty of Sciences, Department of Mathematics, Hongo, Tokyo 113, Japan.

G. Lebeau — Université de Paris Sud, Mathématiques, Batiment 425, 91405 Orsay, France.

G. Métivier — Université de Rennes I, Mathématique et Informatique, Campus de Beaulieu, F 35042 Rennes Cedex, France.

A.J. Milani — Dipartimento di Matematica, Università di Torino, Via Carlo Alberta, 10, I-10123 Torino, Italy.

S. Mizohata — Department of Mathematics, Kyoto University, Kyoto 606, Japan.

Y. Morimoto — Institute of Mathematics, Yoshida College, Kyoto University, 606, Kyoto, Japan.

M. Nacinovich — Università di Pisa, Dipartimento di Matematica, Via Buonarroti 2, 56100 Pisa, Italy.

T. Nishitani — Department of Mathematics, College of General Education, Osaka University, Toyonaka, Osaka, Japan.

C. Parenti — Department of Mathematics, University of Bologna, Piazza di Porta S. Donato, 5, 40127 Bologna, Italy.

J. Persson — Lunds Universitet, Matematiska Institutionen, Box 118, S 22100 Lund, Sweden.

J. Rauch — Department of Mathematics, University of Michigan, Angell Hall, Ann Arbor, MI 48109 USA.

L. Rodino — Università di Torino, Dipartimento di Matematica, Via Carlo Alberto, 10, I-10123 Torino, Italy

X. Saint Raymond	Université de Paris-Sud, Mathématiques, Batiment 425, F 91405 Orsay, France.
J. Sjöstrand	Department of Mathematics, Université de Paris-Sud, F-91405 Orsay, France.
S. Tarama	Department of Applied Mathematics and Physics, Kyoto University, 606 Kyoto, Japan.
J. Vaillant	Unité Associée au CNRS 761, Mathématiques, Tour 45-46, 5 ème étage, Université de Paris VI, 4, place Jussieu, 75252 Paris Cedex 05, France.
C. Wagschal	Laboratoire Central des Ponts et Chaussées, 58 Boulevard Lefebvre, 75732, Paris Cedex 15, France.
S. Wakabayashi	Institute of Mathematics, The University of Tsukuba, Ibaraki 305, Japan.
C. Zuily	Université de Paris Sud, Mathématiques, Batiment 425, F 91405 Orsay, France.

Preface

A semester dedicated to Evolution Equations organized, jointly by the Scuola Normale Superiore and the Department of Mathematics of the Univerisity of Pisa, was held in Pisa during January-June 1987. The semester was financially supported by the Gruppo Nazionale per l'Analisi Funzionale ed Applicazioni (GNAFA) of the Italian National Research Council (CNR) and by the Research Grants from the Italian Education Ministry.

One of the parts of the semester on evolution equations was dedicated to Hyperbolic Equations and this was held during a period of four weeks in the months of March-April 1987, of which the first three weeks were devoted to brief lecture courses and seminars while a workshop was held during the last week.

The present volume collects together the texts of a large part of the lectures delivered during the Workshop and during the previous three weeks. These lectures consisted of some of the most recent contributions of several authors in different sectors of the theory of partial differential equations (mostly that of hyperbolic equations) both linear and nonlinear, in particular: propagation of singularities and wave front sets, well posedness of weakly hyperbolic Cauchy problems, solvability in Gevrey and analytic classes, global solutions for nonlinear equations with small initial data, nonstrictly hyperbolic systems, ramification of solutions to equations of Kowalewsky type.

The entire programme devoted to hyperbolic equations and, in particular, the concluding Workshop was received with great interest; we wish to express our sincere gratitude to all the participants, especially to all the speakers, who, with their contributions of a very high standard, made the conference a success.

J-Y CHEMIN

Bilinear symbolic calculus and controlled interaction in nonlinear strictly hyperbolic equations

INTRODUCTION

We want here to describe the microlocal regularity of a real solution $u \in H^s_{loc}(\Omega)$ of an equation

$$F(x,u,\partial^\alpha u)_{|\alpha| \leqq m} = 0, \qquad (E)$$

where s is big enough, Ω an open subset of $\mathbf{R}^n$, and F a smooth function. In the whole following development, we suppose:

(H_1) $p_m(x,\xi) = \sum_{|\alpha|=m} \partial F/\partial u_\alpha \cdot \xi^\alpha$ is a strictly hyperbolic polynomial for ξ_1.

(H_2) We are in an evolution situation, which means that any null bicharacteristic curve of p_m issued from a point of $\Omega_+ = \Omega \cap (x_1 > 0)$ meets $\Omega_- = \Omega \cap (x_1 < 0)$.

Constructing paradifferential calculus, J-M. Bony proves in [6], that the microlocal singularities H^σ, for $\sigma \leqq 2s-s_0$, have linear properties, considering the geometry of the bicharacteristics of p_m.

For σ bigger than $2s-s_0$, a lot of results have been proved, on the one hand under hypothesis of conormal singularities in the past, by S. Alinhac (see [1], [2], [3]), J-M. Bony (see [7] and [8]), R. Melrose and N. Ritter (see [14]), and the author (see [11]), on the other hand, in one space dimension by P. Godin (see [13]), J. Rauch and M. Reed (see [16]), and the author (see [9]).

Our aim here is the control of the interaction of singularities up to $3s-s_1$, i.e. the proof of an estimate of the H^σ wave front set of a solution of (E), $\sigma \leqq 3s-s_1$, knowing it in the past.

Results about H^σ wave front set have been proved by M. Beals in [4] and [5], where he emphasized the linear propagation of H^σ-microlocal singularities, for $\sigma \leqq 3s-s_1$, in the case of semi-linear equation of order two. With an example, he proved it was an optimal result.

For the study of such singularities in fully non linear equations, we must precise the tools of paradifferential calculus.

The structure of the paper will be as follows:

1. Bilinear symbolic calculus and precised paralinearization.
2. Statements of results.
3. Slow waves.
4. An idea of the proof.

Here, we only give the main steps and ideas of the proof. For further details, see [10].

1. BILINEAR SYMBOLIC CALCULUS AND PRECISED PARALINEARIZATION

The aim of this part is to give improvements of paradifferential calculus adapted to the study of H^{σ}-microlocal singularities, with $\sigma \leq 3s-2s_0$. The basic tool to do this, is the bilinear symbolic calculus. We need to precise the definition of paraproduct as well.

1.1 Paradifferential bilinear symbolic calculus

Here, ρ is a positive real number, not integer.

DEFINITION 1.1.1

(i) Σ^m_{ρ} is the set of functions $a(x,\xi)$ so that, $a(x,\xi) = \sum_{j=0}^{\rho} a_j(x,\xi)$, so that $\| \partial^{\alpha}_{\xi} a_j(\cdot,\xi)\|_{H^{\rho-j+\frac{n}{2}}} \leq C_{j,\alpha}(1+|\xi|)^{m-j-|\alpha|}$.

(ii) $\Sigma^{m,m'}_{\rho}$ is the set of functions $a(x,\xi,\eta)$ so that

$$a(x,\xi,\eta) = \sum_{j=0}^{\rho} a_j(x,\xi,\eta) \text{ so that}$$

$$\cdot \ \| \partial^{\alpha}_{\xi,\eta} a_j(\cdot,\xi,\eta)\|_{H^{\rho-j+\frac{n}{2}}} \leq C_{j,\alpha}(1+|\xi|)^{m}(1+|\xi+\eta|)^{m'-j-|\alpha|}$$

$$\cdot \ \mathrm{Supp}_{\xi,\eta}\, a_j \subset \{(\xi,\eta) \,/\, C\,|\xi| \leq |\eta| \leq C_1\,|\xi|\}$$

For instance, let χ be a C^{∞} function on $\mathbb{R}^{2n}$, let χ be a C^{∞} function on $\mathbb{R}^{2n}$, so that, for any $(\xi,\eta) \in \mathbb{R}^{2n} \cap {}^{c}\{(\xi,\eta)/|\xi+\eta| \geq 1\}$, $|\partial^{\alpha}_{\xi,\eta}\chi(\xi,\eta)| \leq C_{\alpha}(1+|\xi,\eta|)^{-|\alpha|}$, and $\mathrm{supp}\,\chi \cap {}^{c}\{(\xi,\eta)/|\xi+\eta| \geq 1\} \subset \{(\xi,\eta)/C|\xi| \leq |\eta| \leq C_1|\xi|\}$, $\chi \in \Sigma^{0,0}_{\rho}$ and if a in $\underline{\Sigma}_\rho^m$, then $\chi(\xi,\eta)\ a(x,\xi+\eta)$ belongs to $\Sigma_\rho^{o,m'}$.

We must now give a quantization of these symbols.
Let T be a C^∞ map on $\mathbb{R}^{2n}$ so that:

(i) $|\partial_{\xi,\eta}^\alpha\ T(\xi,\eta)| \leqq C_\alpha\ (1+\ |\xi,\eta|)^{-|\alpha|}$

(ii) $\mathrm{Supp}\ T \subset \{(\xi,\eta)\ /\ |\xi| \leqq \varepsilon\ |\eta|\}$

(iii) $\mathrm{Supp}\ (1-T) \setminus B(0,C) \subset \{(\xi,\eta)\ /\ |\xi| \geqq \varepsilon_1 |\eta|\}$,

(ε and ε_1 being two real numbers so that $0 < \varepsilon_1 < \varepsilon < 1$).

Considering such a map, we define a paraproduct of $\mathbb{R}^n$, always called T, by:

$$T_v w(x) = (2\pi)^{-2n} \int e^{ix(\xi+\eta)} T(\xi,\eta)\ \hat{v}(\xi)\ \hat{w}(\eta) d\xi\ d\eta \tag{1.1}$$

v and w belonging to $\mathcal{S}$.
As far as the basic properties of paraproduct are concerned, see [6].

<u>DEFINITION 1.1.2</u>: Let T be a paraproduct on $\mathbb{R}^n$, and $a \in \Sigma_\rho^{m,m'}$; we define, for any v and w in $\mathcal{S}$, $T_a(v,w)$ by

$$T_a(v,w) = \mathcal{F}^{-1}(2\pi)^{-2n} \int T(\zeta-\xi-\eta,\xi+\eta)\mathcal{F}_x a(\zeta-\xi-\eta,\xi,\eta)\ \hat{v}(\xi)\hat{w}(\xi) d\xi\ d\eta.$$

The following theorem gives the basic properties of these operators.

<u>THEOREM 1.1.1</u>: Let T and T' be two paraproducts on $\mathbb{R}^n$, and $a \in \Sigma_\rho^{m,m'}$ then, for any (s,t) so that $s+t \geqq m$,

(i) T_a maps $H^s \times H^t$ into $H^{s+t-m-m'-\frac{n}{2}}$

(ii) $T_a - T'_a$ maps $H^s \times H^t$ into $H^{s+t-m-m'-\frac{n}{2}+\rho}$

The key-tool of the proof of this theorem, is the Littlewood-Paley's theory, which allows the splitting of the symbol as in [12]. See [10] for technical details.

Now, with the same techniques and ideas, we can prove the following symbolic formulas.

DEFINITION 1.1.3: Let $a \in \Sigma_\rho^{m,m'}$, $b \in \underline{\Sigma}_\rho^{m_1}$, $c \in \underline{\Sigma}_\rho^{m_2}$.

(i) $b \# a(x,\xi,\eta) = \sum_{|\alpha|+i+j\leq[\rho]} \partial_\xi^\alpha b_i(x,\xi)|_{\xi+\eta} D_x^\alpha a_j(x,\xi,\eta)$

(ii) $a \# (b,c)(x,\xi,\eta) = \sum_{|\alpha|+|\beta|+i+j+\ell\leq[\rho]} \partial_\xi^\alpha \partial_\eta^\beta a_i(x,\xi,\eta) D_x^\alpha b_j(x,\xi) D_x^\beta c_\ell(x,\eta)$

These definitions are justified by the following theorem:

THEOREM 1.1.2: Let a be in $\Sigma_\rho^{m,m'}$, b in $\underline{\Sigma}_\rho^{m_1}$ and c in $\underline{\Sigma}_\rho^{m_2}$; then:

(i) for any (s,t) so that $s+t \geq m$, $T_b \circ T_a - T_{b\#a}$ maps $H^s \times H^t$ into $H^{s+t-\frac{n}{2}-m-m_1+\rho}$.

(ii) for any (s,t) so that $s + t \geq m + m_1 + m_2$:

$$T_a \circ (I_b \otimes T_c) - T_{a\#(b,c)} \text{ maps } H^s \times H^t \text{ into } H^{s+t-\frac{n}{2}-m-m_1-m_2-\frac{n}{2}+\rho}.$$

The definitions 1.1.1 and 1.1.3 can clearly be localized.

We can now give a local definition of bilinear operators.

DEFINITION 1.1.4: Let Ω be an open subset of $\mathbb{R}^n$; Op $\Sigma_\rho^{m,m'}(\Omega)$ is the set of operators A, so that, for any (s,t) so that $s + t \geq m$:

(i) A maps $H^s_{loc}(\Omega) \times H^t_{loc}(\Omega)$ in $\mathcal{D}'(\Omega)$;

(ii) A is properly supported in a traditional meaning (see [10]);

(iii) a exists belonging to $\Sigma_\rho^{m,m'}(\Omega)$ so that, for any compact subset K of Ω, any $\chi \in C_0^\infty(\Omega)$, and any paraproduct T in $\mathbb{R}^n$:

$$A - \chi T_{\chi a} \text{ maps } H_K^s \times H_K^t \text{ into } H^{s+t-\frac{n}{2}-m-m'+\rho};$$

(iv) when (iii) is verified, we say that a is a symbol of A.

Let us remark that for any a in $\Sigma_\rho^{m,m'}(\Omega)$, A exists in Op $\Sigma^{m,m'}(\Omega)$ so that a is a symbol of A. Like in [6], local versions of Theorems 1.1.1 and 1.1.2 are easily proved in [10].

1.2 Precised localization of a paraproduct

We want to avoid the problem of the invariance of a paradifferential operator.

DEFINITION 1.2:

(i) Let Ω be an open subset of $\mathbb{R}^n$; $P^m_\rho(\Omega)$ is the set of the elements of $\underline{\Sigma}^m_\rho$ which are polynomial in ξ.

(ii) Let T be a paraproduct in $\mathbb{R}^n$, (φ_j,ω_j) a C^∞ partition of the unity of Ω, $(\chi_j)_{j\in\mathbb{N}}$ a sequence of C^∞ functions supported in ω_j so that χ equals 1 near Supp φ_j, and $a \in P^m_\rho(\Omega)$, one defines $OPP_T(a)$ by

$$OPP_T(a) \,.\, v = \sum_{j\in\mathbb{N}} \chi_j\, T_{\chi_j a}\, \varphi_j v$$

$$\text{with } T_{\chi_j a}\, \varphi_j u = \sum_{|\alpha|\leq m} T_{\chi_j a_\alpha} D^\alpha v, \text{ if } a(x,\xi) = \sum_{|\alpha|\leq m} a_\alpha(x)\xi^\alpha .$$

This definition is justified by the following proposition.

PROPOSITION 1.2:

(i) For any compact subset K of Ω, and any $\chi \in C^\infty_0(\Omega)$ which equals 1 near K, $OPP_T(a) - \chi T_{\chi a}$ maps the distributions supported in K, into C^∞_0, for any a belonging to $P^m_\rho(\Omega)$.

(ii) $OPP_T(a)$ is as microlocal as wanted, i.e. : for any $\varepsilon > 0$, ε_0 exists in $\mathbb{R}_+$, so that, if Supp $T \subset \{(\xi,\eta) \,/\, |\xi| \leq \varepsilon_0|\eta|\}$ then

$$WF(OPP_T(a).v) \subset \{(x,\xi) \in T^*\Omega \,/ d(\frac{\xi}{|\xi|}, WF(v) \cap S^{n-1}) \leq \varepsilon\} .$$

It means that $OPP_T(a)$ is in $\underline{\Sigma}^m_\rho$ and that it does not depend on the partition of unity modulo a mollifier.

The proof of this proposition is based on the fact that, if T is a paraproduct of $\mathbb{R}^n$, and v and w two compactly supported distributions, then Supp Sing $T_v w \subset$ Supp v $\cap$ Supp Sing w. This is proved by integration by parts in (1.1).

1.3 Precised paralinearization theorem

The following theorem precises Bony-Meyer's paralinearization theorem.

THEOREM 1.3: Let u be a real solution $H^s_{loc}(\Omega)$ of (E), with $s > \frac{n}{2} + m + 1-d$, d is 2 if (E) is semilinear, 1 if (E) is quasilinear, and 0 if not; then, for any paraproduct T in $\mathbb{R}^n$, R exists in $OP\,\Sigma^{2m-d,0}_{s-\frac{n}{2}-m-1+d}(\Omega)$ so that, if

$p(x,\xi) = \sum_{|\alpha|\leq m} \partial F/\partial u_\alpha . \xi^\alpha$, then:

$$OPP_T(p) . u + R(u,u) \in H^{3s-n-3m-1+2d}_{loc}(\Omega).$$

Let us remark that this theorem is independent of any hypothesis of hyperbolicity on the equation (E). It means that a nonlinear equation can be transformed into a paradifferential one, no longer linear modulo a rest H^{2s-s_0}, as it was in [6], but bilinear modulo a rest H^{3s-2s_0}.

2. STATEMENT OF RESULTS

In everything that follows, u is a real $H^s_{loc}(\Omega)$-solution of (E), with $s > \frac{n}{2} + m + 1-d$, satisfying hypothesis (H_1) and (H_2). Let us define subsets of $T^*\Omega$ describing the propagation of singularities.

DEFINITION 2.1: F_σ is the union of future bicharacteristic strips issued from points of $WF_\sigma(u) \cap \text{Car } p_m \cap (x_1 < 0)$ (Car p_m being the set of the zeros of p_m).

Bony's propagation theorem (see [6]) says, that, for any $\sigma \leq 2s - \frac{n}{2} - m-1+d$, $F_\sigma = WF_\sigma(u)$, and $WF_{2s-\frac{n}{2}-m+d}(u) \subset {}^c\,\text{Car } p_m$.

It is well known, that two microlocal singularities of non opposite directions interact under nonlinear functions by convex closure. M. Beals proved in [4], that, in some cases, the interaction of two microlocal singularities of characteristic opposite directions does not product singularities outside the tangent hyperplane to Car p_m in that direction.

It suggests the following definitions.

DEFINITION 2.2:

(i) $G_\sigma^{(1)}$ is the set of $(x,\xi) \in T^*\Omega$ so that there exist ξ_1 and ξ_2 of non opposite directions so that $\xi \in \mathbf{R}_+\xi_1 + \mathbf{R}_+\xi_2$ and so that, for any (σ_1,σ_2) so that $\sigma_1 + \sigma_2 = \sigma + \frac{n}{2}$, $\xi_1 \in F_{\sigma_1|x}$ or $\xi_2 \in F_{\sigma_2|x}$

(ii) $G_\sigma^{(2)}$ is the set of $(x,\xi) \in T^*\Omega$ so that η exists in $F_{\frac{\sigma}{2}+\frac{n}{4}\big|_x}$ so that $\xi \in T_\eta \text{ Car } p_{m|x}$

(iii) $G_\sigma = G_\sigma^{(1)} \cup G_\sigma^{(2)}$.

Singularities so created by interaction will then propagate.

DEFINITION 2.3: H_σ is the union of future bicharacteristic strips issued from points of $G_\sigma \cap \text{Car } p_m$.

THEOREM 2.4: Let u be a real solution $H^s_{loc}(\Omega)$ of (E), $s > \frac{n}{2} + m+1-d$; then, for any $\sigma \leqq 3s - n - 2m - 2 + 2d$:

- $WF_\sigma(u) \cap \text{Car } p_m \subset F_\sigma \cup H_{\sigma+m+1-d}$
- $WF_{\sigma+1}(u) \cap {}^c\text{ Car } p_m \subset G_{\sigma+m+1-d}$.

3. SLOW WAVES

3.1 Equation of order 2

COROLLARY 3.1: Let u be a solution $H^s_{loc}(\Omega)$, real, of (E), supposed of order 2, $s > \frac{n}{2} + 3-d$; let (x,ξ) be a point of Car p_m and Γ the bicharacteristic issued from (x,ξ); then, for any $\sigma \leqq 3s - n - 6 + 2d$, if u is microlocally H^σ in (x,ξ), then u is microlocally H^σ along Γ.

It is a consequence of Theorem 2, because Car $p_{2|_x}$ is convex, and so if $\sigma \leqq 3s - n - 3 + 2d$, $G_\sigma \cap \text{Car } p_m = \emptyset$. Let us remark that this corollary contains results and conjecture of [5].

3.2 Slow waves

DEFINITION 3.2:

(i) Let Σ be a characteristic hypersurface, it is said to be slow if, for any (x,ξ) belonging to $N^*\Sigma$, $T_\xi \text{ Car } p_m|_x \cap \text{Car } p_m|_x = \mathbb{R}^*\xi$.

(ii) A distribution v is a slow wave of order σ, if $\text{Car } p_m \cap WF_\sigma(v) \subset N^*\Sigma$ and $WF_{\sigma+1}(v) \cap {}^c\text{Car } p_m \subset T_{N^*\Sigma} \text{ Car } p_m$, with $T_{N^*\Sigma} \text{ Car } p_m = \{(x,\xi) \in N^*\Sigma / \exists \eta \in N^*\Sigma|_x / \xi \in T_\eta \text{ Car } p_m|_x\}$, Σ being a slow hypersurface.

Why slow? In fact, let $\square_i = \partial_t^2 - C_i^2\Delta$, with $C_1 > C_2$, if Σ is a characteristic hypersurface for $\square_2$, it is a slow hypersurface for the operator $\square_1 \square_2$.

COROLLARY 3.2.1 (Evolution of a slow wave): Let u be a real solution $H^s_{loc}(\Omega)$ of (E), with $s > \frac{n}{2} + m + 1 - d$, and Σ a slow hypersurface for p_m; for $\sigma \leqq 3s - n - 2m - 2 + d$, if u is a slow wave of order σ in the past, so it is in the future.

3.3 Interaction of two slow waves; an example

Let us consider a real solution $u \in H^s_{loc}(\mathbb{R}^3)$, $s > \frac{n}{2} + 3$, of the equation (E_T) $\square_1 \square_2 u = f(x,u,\partial^\alpha u)_{|\alpha| \leqq 3}$, where f is a smooth function, and suppose now that u is an H^s-slow wave, relatively to the slow cones $C_2^2t^2 = (x \pm a)^2 + y^2$.

The following pictures describe the singularities of u for t = cste.

$t < \frac{a}{c_2}$, nothing occurs $t = \frac{a}{c_2}$, the interaction begins.

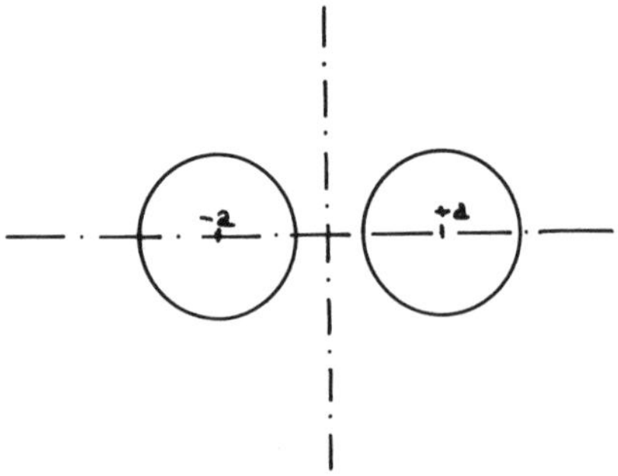

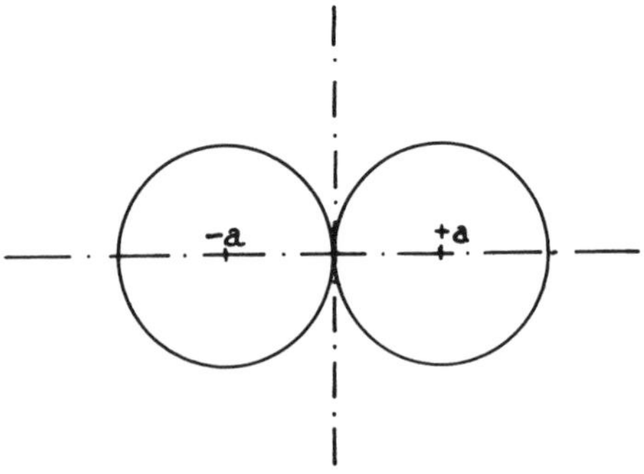

$$\frac{a}{c_2} < t < \frac{a}{c_2} \frac{c_1}{(c_1^2 - c_2^2)^{1/2}}$$

Singularities created by interaction propagate fast

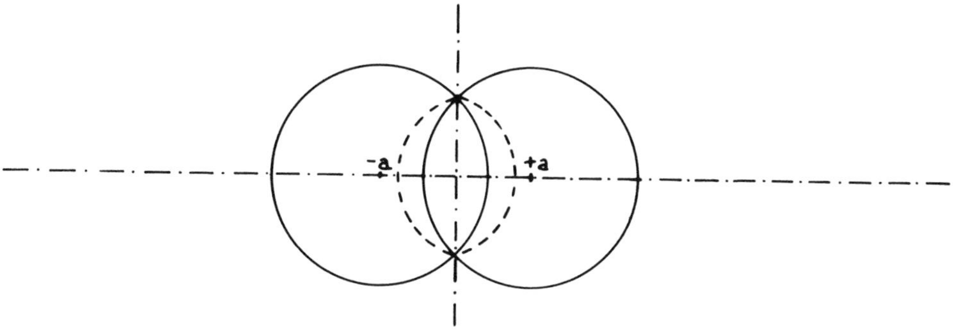

$$t = \frac{a}{c_2} \frac{c_1}{(c_1^2 - c_2^2)^{1/2}}$$, it is the end of the interaction.

When $t \geqq \frac{a}{c_2} \cdot \frac{c_1}{(c_1^2 - c_2^2)^{1/2}}$, each type of singularity propagates at its own speed, there is no more interaction.

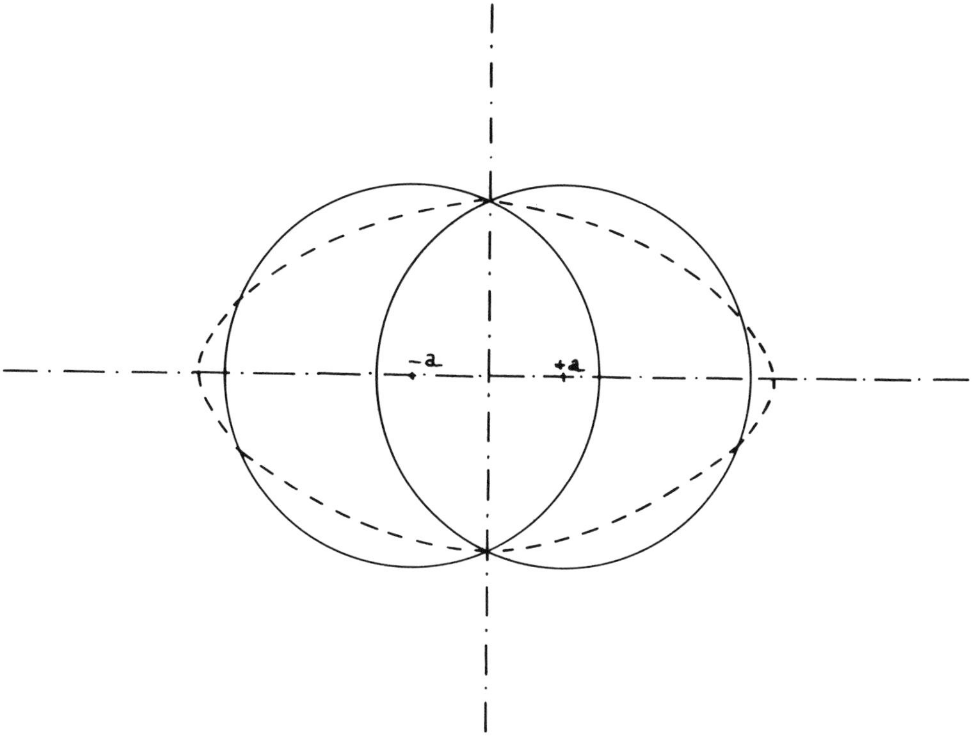

All these results come from the calculus of the sets H; it is left to the reader. Let us just examine the reason why interaction stops.

If $t < \frac{a}{c_2} \frac{c_1}{(c_1^2 - c_2^2)^{1/2}}$, interaction occurs

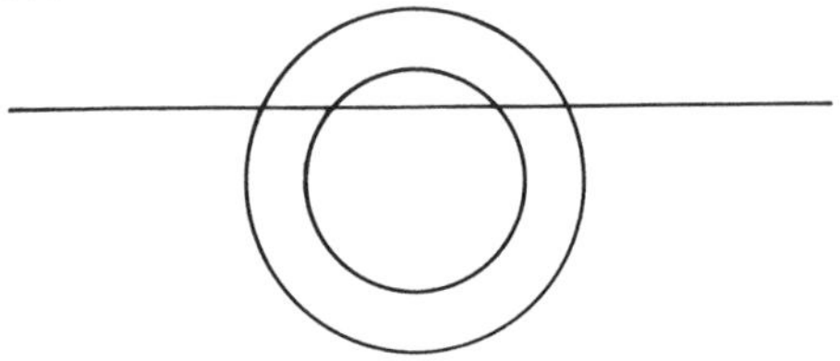

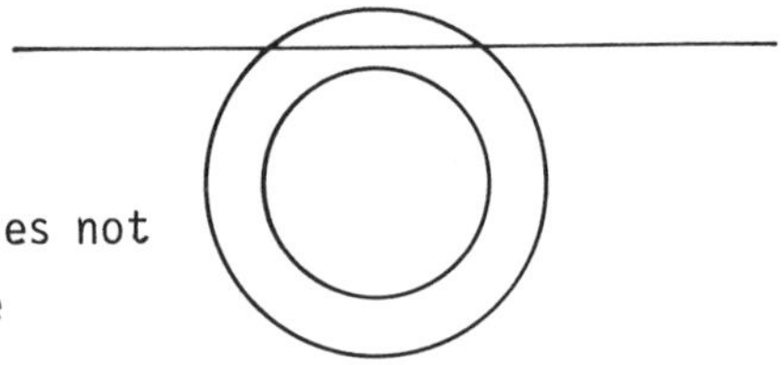

If $t > \frac{a}{c_2} \frac{c_1}{(c_1^2 - c_2^2)^{1/2}}$ interaction does not occur any more

The specificity of slow waves is emphasized by the following fact: we prove in [10] the existence of a solution $u \in H^s_{loc}(R^3)$, $s > \frac{5}{2}$, of the equation $\square_1\square_2 u = \beta u^2$, with $\beta \in C_0^\infty$, so that, in the past, $WF(u) \subset N^*C_1$, and in the future, the regularity of u is as follows.

(1) and (2) u is C^∞
(4) u is not $H^{2s+n/2+7+\varepsilon}$
(3) u is $H^{3s+1+\varepsilon}$

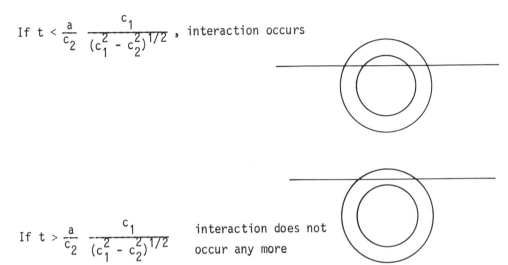

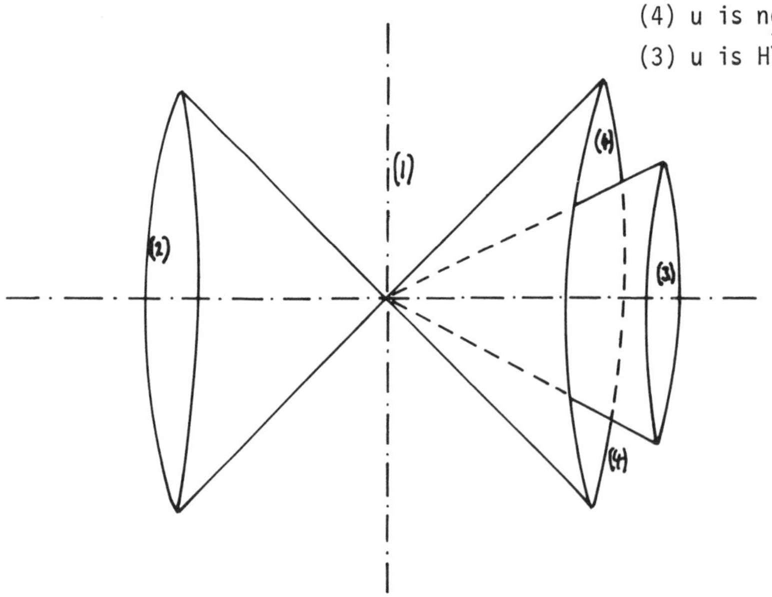

4. AN IDEA OF THE PROOF

The proof is based on the precised paralinearization theorem; the paramount work to do now is the study of microlocal regularity of the rest. To do this, we need spaces describing the regularity of the solution near the characteristic manifold. Thanks to the tangent hyperplan lemma, this description allows us an estimate of the H^{σ} wave front of the rest, for $\sigma \leqq 3s - 2s_1$.

4.1 Algebras associated with a nonlinear equation

DEFINITION 4.1: Let $P \in \mathrm{Op}\ \Sigma^m_\rho(\Omega)$;

$$H^s(P) = \{u \in H^s_{loc}(\Omega)/P^j u \in H^{s-(m-1)j}_{loc}(\Omega),\ \forall j \leqq [\rho];\ P^j \in H^{s+\rho-mj}$$

$$\text{if } j = [\rho] + 1\}$$

The algebras have been introduced by M. Beals in the case when P is a differential operator with smooth coefficients. The paradifferential calculus allows us to prove very easily that $\mathrm{Op}\ \Sigma^{m'}(\Omega).\ H^s(P) \subset H^{s-m}(P)$, that, for any $s \geqq \rho + \frac{n}{2}$, $H^s(P)$ is an algebra, that if Q and Q' are two paradifferential operators so that $P = Q.Q'$, then $H^s(P) \subset H^s(Q)$, and the following proposition as well:

PROPOSITION 4.1: Let u be a real solution $H^s_{loc}(\Omega)$ of (E), $s > \frac{n}{2} + m+1-d$, if $\sigma(P_m)(x,\xi) = \sum_{|\alpha|=m} \partial F/\partial u_\alpha . \xi^\alpha$, u belongs to $H^s(P_m)$.

4.2 Tangent hyperplan lemma

This key lemma will allow us to translate "belonging to $H^s(P_m)$" into terms of microlocal regularity of the rest. As developed in the second section, the problem is the interaction of two microlocal singularities of opposite directions. As p_m is strictly hyperbolic, close to such directions, we consider only the case of order one.

Tangent hyperplane lemma: Let $q_1 \in \mathrm{Op}\ \underline{\Sigma}^1_\rho(\Omega)$ so that, for any $(x,\xi) \in T^*\Omega$, any $\alpha \in \mathbf{R}$, we have $q_1(x,\alpha\xi) = \alpha q_1(x,\xi)$; let $(x_0,\xi_0) \in T^*\Omega$, so that $q_1(x_0,\xi_0) = 0$ and $d_\xi q_1(x_0,\xi_0) \neq 0$; let K and K' two open conic subsets of

R^n, so that:

$$\bar{K}' \subset K; \text{ and } K \cap T_\xi \text{ Car } q_{1|x_0} = \emptyset,$$

it then exists
- an open neighbourhood Ω_0 of x_0
- a conic neighbourhood K_0 of ξ_0
- a pseudo differential symbol λ of order 1, supported

in $\Omega_0 \times K$, and elliptic on $\Omega_0' \times K$, so that: for any $a_\pm \in S^0(\Omega_0)$, supported in $\Omega_0 \times \pm K_0$, it exists $r \in \Sigma_\rho^{0,0}(\Omega_0)$ supported in $\Omega_0 \times K_0 \times -K_0$ and

$$\lambda(x,\xi+\eta)a_+(x,\xi)a_-(x,\eta) = r(x,\xi,\eta)(q_1(x,\xi) + q_1(x,\eta)).$$

PROOF: Considering the localization of K, if δ is small enough, it is easy to see that, if $(\xi,\eta) \in K^\delta \times -K^\delta$ and $\xi + \eta \in K$, then:

$$|\xi + \eta| \leq C\,\delta\,(|\xi| + |\eta|), \quad C_1|\eta| \leq |\xi| \leq C\,|\eta|.$$

(K^δ is the set of any ξ so that $0 \leq |\xi - (\xi|\xi_0)\xi_0| \leq \delta|\xi|$).

Let us now suppose that microlocally near (x_0,ξ_0), $q_1(x,\xi) = \xi_n + s(x,\xi)$ with $d_\xi\, s(x_0,\xi_0) = 0$.

As $s(x,-\xi) = s(x,\xi)$, we have

$$s(x,\xi) + s(x,\eta) = \sum_{j=1}^{n} \int_0^1 (\xi_j+\eta_j)\, \partial_{\xi_j}\, s(x;-\xi + t(\xi+\eta))dt$$

Let $b(x,\xi,\eta) = \dfrac{1}{\xi_n+\eta_n}\,(s(x,\xi) + s(x,\eta))$.

As $M > 0$ exists so that if $(\xi,\eta) \in K^\delta \times -K^\delta$, and if $\xi + \eta \in K$, we have

$\left|\dfrac{\xi_j + \eta_j}{\xi_n + \eta_n}\right| \leq M$, as δ being small enough, we have $|\xi+\eta| \leq \varepsilon\,|\xi|$, $|b(x,\xi,\eta)| \leq \frac{1}{2}$,

as soon as $(\xi,\eta) \in K^\delta \times (-K^\delta)$ and $\xi + \eta \in K$, if δ is small enough, and x near x_0.

Let $\chi \in C^\infty(\mathbf{R}^n)$, homogeneous of degree 0, equal to 1 near $K' \cap (|\xi| \geq 1)$, and supported in $\bar{K} \cap (|\xi| \geq \frac{1}{2})$, we have, for any $a_\pm \in S^0(\Omega_0)$ supported in $\Omega_0 \times \pm K^\delta$, $\chi(\xi+\eta)(\xi_n+\eta_n)a_+(x,\xi)a_-(x,\eta) = r(x,\xi,\eta)(q_1(x,\xi)+q_1(x,\eta))$ with

$$r(x,\xi,\eta) = \frac{\chi(\xi+\eta)a_+(x,\xi)a_-(x,\eta)}{1 - b(x,\xi,\eta)} .$$

It is easy to check that r belongs to $\Sigma_\rho^{0,0}(\Omega_0)$.

PROPOSITION 4.2: Let u be a real solution of (E), $H^s_{loc}(\Omega)$ with $s > \frac{n}{2}+m+1-d$; for any $R \in Op\Sigma^{\mu,\mu'}_{s-\frac{n}{2}-m-1+d}(\Omega)$ with $\mu \leq 2s$, and any $\sigma \leq 3s-n-m-1+d-\mu-\mu'$, $WF_\sigma(R(u,u)) \subset G_{\sigma+\mu+\mu'}$.

COROLLARY 4.2: Let u be as above, then, if $\sigma \geq 3s-2m-n+2d-2$, then $WF_{\sigma-m+1}(OPP_T(p).u) \subset G_{\sigma+m+1-d}$.

To demonstrate this, it is enough to translate the tangent hyperplan lemma in terms of operators, then to repeat the relation we obtain, as long as the regularity of the calculus allows it, and then, to apply the relation through a microlocal partition of unity.

4.3 Proof of Theorem 2: Let $\sigma \in [2s - \frac{n}{2} -m-1+d, 3s-n-2m-2+2d]$ and let $(x_0,\xi_0) \notin Car\ p_m \cup G_{\sigma+m+1-d}$ (resp. C any continuous part of a characteristic Γ so that $F_\sigma \cup H_{\sigma+m+1-d} = \emptyset$); it follows from Bony's theorem of microlocal ellipticity (resp. from Bony's propagation theorem) that a real $\delta > 0$, and two pseudo differential operators of order 0, A and A', exist so that:

- A is elliptic near (x_0,ξ_0) (resp. C)
- $A'u \in H^{2s-\frac{n}{2}-m+d}$ (resp. $H^{2s-\frac{n}{2}-m-1+d}$)
- for any $(x,\xi) \in S\,E\,(A)$, $d(\frac{\xi}{|\xi|}; SE(I-A')|_x \cap S^{n-1}) \geq 2\delta$.

Let T be a paraproduct so that (ii) of Proposition 1.2 can be applied with $\varepsilon = \delta$; if SE(A) is small enough, from Corollary 4.2.2, we suppose that $A.OPP_T(P)u \in H^{\sigma-m+2}$ and then, from Proposition 1.2 (ii), $A.OPP_T(p).A'u \in H^{\sigma-m+1}$. Bony's theorem of microlocal ellipticity (resp. propagation) shows us that u is $H^{\sigma+1}$ microlocally near (x_0,ξ_0) (resp. H^σ microlocally near C).

References

[1] S. Alinhac: Evolution d'une onde simple. Actes des journées E.D.P. de Saint-Jean-de-Monts (1985).

[2] S. Alinhac: Paracomposition et opérateurs paradifférentiels. Comm. in P.D.E. 11 (1) 1986.

[3] S. Alinhac: Interaction d'ondes simples pour des équations complètement non linéaires. Séminaire E.D.P. de l'Ecole Polytechnique 1985-1986.

[4] M. Beals: Self spreading and strength of singularities for solutions of semi linear wave equations. Annals of Math. 118 (1983).

[5] M. Beals: Propagation of smoothness for nonlinear second order strictly hyperbolic differential equations. Proc. of Symp. in Pure Math. 43, 1985.

[6] J-M. Bony: Calcul symbolique et propagation des singularitiés pour des équations aux dérivées partielles non linéaires. Ann. Scient. Ec. Norm. Sup. 1981.

[7] J-M. Bony: Interaction des singularités pour des équations aux dérivées partielles non linéaires. Séminaire Goulaouic-Meyer-Schwartz 1981-1982, n°2.

[8] J-M. Bony: Second microlocalization and propagation for semi linear hyperbolic equations. To appear in Contribution to the workshop and symposium on hyperbolic equations and related topics: Katata and Kyoto, August 25 - Sept. 5 1984.

[9] J-Y. Chemin: Calcul paradifférentiel précisé et applications à des équations aux dérivées partielles non semi linéaires. To appear in Duke Math. Journal.

[10] J-Y. Chemin: Calcul symbolique bilinéaire et interaction contrôlée dans les équations aux dérivées partielles. To appear in Bulletin Soc. Math. France.

[11] J-Y. Chemin: Interaction de trois ondes dans des équations strictement hyperboliques semi linéaires. Note aux C.R.A.S. t. 304 Série I n° 8 (1987).

[12] R. Coifman and Y. Meyer: Au-delà des opérateurs pseudodifférentiels. Astérisque 57 (1978).

[13] P. Godin: Propagation of C^∞ regularity for fully non linear second order strictly hyperbolic equations in two variables. Trans. Amer. Math. Soc. 290 (1985).

[14] R. Melrose and N. Ritter: Interaction of non linear progressing waves. Annals of Math. 121 (1985).

[15] Y. Meyer: Remarque sur un théorème de J-M. Bony. Supp. Rend. Cir. Math. Palermo n° 1, 1981.

[16] J. Rauch and M. Reed: Non linear microlocal analysis of semi linear hyperbolic systems in one space dimension. Duke Math. Journal 49 (1982).

Jean-Yves Chemin
Centre de Mathématiques
de l'Ecole Polytechnique
91 128 Palaiseau Cedex
France.

Y. CHOQUET-BRUHAT

Global existence theorems by the conformal method

INTRODUCTION

The classical field equations of fundamental physical interactions are semi linear partial differential equations on the Minkowski space time. To obey the principle of relativistic causality these equations must be, at least for an appropriate choice of gauge, of hyperbolic type, with characteristic cone the null cone of the Minkowski metric. These conditions are satisfied by the Yang-Mills equations, coupled with spinor and Higgs (i.e. scalar) field equations on Minkowski space-time which are believed to link the three fundamental interactions, electromagnetic, weak and strong with their sources. These equations appear (in an appropriate gauge for the Yang-Mills field) as a semilinear partial differential system with principal part diagonal by blocks, each block being one element, the D'Alembert operator $\square$, or the Dirac operator $D = \Gamma^{\lambda} \partial_{\lambda}$.

The fourth fundamental interaction, and most anciently known, gravitation, does not enter into this scheme. It is represented by a Lorentz metric, i.e. a metric of hyperbolic signature, and obeys Einstein's equations. These equations reduce in an appropriate gauge (for instance appropriate coordinates) to a hyperbolic system, quasi-linear, whose characteristic cone is the null cone of the metric

$$g^{\lambda\mu} \partial^2_{\lambda\mu} g_{\alpha\beta} + f_{\alpha\beta}(g,\partial g) = 0$$

The complete system of fundamental interactions at a classical level could be the Einstein equations for a Lorentz metric with sources the other fields, and the equations for these other fields where the Minkowski metric is replaced by this Lorentz metric.

The physical results pertain only to the quantized fields: it is not known how to quantize the gravitational field, and not even if it should be quantized. However, much work has been devoted, with at least mathematical success, to the quantization of fields in a "curved background", that is on a manifold with a Lorentz metric.

It is an important - essentially open for a curved background - problem to study global existence of solutions of the field equations. A remarkable property of the sourceless Yang-Mills equations in dimension 4, is their conformal invariance. This property is shared, with weights on the fields, by the coupled Yang-Mills, spinor and Higgs equations (cf. [2]). It has been used to transform the problem of existence of global solutions of these equations on Minkowski space time M_4 into a local existence problem on a conformally related manifold, the Einstein cylinder Σ_4, $S^3 \times \mathbb{R}$ with its canonical metric.

In dimension $d > 4$ the Yang-Mills equations are no longer conformally invariant. However, the scalar and spinor (Dirac) wave operators, still have conformal covariance properties. Indeed the conformal mapping from the Minkowski space time M_{n+1} onto a bounded open subset of $\Sigma_{n+1} = S^n \times \mathbb{R}$ has been used by Christodoulou [6] to prove the global existence of solutions of the Cauchy problem for general second order equations of the form $\Box u = f(u, \partial u, \partial^2 u)$, with $f(0,0,0) = f'(0,0,0) = 0$, $\Box$ the usual wave operator on M_{n+1}, and $n \geqq 5$, and odd (result obtained previously, by a different method, by Klainerman [7]), or when $n = 3$ and f satisfies a further condition (suggested in Klainerman [5]).

In this paper we give a brief review of this conformal method. In §1 of this paper we recall the conformal diffeomorphism of M_{n+1} onto a bounded open set of $\mathbb{R}^{n+1}$.

In §2 we explain the transformation by a conformal diffeomorphism of a partial differential operator acting on sections of a vector bundle associated to the orthonormal frame bundle, or to a spin frame bundle of a pseudo Riemannian manifold.

In §3 we expose the proof, following Christodoulou [6], of the global existence theorems for solutions of $\Box u = F(u, \partial u, \partial^2 u)$.

In §4 we indicate how these theorems can be extended to some asymptotically flat manifolds, called here "strongly asymptotically simple". Note that global existence theorems obtained by the conformal method automatically include decay estimates of the solutions (cf. [2], [4]).

1. CONFORMAL MAPPING FROM M_{n+1} INTO Σ_{n+1}

(Penrose [3] treats the case $n = 4$, the generalization is straightforward). We consider the Minkowski space time M_{n+1}, which is a manifold diffeomorphic to $\mathbb{R}^{n+1}$ with a Lorentz metric which reads in canonical coordinates:

$$\eta = dt^2 - \sum_{i=1}^{n} (dx^i)^2$$

$$-\infty < t < +\infty \quad , \qquad -\infty < x^i < +\infty \quad .$$

Or, equivalently

$$\eta = du\, dv - r^2\, dS_{n-1}^2$$

$$u = t + r, \qquad v = t - r, \qquad r = \left\{ \sum_{i=1}^{n} (x^i)^2 \right\}^{1/2}$$

$$-\infty < u < +\infty \, , \quad -\infty < v < +\infty \, , \qquad u - v \geqq 0$$

dS_{n-1}^2: metric of the sphere S_{n-1}, for instance in angular (pseudo) coordinates θ^i, $i = 2,\ldots n-1$.

We define a diffeomorphism from $\mathbb{R}^{n+1}$ onto a bounded open set $\mathcal{U}$ of $R \times S^n$ by setting

$$U = \text{Arctg}\, u \quad , \qquad V = \text{Arctg}\, v$$

$$-\frac{\pi}{2} < U < \frac{\pi}{2} \, , \qquad -\frac{\pi}{2} < V < \frac{\pi}{2} \, , \qquad U - V \geqq 0;$$

$$\Theta^i = \theta^i \quad , \qquad i = 1,\ldots,n-1$$

with U and V linked to the canonical coordinates $T \in \mathbb{R}$, α with $0 \leqq \alpha \leqq \pi$, Θ^i, $i = 2,\ldots,n-1$ angular coordinates on S^{n-1} by:

$$T = U + V, \; \alpha = U - V.$$

Thus, on $\mathcal{U}$:

$$0 \leqq \alpha \leqq \pi, \quad \alpha - \pi < T < \pi - \alpha$$

An easy computation shows that

$$\phi_*^{-1}\eta = \Omega^{-2}\gamma \quad \text{on } U$$

with

$$\Omega = 2 \cos U \cos V$$

$$\gamma = 4\, dU\, dV - \sin^2(U - V) dS_{n-1}^{\ 2}$$

that is

$$\gamma = dT^2 - d\alpha^2 - \sin^2\alpha\, dS_{n-1}^{\ 2}$$

$$\Omega = \cos T + \cos \alpha.$$

We see that γ extends to the whole of $\mathbb{R} \times S^n$ since it is the metric of this cylinder, written above in canonical coordinates. We see that Ω extends also as a C^∞ function to $\mathbb{R} \times S^n$. It is strictly positive on $\mathcal{U}$, but vanishes on ∂U

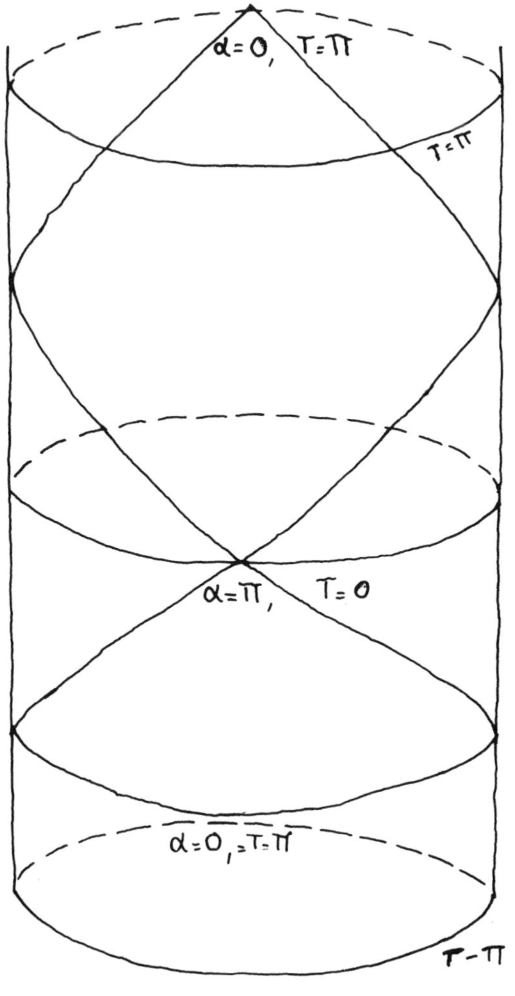

REMARK: The pull back of Ω to $\mathbb{R}^{n+1}$ reads:

$$\Omega \circ \phi = \frac{4}{(1+u^2)(1+v^2)}, \quad u = t + r, \quad v = t - r$$

it decays like t^{-4} in strictly time like directions, like t^{-2} in the null directions $t - r$ = constant or $t + r$ = constant.

In the identification of $\mathcal{U}$ with a subset of $\mathbb{R} \times S^n$, the image of the submanifold $t = 0$ of M_{n+1} is the sphere S^n, submanifold $T = 0$ of $\mathbb{R} \times S^n$ minus a point(1), $\alpha = \pi$, called I_o, which can be considered as the image of the points at space like infinity, $t = 0$, $r = +\infty$. The closure of the image of other space like sections t = constant of $\mathbb{R}^{n+1}$ are space like sections of $\mathbb{R} \times S^n$ which pass also through I_o.

The boundary $\partial\mathcal{U}$ is constituted of two null cones, with vertex I_+, and I_-, points of coordinates respectively $\alpha = 0$, $T = \pi$ and $\alpha = 0$, $T = -\pi$. These points are the closoure of the image in Σ_{n+1} of the time like lines of M_{n+1}, respectively when t tends to $+\infty$ and t tends to $-\infty$.

The other points of $\partial\mathcal{U}$ in Σ_{n+1} are closures of the images of light rays.

2. BASICS OF THE CONFORMAL METHOD

Let (M,g) be a smooth Riemannian manifold of arbitrary signature. Let E and F be two smooth vector bundles over M associated with the orthonormal frame bundle. Denote by (E,m) the vector bundle over M whose fibre at x is $(E,m)_x = E_x \times (T^*_x \otimes E_x) \ldots \times (\otimes^m T^*_x M \otimes E_x)$. A partial differential operator P of order m, mapping sections of E into sections of F is given for each $x \in M$ by a smooth mapping p_x from U_x, an open set of $(E,m)_x$, into F_x. We write

$$p_x(Y) = p(x, y_o,\ldots,y_m) \in F_x, \text{ H} \tag{2.1}$$

$$Y = (y_o,\ldots,y_m) \in (E,m)_x, \text{ i.e. } y_o \in E_x,\ y_1 \in T^*_x M \otimes E_x \ldots y_m \in \otimes^m T^*_x \otimes E_x.$$

(1) Note that $\alpha = 0$ [resp. $\alpha = \pi$] represents, in contradistinction from other values $0 < \alpha < \pi$, only one point of the sphere S_n, the south pole [resp. the north pole].

If v is a smooth section of E the partial differential operator acts on v according to:

$$(Pv)(x) = p(x, (\nabla v)(x),\dots,(\nabla^m v)(x)) \tag{2.2}$$

where ∇ is the covariant derivative in the metric g.

Let $\varphi: M \to \tilde{M}$ by $x \mapsto \tilde{x} = \varphi(x)$ be a diffeomorphism. Let $\tilde{g}$ be a metric on $\tilde{M}$ and M its pull back by φ:

$$g = \varphi_* \tilde{g}\ , \quad \text{equivalently} \quad \tilde{g} = \varphi^* g.$$

Let $\rho = (e_0,\dots,e_n)$ be an orthonormal frame for g in an open set $U \subset M$. Its image by φ is an orthonormal frame for $\tilde{g}$ in $\tilde{U} = \varphi(U) \subset \tilde{M}$;

$$\tilde{\rho}(\tilde{x}) = (\varphi'(x)e_0(x),\dots,\varphi'(x)e_n(x)),\ \tilde{x} = \varphi(x)$$

$$= (\tilde{e}_0(\tilde{x}),\dots,\tilde{e}_n(x))$$

which we denote

$$\tilde{\rho} = \varphi^*\rho\ .$$

Let v be a section over $U \subset M$ of a vector bundle E associated to the principal bundle $\mathcal{O}_g$ of orthonormal frames of g over M by a representation r: the fibre E_x of E is the set of equivalence classes of pairs $(\rho(x), X(x))$ with $\rho(x)$ an orthonormal frame and $X(x) \in V$, some given vector space ($\mathbb{R}^p$ or $\mathbb{C}^p$), with the equivalence relation

$$(\rho_1(x), X_1(x)) \simeq (\rho_2(x), X_2(x))$$

if

$$\rho_2(x) = L(x)\ \rho_1(x), \quad X_2(x) = r(L(x))\ X_1(x)$$

with L(x) any element of the (pseudo) orthogonal group of the metric g. Let $\tilde{E}$ be a vector bundle over $\tilde{M}$ associated to the orthonormal frame bundle $\mathcal{O}_{\tilde{g}}$ by the same representation r. The image of φ of the section v is by definition the section $\tilde{v}$ which has for representant at $\tilde{x} = \varphi(x)$, in the frame $\tilde{\rho}(\tilde{x}) = \varphi'(x)\ \rho(x)$, the element $\tilde{X}(\tilde{x}) = X(x)$, representant of $v(x)$ in the frame $\rho(x)$. We denote this image by $\tilde{v} = \varphi^* v$, equivalently $v = \varphi_* \tilde{v}$.

Case of spinor fields: the spinor fields are sections of a vector bundle associated not to the orthonormal bundle O_g but to a spin bundle S_g, which is a two sheet cover of O_g, when it exists - i.e. if the second Stieffel Whitney class of O_g vanishes. The spin bundle is not uniquely determined, there may even exist inequivalent spin bundles. It is a reason why some ambiguity may appear in the definition of the image of spinor fields by diffeomorphisms. We shall proceed as follows.

Let (M,g) be a (pseudo) Riemannian manifold admitting a spin structure, that is a principal bundle S_g over M (called bundle of spin frames), with typical fibre the spinor group, which projects on O_g:

$$\pi: S_g \to O_g \quad \text{by} \quad s(x) \to \rho(x)$$

with the property that the right action of their respective groups on these bundles, the projection π and the 2 to 1 homomorphism H from the spinor group S_g onto the orthogonal group O_g are such that: if $s_2(x) = s_1(x)\Lambda$ then $\rho_2(x) = \rho_1(x)\ L(x)$, with $L(x) = H\ \Lambda(x)$ and $\rho_i(x) = \pi\ s_i(x)$, $i = 1,2$. Let $\varphi: M \to \tilde{M}$ be a diffeomorphism and $g = \varphi_* \tilde{g}$. Then if (M,g) admits a spin structure the same is true of $(\tilde{M},\tilde{g})$: we choose for spin structure on $(\tilde{M},\tilde{g})$ the image of the bundle of spin frames over (M,g), namely the bundle over $\tilde{M}$ with fibre at $\tilde{x}$ the set $\tilde{S}_{\tilde{x}} = S_x$, $x = \varphi^{-1}(\tilde{x})$: it is a principal bundle with group the spinor group, if we define the right action of the spinor group in $\tilde{S}_{\tilde{x}}$ by its right action on S_x.

We define the projection on the bundle of orthonormal frames $O_{\tilde{g}}$ by

$$\tilde{\pi}: \tilde{s}(\tilde{x}) \to \tilde{\rho}(\tilde{x}) = \varphi'(x)\ \rho(x)$$

Since the linear map $\varphi'(x)$ (left action on the fibre) and the right action of the orthogonal group commute on the fibre O_x: we have

$$\tilde{\rho}_2(\tilde{x}) = \tilde{\rho}_1(\tilde{x})L$$

if and only if

$$\rho_2(x) = \rho_1(x)L.$$

Therefore, by the definition that we have given of the action of the spinor group $\tilde{S}_{\tilde{x}}$ we see that $\tilde{S}_{\tilde{g}}$ is a spin structure over $(\tilde{M},\tilde{g})$.

The image by φ of a spinor field ψ on M is the spinor field on $\tilde{M}$ such that it is represented at $\tilde{x} = \varphi(x)$, in the spin frame $\tilde{s}(\tilde{x}) = s(x)$ by the same element of $\mathbb{C}^p$, $p = [d/2]$, typical fibre of a vector bundle of spinors over a d-dimensional manifold . We denote this image by

$$\tilde{\psi} = \varphi^*\psi .$$

Consider now the covariant derivative ∇v of a section of E, section of the fibred product $T^*M \otimes E$ over M, with fibre $T^*_xM \otimes E_x$. Its image by φ, $\varphi^* \nabla v = \widetilde{\nabla v}$, is by definition such that it has the same representant in the frame $\tilde{\rho}(\tilde{x})$ than ∇v in the frame $\rho(x)$. Since $\tilde{g}$ and g have also the same representant in these frames we have

$$\widetilde{\nabla v} = \tilde{\nabla} \tilde{v}$$

and, analogously

$$\widetilde{\nabla^k v} = \tilde{\nabla}^k \tilde{v} = \varphi^* \nabla^k v = \tilde{\nabla}^k \varphi^* v$$

The image of the operator P on sections of E under the diffeomorphism $\varphi: M \to \tilde{M}$ is by definition the operator $\tilde{P}$ on sections of $(\tilde{E}, \tilde{M})$ defined by

$$\tilde{P} \tilde{v} = \varphi^* P(\varphi_* \tilde{v}).$$

If v and Pv are scalar functions ($E = M \times \mathbb{R}$ and $F = M \times \mathbb{R}$) then

$$\tilde{P} \tilde{v} = (P (\tilde{v} \circ \varphi)) \circ \varphi^{-1}.$$

In the general case, with P defined by (2.2) we have

$$(\tilde{P}\,\tilde{v})(\tilde{x}) = p\,(\varphi^{-1}(\tilde{x})),\ (\varphi_*\,\tilde{v})(\varphi^{-1}(\tilde{x})),\ldots,(\varphi_*\,\tilde{\nabla}^k\,\tilde{v})(\varphi^{-1}(\tilde{x}))$$

EXAMPLE Let P be the Laplace operator on scalar functions in the metric g

$$Pv = \Delta_g v = \mathrm{tr}_g\,\nabla^2 v = \langle g^{\#},\ \nabla_2 v\rangle$$

the mapping p_x is a mapping from $\otimes^2\ T^*_xM$ into $\mathbb{R}$ given by

$$p_x(y_2) = \mathrm{tr}_{g_x}\ y_2,\ y_2 \in \otimes^2\ T^*_xM,$$

i.e. in local coordinates

$$p_x(y_2) = g^{\alpha\beta}(x)\ y_{2\alpha\beta}$$

its image under a diffeomorphism $\varphi : M \to \tilde{M}$ is

$$(\tilde{P}\,\tilde{v})(\tilde{x}) = \langle g^{\#}(\varphi^{-1}(\tilde{x}),\ (\tilde{\nabla}^2\,\tilde{v})(\tilde{x})\rangle$$

thus the image of the Laplace operator Δ_g by a diffeomorphism φ is the Laplace operator in the image metric $\Delta_{\tilde{g}}$.

The Laplacian is invariant under a diffeomorphism $\varphi: M \to M$ is and only if φ is an isometry, i.e. $\tilde{g} = g$.

Suppose that we are given on a manifold N two conformally related metrics h and $\hat{h}$

$$h = \Omega^2\hat{h}$$

with Ω a strictly positive function on N.

The metrics h and $\hat{h}$ have isomorphic bundles of orthonormal frames O_h and $O_{\hat{h}}$ over N, which can be identified through the relation

$$\rho = \Omega^{-1}\ \hat{\rho}.$$

A spin structure on (N,h) defines a spin structure on $(N,\hat{h})$ by $s(x) = \hat{s}(x)$. A vector bundle [associated with O_h or S_h, can be considered as associated with $O_{\hat{h}}$, or $S_{\hat{h}}$.

Let P_h be a partial differential operator on (N,h) from sections of E into sections of F defined by a mapping p which depends also explicitly on h and its Riemann tensor: i.e. p_x is now defined as a mapping

$$P_x : \mathcal{M}_x \times \mathcal{R}_x \times E_x \times (T_x^*M \otimes E_x) \dots \times (\otimes^k T_x^*M \otimes E_x)$$

where $\mathcal{M}_x$ is the open set of metrics at x in the set of symmetric two tensors, and $\mathcal{R}_x$ the fibre of the 4 tensors with the symmetries of the Riemann tensor. The operator P_h is said to be conformally invariant with weight q if there exists numbers q and r such that for any section v of E:

$$P_{\hat{h}}\hat{v} = \Omega^r P_h v, \quad \text{with } \hat{v} = \Omega^q v.$$

The operator P_h is said to be conformally regular if there exists numbers q and r and an operator $Q_{h,\Omega}$ with smooth coefficients for any smooth function Ω such that

$$P_{\hat{h}}\hat{v} = \Omega^r Q_{h,\Omega}v , \quad \hat{v} = \Omega^q v.$$

In physical theories one meets, more generally, systems of partial differential operators $P_{A,g}$ which act on a family v of sections v_I of vector bundles associated with $\mathcal{O}_h$ or $\mathcal{S}_h$. The system is said to be conformally invariant with weight if there exists numbers q_I and r_A such that

$$P_{A,\hat{h}}\, \hat{v} = \Omega^{r_A} P_{A,h}\, v, \text{ with } \hat{v}_I = \Omega^{q_I}\, v_I$$

(no summation) and conformally regular if there exists operators $Q_{A,h,\Omega}$, with smooth coefficients for any smooth function Ω, such that

$$P_{A,h}\, \hat{v} = \Omega^{q_A}\, Q_{A,\Omega,h}\, v, \quad \hat{v}_I = \Omega^{q_I}\, v_I.$$

If a system is conformally invariant with weight [resp. conformally regular] there is an isomorphism between the solutions of the two systems on N

$$P_{\hat{h}}\, \hat{v} = 0 \text{ and } P_h\, v = 0 \quad [\text{resp. } Q_{h,\Omega}\, v = 0]$$

namely

$$\hat{v}_I = \Omega^{q_I} v_I .$$

If on the other hand P_h is the image by a diffeomorphism $\varphi: M \to N = \tilde{M}$ of an operator $R_{\varphi_* \hat{h}}$ on M, then there exists an isomorphism between the solutions of $Q_{h,\Omega}\, u = 0$ on N and the solutions of $R_{\varphi_* \hat{h}}\, v = 0$ on M.

The classical equations of fundamental interactions, massless Maxwell-Dirac equations. Weinberg-Salam theory, chromodynamics, models for unifications, enter in the above category. The unknown are a 1 form A on M, the Yang-Mills potential, with values in a vector space G, Lie algebra of some Lie group G, a spinor multiplet, Ψ_I, and a scalar multiplet Φ_J, also called eventually "Higgs field". The whole system is conformally invariant in dimension n+1 = 4 with weights 0 for A, 1 for Φ and 3/2 for Ψ (cf. [2]), for minimal coupling. It is conformally regular for higher couplings. In other dimensions the Yang-Mills operator is not conformally invariant. The Dirac operator is still conformally invariant, with weight n/2. The Higgs operator has for principal part the wave operator $\Box_g$. We shall use the conformal invariance with weight (n-1)/2 of the operator $\Box_g - \frac{(n-1)}{4n} R(g)$ in the next paragraph.

The conformal method, applied by Christodoulou and myself to solve global Cauchy problems for the fields equations on Minkowski space time, in the case n = 3, uses the diffeomorphism Φ from $\mathbf{R}^{n+1}$ onto the open set U of $S^n \times \mathbf{R}$ of §1 and the conformal property,

$$\gamma = \Omega^2 \varphi^* \eta$$

(η Minkowski metric, γ canonical metric of $S^n \times \mathbf{R}$) to transform a Cauchy problem on Minkowski space time with data on t = 0, for a conformally regular operator with weight, into a Cauchy problem with data on $S^n - \{I_0\}$, on $\mathbf{R} \times S^n$. If the transformed data belong to the relevant functional spaces for this Cauchy problem to have a local solution, and are small enough for the solution on $\mathbf{R} \times S^n$ to exist up to $T = \pi$, then the original problem has a solution on $\mathbf{R}^{n+1}$. Moreover, we can deduce its behaviour at infinity from the regularity properties of the solution of the transformed problem on $\mathfrak{M}$.

NOTE: The transformed data must in general, for hyperbolic problems to have a local solution, belong to some Sobolev spaces $H_s(S^n)$: this imposes on the original data properties of fall off at infinity of $\mathbb{R}^n$ - depending also on the weights.

3. NONLINEAR WAVE EQUATIONS IN DIMENSIONS $n+1 \geq 6$, $n+1 = 4$ (Christodoulou [6])

We consider the equation on $\mathbb{R}^{n+1}$ with unknown a scalar function u

$$\Box u + F(u, \partial_\alpha u, \partial^2_{\alpha\beta} u) = 0 \tag{3.1}$$

where in the Cartesian coordinates x of $\mathbb{R}^{n+1}$ with $x^0 = t$, x^i, $i = 1,\ldots,n$

$$\Box = \frac{\partial^2}{\partial t^2} - \sum_{i=1}^{n} \frac{\partial^2}{(\partial x^i)^2}$$

$\partial_\alpha u$, $\partial^2_{\alpha\beta} u$ denotes the set of partial derivatives $\frac{\partial u}{\partial x^\alpha}$, $\frac{\partial^2 u}{\partial x^\alpha dx^\beta}$; F is a smooth mapping from an open set of $\mathbb{R}^{n+1} \times \mathbb{R}^{(n+1)(n+2)/2}$ into $\mathbb{R}$. This equation corresponds to an operator P on Minkowski space time $M_{n+1} = (\mathbb{R}^{n+1}, \eta)$ with

$$p_\eta(x, y^{(0)}, \ldots, y^{(2)}) = \langle \eta^\#, y^{(2)} \rangle + q(x, y^{(0)}, y^{(1)}, y^{(2)}) \tag{3.2}$$

where $q_x \equiv q(x,.,.,.)$ is the mapping from $\mathbb{R} \times T^*_x \mathbb{R}^{n+1} \times (\otimes^2 T^*_x \mathbb{R}^{n+1})$ into $\mathbb{R}$ represented by F(.,.,.) in the natural coframe of Cartesian coordinates in $\mathbb{R}^{n+1}$. The transformed of equation (3.1) by a diffeomorphism $\varphi : M = \mathbb{R}^{n+1} \to \tilde{M} = \mathcal{U}$ is the equation

$$\tilde{P}_\eta \tilde{u} \equiv \varphi_* P_\lambda \varphi^* \tilde{u} = 0, \quad \varphi^* \tilde{u} = \tilde{u} \circ \varphi \tag{3.3}$$

If we take coordinates (X^a) on $\mathcal{U}$ we find that the operator $\tilde{P}_\eta$ is represented in these coordinates by:

$$\Box_{\tilde{\eta}} \tilde{u} + F(\tilde{u}, \frac{\partial X^a}{\partial x^\alpha} \tilde{\nabla}_a \tilde{u}, \frac{\partial X^a}{\partial X^\lambda} \frac{\partial X^b}{\partial x^\mu} \tilde{\nabla}^2_{a\,b} \tilde{u}) \tag{3.4}$$

The function $\tilde{u}$ on $\tilde{M} = \mathcal{U}$ satisfies this equation if and only if $u = \tilde{u} \circ \varphi$ satisfies (3.1) on $M = \mathbb{R}^{n+1}$.

By the §2 we know that, on U

$$\eta = \Omega^{-2}\gamma , \qquad \Omega = \cos T + \cos \alpha$$

$$\gamma = dT^2 - d\alpha^2 - \sin^2\alpha\,(\sin^2\theta\, d\varphi^2 + d\theta^2).$$

If $\hat{g}$ and g are two conformal metrics and v a scalar function on an n+1 dimensional manifold the following identity holds:

$$(\Box_{\hat{g}} - \frac{(n-1)}{4n}\hat{R})\,\hat{v} \equiv \Omega^{(3+n)/2}\,(\Box_g v - \frac{n-1}{4n} R\, v) \tag{3.5}$$

with

$$\hat{g} = \Omega^{-2} g , \qquad \hat{v} = \Omega^{(n-1)/2} v,$$

R [resp. $\hat{R}$] scalar curvature of g [resp. $\hat{g}$].

The Minkowski metric η, as well as its image by a diffeomorphism, has a zero curvature, while γ has scalar curvature -n(n-1). Therefore the identity (3.5) applied to (3.4) gives, by setting

$$\tilde{u} = \Omega^{(n-1)/2} v$$

$$\Omega^{(3+n)/2}\left(\Box_\gamma v + \frac{(n-1)^2}{4} v\right) + F\left(\Omega^{(n-1)/2} v, \frac{\partial X^a}{\partial x^\alpha}\tilde{\nabla}_a(\Omega^{(n-1)/2} v), \frac{\partial X^a}{\partial x^\alpha}\frac{\partial X^b}{\partial x^\beta}\tilde{\nabla}^2_{a\,b}(\Omega^{(n-1)/2} v)\right) = 0. \tag{3.6}$$

We make on F the hypothesis that it is linear in its last argument:

$$F(u, \partial_\alpha u, \partial^2_{\alpha\,\beta} u) \equiv h^{\alpha\beta}\, u, \partial_\lambda\, u)\, \partial^2_\alpha\, \partial_\beta u + f(u, \partial_\lambda u) \tag{3.7}$$

then

$$F(\Omega^{(n-1)/2} v, \frac{\partial X^a}{\partial x^\alpha}\tilde{\nabla}_a\,(\;^{(n-1)/2} v), \frac{\partial X^a}{\partial x^\alpha}\frac{X^b}{\partial x^\beta}\tilde{\nabla}_{ab}\,(\Omega^{(n-1)/2} v)$$

$$\equiv h^{\alpha\beta}(\Omega^{(n-1)/2} v, \frac{\partial X^a}{\partial x^\alpha}\tilde{\nabla}_a(\Omega^{(n-1)/2} v))\,\frac{\partial X^a}{\partial x^\alpha}\frac{\partial X^b}{\partial x^\beta}\tilde{\nabla}_{ab}(\Omega^{(n-1)/2} v)$$

$$+ f(\Omega^{(n-1)/2} v, \frac{\partial X^a}{\partial x^\alpha}\tilde{\nabla}_a(\Omega^{(n-1)/2} v)) \tag{3.8}$$

we express this operator on $(\mathfrak{U},\gamma)$ by using the well known relation between the connections of two conformal metrics, here $\tilde{\eta}$ and γ, namely

$$\tilde{\Gamma}_{a\ b}^{\ c} - \Gamma_{a\ b}^{\ c} = \Omega^{-1}(\delta_a^c \nabla_b \Omega + \delta_b^c \nabla_a \Omega - \gamma^{cd} \gamma_{ab} \nabla_d \Omega)$$

For a scalar function v we have

$$\tilde{\nabla}_a v = \nabla_a v = \partial_a v$$

$$\tilde{\nabla}_{ab} v = \nabla_{ab} v - \Omega^{-1} (\nabla_a v \nabla_b \Omega + \nabla_b v \nabla_a \Omega - \gamma^{cd} \gamma_{ab} \nabla_c v \nabla_d \Omega)$$

while

$$\nabla_a (\Omega^{(n-1)/2} v) = \Omega^{(n-1)/2} \left(\nabla_a v + \frac{(n-1)}{2} \Omega^{-1} \nabla_a \Omega v\right)$$

and

$$\nabla_{ab} (\Omega^{(n-1)/2} v) = \Omega^{(n-1)/2} \left(\nabla_{ab} v + \frac{(n-1)(n-3)}{2} \Omega^{-2} \nabla_a \Omega \nabla_b \Omega v\right.$$

$$\left. + \frac{(n-1)}{2} \Omega^{-1} \nabla^2_{ab} \Omega v + \frac{(n-1)}{2} \Omega^{-1} \nabla_a \Omega \nabla_b v\right)$$

For the particular function $\Omega = \cos T + \cos \alpha$ on $S^n \times \mathbf{R}$ and the diffeomorphism ϕ we have the property that $\Omega^{-1} \frac{\partial X^a}{\partial x^\alpha} \nabla^2_{a\ b} \Omega$ are smooth functions on $\mathfrak{U}$, which extend to smooth functions on $S^n \times \mathbf{R}$. Indeed $\frac{\partial X^a}{\partial t}$ and $\frac{\partial X^a}{\partial r}$ extend to smooth functions on $S^n \times \mathbf{R}$, as well as $\frac{\partial X^a}{\partial \theta^i} = 0$, moreover

$$\frac{\partial X^a}{\partial t} \partial_a = - 2 \frac{\partial U}{\partial t} \sin U \cos V - 2 \frac{\partial V}{\partial t} \sin V \cos U$$

with

$$\frac{\partial U}{\partial t} = \cos^2 U, \quad \frac{\partial V}{\partial t} = \cos^2 V$$

thus

$$\frac{\partial X^a}{\partial t} \nabla_a \Omega = - \Omega(\sin 2 U + \sin 2 V) = - 2 \Omega \sin T \cos \alpha$$

also

$$\frac{\partial X^a}{\partial r} \nabla_a \Omega = - 2 \Omega \cos T \sin \alpha$$

while

$$\frac{\partial X^a}{\partial \theta^i} \partial_\rho \Omega = 0.$$

Analogously we find, with $x^{\alpha'} = (t,r,\theta^i)$, $u = t + r$, $v = t - r$:

$$\frac{\partial X^a}{\partial x^{\alpha'}} \frac{\partial X^b}{\partial x^{\beta'}} \tilde{\nabla}_{ab}\Omega = - 4 \Omega(\cos^2 U + 2 \sin^2 U) \partial_{\alpha'} u \partial_{\beta'} u$$
$$-4 (\cos^2 V + 2 \sin^2 V) \partial_{\alpha'} v \partial_{\beta'} v$$

from which the conclusion follows easily when x^α are the Cartesian coordinates of $\mathbb{R}^{n+1}$.

We conclude that (3.6) may be written

$$\Omega^{(n-1)/2} h^{\alpha\beta}\left(\Omega^{(n-1)/2} v, \Omega^{(n-1)/2}\left(\frac{\partial X^a}{\partial x^\alpha} \nabla_a v + a_\alpha v\right)\right) \tag{3.9}$$
$$\left(\frac{\partial X^a}{\partial x^\alpha}\frac{\partial X^b}{\partial x^\beta} \nabla_{ab} v + a_\alpha{}^a{}_\beta \nabla_\alpha v + a_{\alpha\beta} v\right) + f\left(\Omega^{(n-1)/2} v, \Omega^{(n-1)/2}\left(\frac{\partial X^a}{\partial x^a} \nabla_a v + b_\alpha v\right)\right)$$

where a_α, $a_\alpha{}^c{}_\beta$ and b_α are smooth functions on $\mathcal{U}$ which extend as smooth functions to $S^n \times \mathbb{R}$.

We now make the following hypothesis:

HYPOTHESIS: $h^{\alpha\beta}$ vanishes for the zero sections, f and its first derivative vanish for the zero section:

$$h^{\alpha\beta}(0,0) = 0$$

$$f(0,0) = 0 \text{ and } f'(0,0) = 0$$

(hence F and its first derivative vanish for these zero sections). Then, since these functions are smooth, we have, in a neighbourhood of the zero sections:

$$h^{\alpha\beta}(\lambda y, \lambda z) = \lambda k^{\alpha\beta}(\lambda, y, z)$$
$$f(\lambda y, \lambda z) = \lambda^2 K(\lambda, y, z)$$

with $k^{\alpha\beta}$ and K smooth functions of their arguments.

We can therefore write (3.8) as:

$$\Omega^{n-1}\{\ell^{ab}(X^c, v, \nabla_c v)\nabla^2_{a\,b} v + m(X^c, v, \nabla_a v)\} \tag{3.10}$$

where $\ell^{ab} = \frac{\partial X^a}{\partial x^\alpha}\frac{\partial X^b}{\partial x^\beta} h^{\alpha\beta}$ and m are smooth in all their arguments, and vanish for $v = \nabla v = 0$.

We return now to the equation (3.6) and we see that it extends to a smooth equation on $S^n \times \mathbf{R}$ if n is odd and $n-1 > \frac{3+n}{2}$, i.e. $n \geq 5$. Setting $n-5 = p$ this equation can indeed be written:

$$\Box_\gamma v + \frac{(n-1)^2}{4} v + \Omega^p(\ell^{ab}(X^c,v,\nabla_c v)\nabla_a \partial_b v + m(X^c,v,\nabla_c v)) \tag{3.11}$$

It is a quasi linear second order equation on $S^n \times \mathbf{R}$, with smooth coefficients, if n is odd and $p \geq 0$, which is hyperbolic in a neighbourhood of the zero section $v = 0$.

The equation (3.10) admits a solution on $S^n \times [0,\pi]$ for small Cauchy data $(v|_{T=0}, \nabla v|_{T=0})$ in the space $H_s(S^n) \times H_{s-1}(S^n)$, $s > \frac{n}{2} + 2$ (general case).

The original equation (3.1) admits therefore a global solution on $\mathbf{R}^{n+1}$ for small data whose weighted image is in this space - which means for the original data fall off properties at infinity: a sufficient but not necessary condition is that these data belong to the weighted Sobolev space $H_{k,k-1}(\mathbf{R}^n) \times H_{k-1,k}$, $k = \frac{n+1}{2} + 2$ (for details see [2] and [6]).

Case n = 3

In the case n = 3 Christodoulou proves the global existence for solutions of (3.1) on $\mathbf{R}^{n+1}$ when the following hypothesis(1) is made in addition to the previous ones.

Null condition: The mapping defined by F is said to satisfy the null condition if the quadratic form given by its second derivative at the zero section is such that:

(1) Already suggested by Klainerman in [7], see also Hörmander [8].

$$F''(0,0,0)(u, y, z) = 0$$

whenever y and z are sections of $T^* \mathbb{R}^{n+1}$ and $\otimes^2 T^* \mathbb{R}^{n+1}$ respectively such that

$$y_\alpha = a\, \ell_\alpha \text{ and } z_{\alpha\beta} = b\, \ell_\alpha\, \ell_\beta$$

where ℓ_α is a null vector of the Minkowski metric, i.e. such that

$$\eta^{\alpha\beta}\, \ell_\alpha\, \ell_\beta = 0.$$

The proof, in the conformal method, is straightforward, though lengthy. It proceeds as follows.

For a mapping F of the quasilinear form (3.7) the null condition reads

$$h'^{\alpha\beta}(0,0)(u,y) z_{\alpha\beta} = 0$$

which implies that $h'^{\alpha\beta}_u(0,0)\, z_{\alpha\beta}$ vanishes for each 2-tensor $z_{\alpha\beta} = \ell_\alpha\, \ell_\beta$, with ℓ_α a null covector of the Minkowski metric, therefore there exists a scalar μ such that

$$h'^{\alpha\beta}_u(0,0) = \mu\, \eta^{\alpha\beta}$$

and

$$h'^{\alpha\beta}_{y\lambda}(0,0) y_\lambda\, z_{\alpha\beta} = 0,\ y_\lambda = \ell_\lambda,\ z_{\alpha\beta} = \ell_\alpha \ell_\beta$$

This implies that the symmetrized 3-tensor

$$s^{\alpha\beta} = \frac{1}{3}\,(h'^{\alpha\beta}_{y\lambda}(0,0) + h'^{\beta\lambda}_{y_\alpha}(0,0) + h'^{\lambda\alpha}_{y_\beta}(0,0))$$

is of the form

$$s^\alpha\, \eta^{\beta\lambda} + s^\lambda\, \eta^{\beta\alpha} + s^\beta\, \eta^{\alpha\lambda}$$

with s^α an arbitrary vector.

And

$$f''(0,0)(u,y) = 0$$

which implies

$$f''_{uu}(0,0) = 0$$

$$f''_{uy_\lambda}(0,0)y_\lambda = 0 \ , \ \forall y_\lambda = \ell_\lambda, \text{ i.e. } f''_{uy_\lambda}(0,0) = 0$$

and

$$f''_{y_\lambda y_\mu}(0,0)\, y_\lambda\, y_\mu = 0, \ \ \forall y_\lambda = \ell_\lambda$$

therefore

$$f''_{y_\lambda y_\mu}(0,0) = \nu\, \eta^{\lambda\mu} \ .$$

The mapping

$$\chi^{\alpha\beta}(u,y) \equiv h^{\alpha\beta}(u,y) - h'^{\alpha\beta}(0,0)(u,y)$$

vanishes for $u = y = 0$ as well as its first derivative, while

$$\varphi(u,y) \equiv f(u,y) - \frac{1}{2} f''(0,0)(u,y)$$

vanishes together with its first and second derivative. Therefore there exist smooth functions $\check{\chi}^{\alpha\beta}$ and $\check{\varphi}$ of λ, u, y such that

$$\chi^{\alpha\beta}(\lambda u,\lambda y) = \lambda^2\, \check{\chi}^{\alpha\beta}(\lambda,u,y)$$

$$\varphi(u,y) = \lambda^3\, \check{\varphi}(u,y)$$

if we replace in (3.9) the functions $h^{\alpha\beta}$ and f respectively by $\chi^{\alpha\beta}$ and φ we obtain an expression which reads

$$\Omega^{3(n-1)/2} \left\{ \check{\chi}^{\alpha\beta}\left(\Omega^{(n-1)/2},\ v,\ \frac{\partial X^a}{\partial x^\alpha}\nabla_a v + a_\alpha v\right) \right.$$
$$\left. + \check{\varphi}\left(\Omega^{(n-1)/2},\ v,\ \frac{\partial X^a}{\partial x^\alpha}\nabla_a v + b_\alpha v\right)\right\} . \qquad (3.12)$$

The product of (3.12) by $\Omega^{-(3+n)/2}$ is smooth (for n odd) if $n - 3 \geqq 0$.

We have now to prove that the expression (3.9) with $h^{\alpha\beta}(u,y)$ replaced by $h'^{\alpha\beta}(0,0)(u,y)$ and $f(u,y)$ by $f''(0,0)(u,y)$ has the same property, under the null condition. This expression reads:

$$\Omega^{n-1}\left\{\mu\,\eta^{\alpha\beta}v + h'^{\alpha\beta}_{y_\lambda}(0,0)\left(\frac{\partial X^c}{\partial x^\alpha}\nabla_c v + a_\alpha v\right)\right\}\left(\frac{\partial X^a}{\partial x}\frac{\partial X^b}{\partial x}\nabla^2_{a\,b}v + a_{\alpha\beta}v\right) \tag{3.13}$$

$$+\,\Omega^{n-1}\,f''_{y_\alpha y_\beta}(0,0)\left(\frac{\partial X^a}{\partial x^\alpha}\nabla_a v + b_\alpha v\right)\left(\frac{\partial X^b}{\partial x^\beta} + b_\alpha v\right)\Bigg\}$$

We have

$$\eta^{\alpha\beta}\,\frac{\partial X^a}{\partial x^\alpha}\frac{\partial X^b}{\partial x^\beta} = \tilde{\eta}^{ab} = \Omega^2\,\gamma^{ab} \tag{3.14}$$

Returning to the formula (3.9) we see that $a_{\alpha\,\beta}^{\;\;a}$ was of the form, up to numerical factors:

$$\frac{\partial X^a}{\partial x^\alpha}\frac{\partial X^b}{\partial x^\beta}\,\Omega^{-1}\,\nabla_b\,\Omega\,\nabla_a\,v \tag{3.15}$$

we had previously used the fact that $\Omega^{-1}\dfrac{\partial X^b}{\partial x^\beta}\nabla_b\,\Omega$ extends to a smooth function on $S^n \times \mathbf{R}$. We now use (3.14) to see that the product by Ω^{-1} of $\eta^{\alpha\beta}\,a_{\alpha\,\beta}^{\;\;a}$ is also such a smooth function.

In $a_{\alpha\beta}$ we have terms of the form

$$\frac{\partial X^a}{\partial x^\alpha}\,\frac{\partial X^b}{\partial x^\beta}\,\Omega^{-2}\,\nabla_a\,\Omega\,\nabla_b\,\Omega$$

and

$$\frac{\partial X^a}{\partial x^\alpha}\frac{\partial X^b}{\partial x^\beta}\,\Omega^{-1}\,\nabla^2_{a\,b}\,\Omega$$

whose contraction with $\eta^{\alpha\beta}$ gives

$$\gamma^{ab}\,\nabla_a\,\Omega\,\nabla_b\,\Omega \equiv \Omega\,(\cos\alpha - \cos T)$$

<u>REMARK:</u> $\nabla\Omega$ is a null vector when $\Omega = 0$, in particular on $\partial\mathfrak{U}$. We also have:

$$\Omega\,\gamma^{ab}\,\nabla^2_{a\,b}\,\Omega = \Omega(\cos\alpha - \cos T).$$

It is also easy to check that the product of Ω^{-1} by the term in $F''_{y_\alpha y_\beta}(0,0)$ is smooth on $S^n \times \mathbf{R}$. The same result is proved in detail in Christodoulou [6] for the most difficult term, involving $h'^{\alpha\beta}_{y_\lambda}(0,0)$, through the decomposition of this tensor into symmetrized and antisymmetrized parts.

4. EXTENSION TO ASYMPTOTICALLY FLAT MANIFOLDS

Let $\hat{g}$ be a hyperbolic metric on $\mathbb{R}^{n+1}$ which can be written

$$\hat{g} = f\, du\, dv - h\, r^2\, dS^2_{n-1}$$

with $r = (u-v)/2$, and f, h smooth functions of u, v, and admissible coordinates on S^{n-1}.

We consider again the diffeomorphism from $\mathbb{R}^{n+1}$ onto the open bounded set $\mathcal{U}$, $\alpha - \pi < T < \alpha + \pi$ of $\mathbb{R} \times S^n$

$$\Phi : U = \text{Arctg}\, u, \quad V = \text{Arctg}\, v$$

where U and V are linked to the usual coordinates T, α of $\mathbb{R} \times S^n$ by

$$U = \frac{1}{2}(T + \alpha), \qquad V = \frac{1}{2}(T - \alpha)$$

The image of $\hat{g}$ by Φ is the metric on U

$$\tilde{g} = \Phi^* \hat{g} = \Omega^{-2} g$$

with as before

$$\Omega = 2 \cos U \cos V = \cos T + \cos \alpha$$

while now

$$g = 4\, f \circ \Phi^{-1}\, dU\, dV - h \circ \Phi^{-1} \sin^2(U - V)\, dS^2_{n-1}$$

equivalently:

$$g = (f \circ \Phi^{-1})(dT^2 - d\alpha^2) - h \circ \Phi^{-1} \sin^2\alpha\, dS^2_{n-1}.$$

The metric $\hat{g}$ is said <u>strongly asymptotically flat (or simple)</u>(1) if the metric

(1) A weaker notion of an asymptotically simple metric is introduced through geometric notions, and the conformal mapping idea, in Penrose [3].

g extends to a smooth globally hyperbolic metric on $\mathbb{R} \times S^n$. It will be the case if the functions $f \circ \Phi^{-1}$ and $h \circ \Phi^{-1}$ extend to smooth strictly positive functions on $\mathbb{R} \times S^n$, therefore, in particular, if f-1 and h-1 are functions with compact support on $\mathbb{R}^{n+1}$ or, more generally, tend to zero together with enough derivatives of $\mathbb{R}^{n+1}$, in an appropriate sense. We consider on $\mathbb{R}^{n+1}$ the equation:

$$\square_{\hat{g}} v + F(\hat{v}, \hat{\nabla} v, \hat{\nabla}^2 \hat{v}) = 0 \tag{4.1}$$

with

$$F(\hat{v}, \hat{\nabla} v, \hat{\nabla}^2 v) = h^{\alpha\beta}(\hat{v}, \hat{\nabla}\hat{v}) \, \hat{\nabla}^2_{\alpha\beta} \hat{v} + f(\hat{v}, \hat{\nabla} v)$$

$$h^{\alpha\beta}(0,0) = 0$$

$$f(0,0) = 0, \quad f'(0,0) = 0.$$

By the diffeomorphism Φ and the identity (3.5) we transform it into a formula analogous to (3.6)

$$\square_g v - \frac{n-1}{4n} R(g)v + \Omega^{-2} \tilde{R}(\tilde{g})v + \Omega^{-(3+n)/2} F\Big(\Omega^{(n-1)/2} v, \frac{\partial X^b}{\partial x^\alpha} \tilde{\nabla}_b (\Omega^{(n-1)/2} v),$$
$$\frac{\partial X^a}{\partial x^\alpha} \frac{\partial X^b}{\partial x^\beta} \tilde{\nabla}^2_{a\,b} (\Omega^{(n-1)/2} v)\Big) = 0.$$

This equation extends to an equation on $S^n \times \mathbb{R}$, with smooth coefficients if n is odd, $n \geqq 5$ and $\Omega^{-2} \tilde{R}(\tilde{g})$ extends to a smooth function: it will be the case if $\Omega^{-2} \tilde{R}(\tilde{g})$ vanishes on $\partial\mathfrak{U}$ with as many of its derivatives as required for the smoothness.

NOTE: if the operator $\square_{\hat{g}}$ in (4.1) is replaced by the conformally invariant one $\square_{\hat{g}} - \frac{n-1}{4n} \hat{R}(\hat{g})$, this last condition is not necessary.

References

[1] D. Christodoulou, Comptes Rendus Ac. Sc. Paris 293 (1981), p. 39-42.

[2] Y. Choquet-Bruhat and D. Christodoulou, Ann. E.N.S. 1981 14 p.481-506.

[3] R. Penrose, in "Battelle rencontres", C. DeWitt and J. Wheeler eds., Benjamin 1967.

[4] Y. Choquet-Bruhat and I.E. Segal, J. Funct. Analysis, 53, n°2 (1983) p. 112-150.

[5] S. Klainerman, Proc. of I.C.M. Warsaw 1983.

[6] D. Christodoulou, Comm. in pure and applied math. 39 (1986) p. 267-282.

[7] S. Klainerman, Comm. in pure and applied math., 33, (1980) p. 43-101.

[8] L. Hörmander, Report 9, 1985 Institut Mittag-Leffler.

Yvonne Choquet-Bruhat
Institut de Mécanique,
Université Paris VI,
4 Place Jussieu
75252 Paris
France.

M. CICOGNANI

The propagation of Gevrey singularities for some hyperbolic operators with coefficients Hölder continuous with respect to time

We describe here a result on the propagation of Gevrey singularities for the solution of the Cauchy problem

$$\begin{cases} P(t,x,D_t,D_x)u(t,x) = 0 & (t,x) \in [0,T] \times R^n \\ D_t^j u(0,x) = g_j(x) & x \in R^n, \quad j = 0,1, \end{cases} \qquad \text{(C.P.)}$$

$T > 0$, for a strictly hyperbolic operator

$$(*) \quad P(t,x,D_t,D_x) = D_t^2-\alpha^2(t) \sum_{j,k=1}^{n} a_{j,k}(x)D_{x_j}D_{x_k}+b(t,x)D_t + \sum_{\ell=1}^{n} c_\ell(t,x)D_{x_\ell}+d(t,x)$$

$$D_t = -i\partial/\partial t,\ x = (x_1,\dots,x_n),\ D_x = (D_{x_1},\dots,D_{x_n}),\ D_{x_h} = -i\partial/\partial x_h,\ h=1,\dots,n,$$

assuming that $\alpha(t)$ is a Hölder continuous function of exponent χ, $0 < \chi < 1$, on $[0,T]$, $a_{j,k}(x)$ and $b(t,x)$, $c_\ell(t,x)$, $d(t,x)$ for any fixed $t \in [0,T]$, are Gevrey functions of type $\sigma \in]1,1/(1-\chi)]$ on R^n, $j,k,\ell = 1,\dots,n$.

Hyperbolic operators with coefficients which are in a Gevrey class with respect to space variable and Hölder continuous with respect to time, have been considered in [1], [2], [3], [4], [5], [6] and the well-posedness of the related Cauchy problem in some Gevrey spaces of functions of ultra-distributions has been proved. On the other hand the propagation of Gevrey singularities of the solution has not been studied till now.

We reduce the problem (C.P.) to the problem

$$(\text{C.P.})_S \begin{cases} L(t,x,D_t,D_x)U(t,x) = 0 & (t,x) \in [0,T] \times R^n \\ U(0,x) = G(x) & x \in R^n \end{cases}$$

for a strictly hyperbolic system

$$(*)_S \quad L(t,x,D_t,D_x) = D_t-\alpha(t)\begin{bmatrix} \mu(x,D_x) & 0 \\ 0 & -\mu(x,D_x) \end{bmatrix} + (b_{j,k}(t,x,D_x))_{j,k=1,2}$$

where $\mu(x,D_x)$ and $b_{j,k}(t,x,D_x)$ are pseudo-differential operators of order 1 and $1-\chi$ respectively of the type studied in [7] and [8], then we construct a parametrix for the problem $(C.P.)_S$ represented as a matrix of Fourier integral operators. From the particular form of the characteristic roots of L in $(*)_S$ we obtain a commutation law for products of phase functions (see Section 3), and determine the amplitudes as asymptotic sums in the classes of symbols of infinite order defined in [9], [10] and [11] by solving transport equations. By means of this parametrix we obtain results on the propagation of Gevrey singularities for the solutions of problems (C.P.) and $(C.P.)_S$. We shall apply similar techniques to operators of a more general form than (*) in a future work.

Section 1 contains the statement of our main result (Theorem 1.3); in Section 2 we recall the results of [9], [10] on Fourier integral operators with amplitude of infinite order we shall use here. Section 3 is devoted to products of phase functions and in Section 4 we construct a parametrix for the problem $(C.P.)_S$ after we have reduced problem (C.P.) to the system form. In Section 5 we study the propagation of the Gevrey wave front set of the solutions of (C.P.) and $(C.P.)_S$ by using the parametrix we have constructed in Section 4.

The author wishes to thank Professor L. Cattabriga for his constant help and Professor Y. Morimoto for many useful discussions in Bologna and here in Pisa.

0. MAIN NOTATIONS

For $x = (x_1,\ldots,x_n) \in R^n$ we set $D_x = (D_{x_1},\ldots,D_{x_n})$, $D_{x_j} = -i\partial/\partial x_j$, $j=1,\ldots,n$ and for $\alpha = (\alpha_1,\ldots,\alpha_n) \in Z^n_+$, Z_+ the set of all non negative integers, let $D^\alpha_x = D^{\alpha_1}_{x_1} \ldots D^{\alpha_n}_{x_n}$, $|\alpha| = \sum_{j=1}^n \alpha_j$, $\alpha! = \alpha_1! \ldots \alpha_n!$. For $(x,\xi) \in R^n \times R^n$ we shall also write $\langle x,\xi\rangle = \sum_{j=1}^n x_j\xi_j$ and $\langle\xi\rangle = (1 + |\xi|^2)^{1/2}$.

Let $\Omega \subset R^n$ be an open set. For $\sigma > 1$, $A > 0$ we denote by $\gamma^{(\sigma)}_A(\Omega)$ the Banach space of all complex valued functions $u \in C^\infty(\Omega)$ such that:

$$\|u\|_{\Omega,A} = \sup_{\substack{x\in\Omega \\ \alpha\in Z_+^n}} A^{-|\alpha|} \alpha!^{-\sigma}|D_x^\alpha u(x)| < +\infty \tag{0.1}$$

and set

$$\gamma^{(\sigma)}(\Omega) = \lim_{\substack{\longrightarrow \\ A\to+\infty}} \gamma_A^{(\sigma)}(\Omega), \quad G^{(\sigma)}(\Omega) = \lim_{\substack{\longleftarrow \\ \Omega'\subset\subset\Omega}} \gamma^{(\sigma)}(\Omega')$$

$$G_0^{(\sigma)}(\Omega) = \lim_{\substack{\longrightarrow \\ \Omega'\subset\subset\Omega}} \lim_{\substack{\longrightarrow \\ A\to+\infty}} \gamma_A^{(\sigma)}(\Omega') \cap C_0^\infty(\Omega')$$

where Ω' are relatively compact open subsets of Ω.

The dual spaces of $G^{(\sigma)}(\Omega)$ and $G_0^{(\sigma)}(\Omega)$, called spaces of ultradistribution of Gevrey type σ, will be denoted by $G^{(\sigma)'}(\Omega)$ and $G_0^{(\sigma)'}(\Omega)$ respectively. As is well known, the former can be identified with the subspace of ultra-distributions of $G_0^{(\sigma)'}(\Omega)$ with compact support. For $u \in G^{(\sigma)'}(\Omega)$ the Fourier transform $\tilde{u}$ is defined by $\tilde{u}(\xi) = u(e^{-i\langle\cdot,\xi\rangle})$. We shall also denote by $WF_{(\sigma)}(u)$ the σ-wave front set of $u \in G_0^{(\sigma')}(\Omega)$ defined as the complement in $\Omega \times R^n\setminus\{0\}$ of the set of all (x_0,ξ_0) such that there exist a conic neighbourhood Γ of ξ_0 in $R^n\setminus\{0\}$ and a function $\phi \in G_0^{(\sigma)}(\Omega)$ with $\phi(x_0) \neq 0$ such that

$$|(\tilde{\phi u})(\xi)| \leq C \exp(-h\langle\xi\rangle^{1/\sigma}), \quad \xi \in \Gamma \tag{0.2}$$

for some positive constants C and h.

For a given X subset of $T^*(R^n)\setminus\{0\}$, X^{con} will denote the conic hull of X.

If V is a topological vector space, $A \subset R^k$, $m \in Z_+$, $M^m(A,V)$ shall denote the set of all functions defined in A with values in V which are continuous and bounded together with their derivatives up to order m. We shall also write $M^\infty(A,V) = M(A,V) = \bigcap_{m\geq 0} M^m(A,V)$ and sometimes $M_z^m(A,V)$ and $M_z(A,V)$ respectively to underline the independent variable z.

1. THE MAIN RESULT

As we mentioned in the introduction, we consider here a second order differential operator of the form

$$P(t,x,D_t,D_x) = D_t^2 - \alpha^2(t) \sum_{j,k=1}^{n} a_{j,k}(x) D_{x_j} D_{x_k} + b(t,x) D_t + \\ + \sum_{\ell=1}^{n} c_\ell(t,x) D_{x_\ell} + d(t,x). \tag{1.1}$$

We assume the strict hyperbolicity of P in the following sense:

$$\alpha(t) \geqq \alpha_0 > 0, \qquad \sum_{j,k=1}^{n} a_{j,k}(x)\xi_j\xi_k \geqq a_0|\xi|^2, \quad a_0 > 0, \tag{1.2}$$

for every $(t,x,\xi) \in [0,T] \times T^*(R^n)$. Concerning the coefficients of P we assume

$$\alpha(t) \in C^{0,\chi}([0,T]), \ 0 < \chi < 1, \quad a_{j,k}(x) \in \gamma^{(\sigma)}(R^n), \ \sigma > 1 \tag{1.3}$$

$$b(t,x),\ c_\ell(t,x),\ d(t,x) \in M^0([0,T],\ \gamma^{(\sigma)}(R^n)) \tag{1.4}$$

where the exponents χ and σ satisfy

$$1 < \sigma < 1/(1-\chi). \tag{1.5}$$

Here and in the following we shall denote by $\lambda_1(t,x,\xi) = \alpha(t)\mu(x,\xi) = \alpha(t)(\sum_{j,k=1}^{n} a_{j,k}(x)\xi_j\xi_k)^{1/2}$ and $\lambda_2(t,x,\xi) = -\alpha(t)\mu(x,\xi) = -\alpha(t)(\sum_{j,k=1}^{n} a_{j,k}(x)\xi_j\xi_k)^{1/2}$ the characteristic roots of P. The strict hyperbolicity of P yields

$$\lambda_1(t,x,\xi) - \lambda_2(t,x,\xi) \geqq \lambda_0|\xi|, \quad \lambda_0 > 0 \tag{1.6}$$

for every $(t,x,\xi) \in [0,T] \times T^*(R^n)$.

DEFINITION 1.1: Let $j \in \{1,2\}$. We say that a curve $\{t, q_j(t,s;y,\eta), p_j(t,s;y,\eta)\} \subset [0,T] \times T^*(R^n)$ is the bicharacteristic curve with respect to λ_j through (s,y,η) if (q_j,p_j) satisfies the equation

$$\begin{cases} dq_j/dt = -\nabla_\xi \lambda_j(t,q_j,p_j), \quad dp_j/dt = \nabla_x \lambda_j(t,q_j,p_j) \\ (q_j,p_j)|_{t=s} = (y,\eta) \end{cases} \tag{1.7}$$

We denote by $C_j(t,s)$ the transformation

$$T^*(R^n)\setminus\{0\} \ni (y,\eta) \to C_j(t,s)(y,\eta) = (q_j(t,s;y,\eta), p_j(t,s;y,\eta)) \in T^*(R^n)\setminus\{0\} \tag{1.8}$$

<u>DEFINITION 1.2</u>: Let $t_o \in]0,T]$, $\nu \geq 1$, and $t_j \in]0,t_o]$, $j = 1,\dots,\nu$ be such that $0 = t_{\nu+1} < t_\nu < t_{\nu-1} < \dots < t_1 < t_o$. A curve $\{(t,q(t),p(t))\} \subset [0,t_o] \times T^*(R^n)$ is called a trajectory of step ν issuing from (y,η) with jointing points t_j, $j = 1,\dots,\nu$, if $(q,p)|_{t=0} = (y,\eta)$ and there exists a $(\nu+1)$-repeated permutation $(h_o,\dots,h_\nu)$ with $h_k \in \{1,2\}$, $h_k \neq h_{k+1}$, $k = 0,\dots,\nu-1$, such that $\{(t,q(t),p(t))\}$ is the bicharacteristic curve with respect to λ_{h_k} through $(t_{k+1},q(t_{k+1}), p(t_{k+1}))$ when $t \in [t_{k+1},t_k]$, $k = 0,\dots,\nu$.

A point

$$C_{h_o}(t_o,t_1)C_{h_1}(t_1,t_2)\dots C_{h_\nu}(t_\nu,0)(y,\eta)$$

is called an end point at $t = t_o$ of the trajectory. For a set $F \subset [0,T]$, $F \cap]0,t_o[\neq \emptyset$, and V a conic set in $T^*(R^n)\setminus\{0\}$, $\Gamma_\nu(t_o,F,V)$ will denote the conic hull of the set of end points at $t = t_o$ of all trajectories of step ν issuing from a $(y,\eta) \in V$ for large $|\eta|$, with jointing points $t_j \in F$, $j = 1,\dots,\nu$. If $F \cap]0,t_o[= \emptyset$ we put $\Gamma_\nu(t_o,F,V) = \emptyset$, $\nu \geq 1$.

A bicharacteristic curve through $(0,y,\eta)$ will also be called a trajectory of step 0 and, to have uniform notation, we denote by $\Gamma_o(t_o,F,V)$ the set $\{C_j(t_o,0)(y,\eta);(y,\eta) \in V,\ j \in \{1,2\},\ |\eta| \text{ large}\}^{con}$ for every $F \subset [0,T]$ and conic $V \subset T^*(R^n)\setminus\{0\}$, even if this set is actually independent of F.

Let us consider the Cauchy problem

$$\begin{cases} P(t,x,D_t,D_x)u(t,x) = 0 & (t,x) \in [0,T] \times R^n \\ D_t^j u(0,x) = g_j(x) & x \in R^n,\ j = 0,1, \end{cases} \tag{1.9}$$

for the operator given by (1.1), and denote by F the complement in $[0,T]$ of the largest open set I in $[0,T]$ such that $\alpha(t) \in G^{(\sigma)}(I)$ and

$b(t,x)$, $c_\ell(t,x)$, $d(t,x) \in \gamma^{(\sigma)}(I' \times R^n)$ for every relatively compact open $I' \subset I$. We have

THEOREM 1.3: Assume for P conditions (1.1)-(1.5). Then for a solution $u(t,x)$ of the problem (1.9) with initial data $g_0, g_1 \in G^{(\sigma')}(R^n)$ the following estimate holds at a fixed $t_0 \in]0,T]$:

$$\bigcup_{j=0}^{1} WF_{(\sigma)}(D_t^j j(t_0,\cdot)) \subset \text{the closure of } \bigcup_{\nu=0}^{\infty} \Gamma_\nu(t_0, F, \bigcup_{j=0}^{1} WF_{(\sigma)}(g_j)) \tag{1.10}$$

Moreover there exists $F_0 \subset]0,T]$ depending only on F and the function $\alpha(t)$ such that

$$\bigcup_{\nu=1}^{\infty} \Gamma_\nu(t_0, F, \bigcup_{j=0}^{1} WF_{(\sigma)}(g_j)) = \Gamma_1(t_0, F_0, \bigcup_{j=0}^{1} WF_{(\sigma)}(g_j)). \tag{1.11}$$

We shall describe the set F_0 in Section 3.

REMARK 1.4: By the results proved in [1], [3] and [4] condition (1.5) is sufficient for the well-posedness of problem (1.9) in Gevrey classes of functions and ultradistributions of exponent σ. On the other hand from Theorem 10 in [1] for given σ and χ, $\sigma > 1/(1-\chi) > 1$ one can find $\alpha(t) \in C^{0,\chi}([0,T])$, $\alpha(t) \geq \alpha_0 > 0$, and initial data $g_0, g_1 \in G^{(\sigma)}(R)$ such that problem (1.9) for $P(t,x,D_t,D_x) = D_t^2 - \alpha^2(t)D_x^2$, (here $n = 1$), is not well-posed in $G^{(\sigma)}(R)$.

REMARK 1.5: If the coefficients of P are in $\gamma^{(\sigma)}([0,T] \times R^n)$ then $F = \emptyset$ and (1.10) reduces to a particular case of the estimate obtained in [12] and [13] for the solution $u(t,x)$ of a Cauchy problem for a differential operator with coefficients in a Gevrey class with respect to (t,x) and characteristic roots of constant multiplicity. If $F \neq \emptyset$ one can see by means of simple examples that points of $\Gamma_\nu(t_0, F, \bigcup_{j=0}^{1} WF_{(\sigma)}(g_j))$ for $\nu \geq 1$ really appear in $WF_{(\sigma)}(u(t_0,\cdot))$, so strict hyperbolicity is not sufficient, in this case, to have the propagation of singularities of the solution along bicharacteristic curves through $(0,y,\eta)$, $(y,\eta) \in \bigcup_{j=0}^{1} WF_{(\sigma)}(g_j)$ only.

REMARK 1.6: Estimate (1.10) is similar to that one proved in [14] for a hyperbolic Cauchy problem with the coefficients of the equation in a Gevrey class with respect to (t,x) and characteristic roots of variable multiplicity. Here the condition $t_j \in F$ for the jointing points replaces the so called ε-admissibility of the trajectories that is $|\lambda_{h_j}(t_j,q(t_j),p(t_j))-\lambda_{h_{j-1}}(t_j,q(t_j), p(t_j))| \leq \varepsilon\, |\eta|$ for small positive ε. (See also [15] for the case of C^∞-coefficients).

2. SYMBOLS OF INFINITE ORDER AND FOURIER INTEGRAL OPERATORS IN GEVREY CLASSES

The following classes of infinite order symbols of Gevrey type have been introduced in [9], [10] and [11].

DEFINITION 2.1: Let $\sigma > 1$, $\mu \in [1,\sigma]$, $A > 0$, $B_0 > 0$, $B \geq 0$ be real constants and let us set $T^n(C) = \{(x,\xi)\ T^*(R^n);\ \ |\xi| \geq C\}$. We denote by $S^{\infty,\mu}_{G^{(\sigma)}}(A,B_0,B)$ the space of all complex valued functions $a(x,\xi) \in C^\infty(R^{2n}) \cap M°(Y;G^{(\sigma)}(R^n_x))$, for every relatively compact open set $Y \subset R^n_\xi$, such that for every $\varepsilon > 0$

$$\|a\|_\varepsilon^{A,B_0,B} = \sup_{\alpha,\beta\in Z^n_+} \ \sup_{(x,\xi)\in T^n(B|\alpha|^\sigma+B_0)} A^{-|\alpha|-|\beta|}\alpha!^{-\mu}\beta!^{-\sigma}\langle\xi\rangle^{|\alpha|} \tag{2.1}$$

$$\exp(-\varepsilon\langle\xi\rangle^{1/\sigma})\cdot|D^\alpha_\xi D^\beta_x a(x,\xi)| < +\infty$$

and set

$$S^{\infty,\mu}_{G^{(\sigma)}} = \varinjlim_{A,B_0,B\to+\infty} S^{\infty,\mu}_{G^{(\sigma)}}(A,B_0,B) \tag{2.2}$$

$$\tilde{S}^{\infty,\mu}_{G^{(\sigma)}} = \varinjlim_{A,B_0\to+\infty} S^{\infty,\mu}_{G^{(\sigma)}}(A,B_0,0).$$

The asymptotic sums in $S^{\infty,\mu}_{G^{(\sigma)}}$ are defined by:

DEFINITION 2.2: We say that $a(x,\xi) \in S^{\infty,\mu}_{G^{(\sigma)}}$ has an asymptotic expansion $a(x,\xi) \sim \sum_{j\geq0} a_j(x,\xi)$ if there exist constants $A > 0$, $B \geq 0$, $B_0 > 0$ such that for every $\varepsilon > 0$ the sequence $\{a_j(x,\xi)\}_{j\geq0} \subset S^{\infty,\mu}_{G^{(\sigma)}}$ satisfies

$$\sup_{j\in Z_+}\ \sup_{\alpha,\beta\in Z_+^n}\ \sup_{(x,\xi)\in T^n(B|\alpha|+j)^\sigma+B_0)} A^{-|\alpha|-|\beta|-j}\alpha!^{-\mu}(j!\beta!)^{-\sigma}$$

$$\langle\xi\rangle^{|\alpha|+j}\exp(-\varepsilon\langle\xi\rangle^{1/\sigma})|D_\xi^\alpha D_x^\beta\ a_j(x,\xi)| < +\infty \tag{2.4}$$

and

$$\sup_{s\in Z_+}\ \sup_{\alpha,\beta\in Z_+^n}\ \sup_{(x,\xi)\in T^n(B|\alpha|+j)^\sigma+B_0)} A^{-|\alpha|-|\beta|-s}\alpha!^{-\mu}(s!\beta!)^{-\sigma}\langle\xi\rangle^{|\alpha|+s}$$

$$\exp(-\varepsilon\langle\xi\rangle^{1/\sigma})\ |D_\xi^\alpha D_x^\beta\ \sum_{j<s}\ (a(x,\xi)-a_j(x,\xi))| <+\infty.$$

holds.

THEOREM 2.3 ([9]): For every sequence $\{a_j(x,\xi)\}_{j\geq 0}\subset S^{\infty,\mu}_{G^{(\sigma)}}$ satisfying (2.4) there exists $a(x,\xi)\in S^{\infty,\mu}_{G^{(\sigma)}}$ such that $a(x,\xi)\sim\sum_{j\geq 0} a_j(x,\xi)$.

We shall also need symbols of finite order ($\subset S^{\infty,\mu}_{G^{(\sigma)}}$) of the type studied in [7] and [8].

DEFINITION 2.4: For $\sigma>1$, $\mu\in[1,\sigma]$, $A>0$, $B\geq 0$, $B_0>0$, $m\in R$ we denote by $S^{m,\mu}_{G^{(\sigma)}}(A,B_0,B)$ the Banach space of all complex valued functions $a(x,\xi)\in C^\infty(R^{2n})\cap M^\circ(Y;G^{(\sigma)}(R^n_x))$, for every relatively compact open $Y\subset R^n_\xi$, with the norm

$$\|a\|_m^{A,B_0,B} = \sup_{\alpha,\beta\in Z_+^n}\ \sup_{(x,\xi)\in T^n(B|\alpha|^\sigma+B_0)} A^{-|\alpha|-|\beta|}\alpha!^{-\sigma}\beta!^{-\sigma}\langle\xi\rangle^{-m+|\alpha|}$$

$$\cdot|D_\xi^\alpha D_x^\beta a(x,\xi)| < +\infty \tag{2.6}$$

and define

$$S^{m,\mu}_{G^{(\sigma)}} = \varinjlim_{A,B_0,B\to+\infty} S^{m,\mu}_{G^{(\sigma)}}(A,B_0,B) \tag{2.7}$$

$$\tilde{S}^{m,\mu}_{G^{(\sigma)}} = \varinjlim_{A,B_0\to+\infty} S^{m,\mu}_{G^{(\sigma)}}(A,B_0,0). \tag{2.8}$$

DEFINITION 2.5: Let A be an open set in R^k. We say that $a(z,x,\xi)$ belong to a class $G^{(\sigma)}(A;S^{\infty,\mu}_{G^{(\sigma)}})$ if $a(z,\cdot,\cdot) \in S^{\infty,\mu}_{G^{(\sigma)}}$ for every $z \in A$ and for every relatively compact open $A' \subset A$ there exist $A > 0$, $B_0 > 0$, $B \geq 0$ such that

$$\sup_{\gamma \in Z^k_+} \sup_{z \in A'} A^{-\gamma} \gamma!^{-\sigma} \|D^\gamma_z a(z,\cdot,\cdot)\|_\varepsilon^{A,B_0,B} < +\infty \tag{2.9}$$

for every $\varepsilon > 0$. In a similar way we define the classes $G^{(\sigma)}(A;\tilde{S}^{\infty,\mu}_{G^{(\sigma)}})$, $G^{(\sigma)}(A;S^{m,\mu}_{G^{(\sigma)}})$, $G^{(\sigma)}(A;\tilde{S}^{m,\mu}_{G^{(\sigma)}})$ replacing $\|D^\gamma_z a(z,\cdot,\cdot)\|_\varepsilon^{A,B_0,B}$ with $\|D^\gamma_z a(z,\cdot,\cdot)\|_\varepsilon^{A,B_0,0}$, $\|D^\gamma_z a(z,\cdot,\cdot)\|_m^{A,B,B_0}$, $\|D^\gamma_z a(z,\cdot,\cdot)\|_m^{A,B_0,0}$ respectively.

DEFINITION 2.6: Let $\phi(x,\xi) \in \tilde{S}^{1,\mu}_{G^{(\sigma)}}$ be real valued and $a(x,\xi) \in S^{\infty,\mu}_{G^{(\sigma)}}$. We define the Fourier integral operator $a_\phi(x,D_x)$ on $G^{(\sigma)}_0(R^n)$ by

$$a_\phi(x,D_x)u(x) = \int e^{i\phi(x,\xi)} a(x,\xi)\tilde{u}(\xi)đ\xi, \quad u \in G^{(\sigma)}_0(R^n), \quad đ\xi = (2\pi)^{-n}d\xi \tag{2.10}$$

$\phi(x,\xi)$ and $a(x,\xi)$ will be called the phase function and the amplitude of $a_\phi(x,D_x)$ respectively. If $\phi(x,\xi) = \langle x,\xi\rangle$ we shall write $a(x,D_x)$ instead of $a_\phi(x,D_x)$ and we shall call $a(x,D_x)$ a pseudo-differential operator with symbol $a(x,\xi)$.

THEOREM 2.7: ([10]): Let $\phi(x,\xi)$ and $a(x,\xi)$ be as in Definition 2.6. Then (2.10) defines a continuous linear map from $G^{(\sigma)}_0(R^n)$ to $G^{(\sigma)}(R^n)$ which extends to a continuous linear map from $G^{(\sigma)'}(R^n)$ to $G^{(\sigma)'}_0(R^n)$. If $\phi(x,\theta\xi) = \theta\phi(x,\xi)$ for $\theta > 0$ and large $|\xi|$, then for every $u \in G^{(\sigma)}(R^n)$ we have

$$WF_{(\sigma)}(a_\phi(x,D_x)u(x)) \subset \{T_\phi(y,\eta);(y,\eta) \in WF_{(\sigma)}(u),\ |\eta| \text{ large}\}^{con} \tag{2.11}$$

where $T_\phi : T^*(R^n) \to T^*(R^n)$, called the transformation generated by ϕ, is defined by

$$T_\phi(y,\eta) = (x,\xi) \text{ if and only if } y = \nabla_\xi \phi(x,\eta) \text{ and } \xi = \nabla_x\phi(x,\eta). \quad (2.12)$$

DEFINITION 2.8: A continuous linear map from $G_0^{(\sigma)}(R^n)$ to $G^{(\sigma)}(R^n)$ is said to be a σ-regularizing operator if it extends to a continuous linear map from $G^{(\sigma)'}$ to $G^{(\sigma)}$. The class of σ-regularizing operators will be denoted by $\mathcal{R}_{G^{(\sigma)}}$.

THEOREM 2.9: Let $a(x,\xi) \in S^{\infty,\mu}_{G^{(\sigma)}}$, $a(x,\xi) \sim 0$ according to Definition 2.2. Then $a_\phi(x,D_x) \in \mathcal{R}_{G^{(\sigma)}}$ for every phase function ϕ.

The following result on composition of operators of type (2.10) will be used later.

THEOREM 2.10: ([10]: Let $p^1(x,D_x)$ be a pseudo-differential operator with symbol $p^1(x,\xi) \in \tilde{S}^{\infty,1}_{G^{(\sigma)}}$ and $p^2_\phi(x,D_x)$ a Fourier integral operator with amplitude $p^2(x,\xi) \in S^{\infty,\mu}_{G^{(\sigma)}}$. Then

$$p^1(x,D_x) \circ p^2_\phi(x,D_x) = q_\phi(x,D_x) + R$$

where $R \in \mathcal{R}_{G^{(\sigma)}}$ and the amplitude $q(x,\xi) \in S^{\infty,\mu}_{G^{(\sigma)}}$ has an asymptotic expansion $q(x,\xi) \sim \sum_{j\geq 0} q_j(x,\xi)$ given by

$$q_j(x,\xi) = \sum_{|\alpha|=j} \alpha!^{-1}\, D_y[D_\xi^\alpha p^1(x,\tilde{\nabla}_x\phi(x,y,\xi))p^2(y,\xi)]_{y=x} \quad (2.13)$$

where $\tilde{\nabla}_x\phi(x,y,\xi) = \int_0^1 \nabla_x\phi(y+\theta(x-y),\theta)d\theta.$

3. PRODUCTS OF PHASE FUNCTIONS

Let $\lambda_j'(t,x,\xi) = (-1)^{j-1}\alpha(t)\mu'(x,\xi) \in M^0([0,T]; \tilde{S}^{1,1}_{G^{(\sigma)}}) \cap G^{(\sigma)}(I;\tilde{S}^{1,1}_{G^{(\sigma)}}), j = 1,2,$ be real valued symbols such that $\mu'(x,\xi) = \mu(x,\xi)$ for $|\xi| \geq B_0 > 0$ where, as we said in Section 1, $\lambda_j(t,x,\xi) = (-1)^{j-1}\alpha(t)\mu(x,\xi)$, $j = 1,2$, are the characteristic roots of P given by (1.1) and I denotes an open set in [0,T] such that $\alpha(t) \in G^{(\sigma)}(I)$. Consider the solutions $\phi_j(t,s;x,\xi)$ of the eikonal equations

$$\begin{cases} \partial_t\phi_j(t,s;x,\xi) = \lambda_j'(t,x,\nabla_x\phi_j(t,s;x,\xi)) \\ \phi_j(s,s;x,\xi) = \langle x,\xi\rangle \qquad j = 1,2 \end{cases} \tag{3.1}$$

then we have:

<u>THEOREM 3.1</u>: ([13], [14]): There exists T_0, $0 < T_0 \leqq T$ such that $\phi_j \in M^1([0,T_0]^2; \tilde{S}^{1,\sigma}_{G(\sigma)} \cap G^{(\sigma)}(I^2; \tilde{S}^{1,\sigma}_{G(\sigma)})$. Moreover $T_{\phi_j(t,s)}(y,\eta) = C_j(y,\eta)$ for large $|\eta|$ where $T_{\phi_j(t,s)}$ is the transformation generated by $\phi_j(t,s)$ (see (2.12)) and $C_j(t,s)$ is the transformation defined by (1.8).

<u>DEFINITION 3.2</u>: We define the product $\phi_k(t,t_1) \# \phi_\ell(t_1,s) = \Phi_{k,\ell}(t,t_1,s)$, $k, \ell \in \{1,2\}$, as the solution of the equation

$$\begin{cases} \partial_t\Phi_{k,\ell}(t,t_1,s;x,\xi) = \lambda_k'(t,x,\nabla_x\Phi_{k,\ell}(t,t_1,s;x,\xi)) \\ \Phi_{k,\ell}(t_1,t_1,s) = \phi_\ell(t_1,s). \end{cases} \tag{3.2}$$

Equations (3.1) and (3.2) can be solved by similar arguments. We obtain, with a smaller T_0 if necessary:

<u>THEOREM 3.3</u>: $\Phi_{k,\ell}(t,t_1,s) \in M^1([0,T_0]^3; \tilde{S}^{1,\sigma}_{G(\sigma)}) \cap G^{(\sigma)}(I^3, \tilde{S}^{1,\sigma}_{G(\sigma)})$ and the transformation $T_{\Phi_{k,\ell}(t,t_1,s)}$ generated by $\Phi_{k,\ell}(t,t_1,s)$ admits the decomposition

$$T_{\Phi_{k,}}(t,t_1,s) = T_{\phi_k(t,t_1)} \circ T_{\phi_\ell(t_1,s)} = C_k(t,t_1) \circ C_\ell(t_1,s). \tag{3.3}$$

For $(y,\eta) \in T^*(R^n)\setminus\{0\}$ and $0 < t_1 < t_0$, $(x,\xi) = T_{\Phi_{k,\ell}(t_0,t_1,0)}(y,\eta)$ is the end point at $t = t_0$ of a trajectory of step 1 issuing from (y,η) with t_1 as jointing point.

The use of η instead of ξ to denote the dual variable of x in $\Phi_{k,\ell}$ will make the following definition and proposition clearer:

<u>DEFINITION 3.4</u>: For $(x,\eta) \in T^*(R^n)$ set $y = y(t,t_1,s;x,\eta) = \nabla_\xi\Phi_{k,\ell}(t,t_1,s;x,\eta)$.

We call $(X_{k,\ell}, \Xi_{k,\ell}) = (X_{k,\ell}, \Xi_{k,\ell})(t,t_1,s;x,\eta) = C\,(t_1,s;y,\eta)$ the critical point of $\Phi_{k,\ell}(t,t_1,s)$.

PROPOSITION 3.5: The following statements hold:

(i) $\Phi_{k,\ell}(t,t,s) = \phi_\ell(t,s)$, $\Phi_{k,\ell}(t,s,s) = \phi_k(t,s)$.

(ii) $X_{k,\ell}, \Sigma_{k,\ell}\langle\eta\rangle^{-1} \in M^1([0,T_0]^3; \tilde{S}^{0,\sigma}_{G(\sigma)}) \cap G^{(\sigma)}(I^3; \tilde{S}^{0,\sigma}_{G(\sigma)})$ and

$|\Xi_{k,\ell}| \geq \delta\,|\eta|$, $\delta > 0$, for large $|\eta|$.

(iii) $(X_{k,\ell}, \Xi_{k,\ell})$ satisfies the equation

$$\begin{cases} X_{k,\ell} = \nabla_\xi \phi_k(t,t_1;x,\Xi_{k,\ell}) \\ \Xi_{k,\ell} = \nabla_x \phi_\ell(t_1,s;X_{k,\ell},\eta). \end{cases} \tag{3.4}$$

(iv) If $\xi = \nabla_x \Phi_{k,\ell}(t,t_1,s;x,\eta)$ then

$\xi = \nabla_x \phi_k(t,t_1;x,\Xi_{k,\ell})$, $y = \nabla_\xi \phi_\ell(t_1,s;X_{k,\ell},\eta)$.

(v) $\Phi_{k,\ell}(t,t_1,s;x,\eta) = \phi_k(t,t_1;x,\Xi_{k,\ell}) - \langle X_{k,\ell}\, \Xi_{k,\ell}\rangle + \phi_\ell(t_1,s;X_{k,\ell},\eta)$.

(vi) $\partial_{t_1}\Phi_{k,\ell}(t,t_1,s;x,\eta) = \lambda'_\ell(t_1,X_{k,\ell}\,\Xi_{k,\ell}) - \lambda'_k(t_1,X_{k,\ell}\,\Xi_{k,\ell})$.

REMARK 3.6: In [13], following [15], the product of phase functions is defined by $\phi_k \# \phi_\ell(x,\eta) = \phi_k(x,\Xi_{k,\ell}) - \langle X_{k,\ell}, \Xi_{k,\ell}\rangle + \phi_\ell(X_{k,\ell}\,\eta)$ after solving the equation of the critical point (3.4). Moreover the composition formula $a_{\phi_k}(x,D_x) \circ b_{\phi_\ell}(x,D_x) = c_{\phi_k\#\phi_\ell}(x,D_x)$ + regularizing operator is proved in the Gevrey classes. Here we shall not use composition of Fourier integral operators, so we have given an equivalent definition of products of phase functions that points out their geometrical meaning.

By using statements (i), (ii), (vi) in Proposition 3.5, property (1.6) and Theorems 2.3 and 2.7 we can prove the following result with iterated integrations by parts, noting that $(i\partial_{t_1}\Phi_{k,\ell})^{-1}\,\partial_{t_1} e^{i\Phi_{k,\ell}} = e^{i\Phi_{k,\ell}}$.

THEOREM 3.7: For $u \in G^{(\sigma)'}(R^n)$ consider

$$E(t)u = \int_0^t p_{\Phi_{k,\ell}(t,t_1,0)}(t,t_1;x,D_x)u(x)dt_1, \quad 0 \leq t \leq T_0$$

with the amplitude $p(t,t_1) \in M^\circ_{t_1}([0,T];S^{\infty,\mu}_{G^{(\sigma)}}) \cap G^{(\sigma)}_{t_1}(I \cap [0,t]; S^{\infty,\mu}_{G^{(\sigma)}})$.

Then

$$WF_{(\sigma)}(E(t)u) \subset \bigcup_{\nu=0}^{1} \Gamma_\nu(t,F,WF_{(\sigma)}(u)), \quad F = [0,T_0] \setminus I.$$

Now we prove a suitable commutation law for products of phase functions. Set

$$A(t) = \int_0^t \alpha(\tau)d\tau \tag{3.5}$$

and

$$\theta = \theta(t,t_1,s) = A^{-1}(A(t) - A(t_1) + A(s)). \tag{3.6}$$

We have

$$\partial_t \Phi_{1,2}(t,\theta,s) = (\partial_t\Phi_{1,2})(t,\theta,s) + (\partial_{t_1}\Phi_{1,2})(t,\theta,s)\partial\theta/\partial t =$$

$$= \alpha(t)\mu'(x,\nabla_x\Phi_{1,2}(t,\theta,s)) - 2\,\alpha(\theta)\,\mu'(X_{1,2}, \Xi_{1,2})\alpha(t)\alpha(\theta)^{-1} =$$

$$= -\alpha(t)\mu'(x,\nabla_x\Phi_{1,2}(t,\theta,s)) = \lambda_2'(t,x,\nabla_x\Phi_{1,2}(t,\theta,s)),$$

using Proposition 3.5 and the fact that μ' is constant on bicharacteristic curves. This and $\Phi_{1,2}(t,\theta,s)|_{t=t_1} = \Phi_{1,2}(t_1,s,s) = \phi_1(t_1,s)$ yield

$$\Phi_{2,1}(t,t_1,s) = \Phi_{1,2}(t,\theta,s) \tag{3.7}$$

and since $\theta(t,\theta,s) = t_1$ we have also

$$\Phi_{1,2}(t,t_1,s) = \Phi_{2,1}(t,\theta,s). \tag{3.8}$$

We can now describe the set F_o in (1.11). Let

$$(x,\xi) = C_{h_o}(t_o,t_1)\, C_{h_1}(t_1,t_2)\ldots C_{h_\nu}(t_\nu,0)(y,\eta)$$

be an end point at $t = t_o$ of a trajectory of step ν, $h_j \in \{1,2\}$, $h_j \neq h_{j+1}$, $j = 0,\ldots,\nu-1$, and put $\theta_\nu(t_1,t_2,\ldots,t_\nu) = A^{-1}(\sum_{j=0}^{\nu-1} (-1)^j A(t_{\nu-j}))$, $\theta^\nu(t_o,t_1,\ldots,t_\nu) = \theta(t_o,\theta_\nu,0) = A^{-1}(A(t_o)-A(\theta_\nu))$ (note that $A(0) = 0$). From Theorem 3.3 and commutation laws (3.7), (3.8) we get

$$(x,\xi) = C_{h_o}(t_o,\theta_\nu)\, C_{h_1}(\theta_\nu,0)(y,\eta) = C_{h_1}(t_o,\theta^\nu)C_{h_o}(\theta^\nu,0)(y,\eta),$$

therefore (1.11) holds with

$$F_o = F_o(t_o) = \{\theta_\nu(t_1,\ldots,t\ ),\ \theta^\nu(t_o,t_1,\ldots,t\ ); \tag{3.9}$$

$$0 < t_\nu < \ldots < t_1 < t_o,\quad t_j \in F\quad j = 1,\ldots,\nu,\ \nu \geq 1\}.$$

4. TRANSPORT EQUATIONS

In this section we reduce the Cauchy problem (1.9) to the system form

$$\begin{cases} L(t,x,D_t,D_x)U = 0 & (t,x) \in [0,T] \times R^n \\ u(0,x) = G(x) & x \in R^n \end{cases} \tag{4.1}$$

for

$$L(t,x,D_t,D_x) = D_t-\alpha(t)\begin{bmatrix} \mu'(x,D_x) & 0 \\ 0 & -\mu'(x,D_x) \end{bmatrix} + (b_{j,k}(t,x,D_x))_{j,k=1,2}$$

with $b_{j,k}(t,x,\xi) \in M^\circ([0,T];\ \tilde{S}^{1-\chi,\sigma}_{G(\sigma)}) \cap G^{(\sigma)}(I;\ \tilde{S}^{1-\chi,\sigma}_{G(\sigma)})$ and we construct a parametrix for the problem (4.1).

Let $\rho(\tau) = c^{-1}\exp(-\langle\tau\rangle)$, $c = \int_{-\infty}^{+\infty} \exp(-\langle\tau\rangle)d\tau$ and $\tilde{\alpha}(\tau) = \alpha(\tau)$ for

$\tau \in [0,T]$, $\tilde{\alpha}(\tau) = \alpha(0)$ for $\tau < 0$, $\tilde{\alpha}(\tau) = \alpha(T)$ for $\tau > T$. Setting

$$\alpha_r(t,\xi) = \int \tilde{\alpha}(\tau)\rho((t-\tau)\langle\xi\rangle)\langle\xi\rangle d\tau \tag{4.3}$$

we have $\alpha_r \in M^\circ([0,T];\ \tilde{S}^{0,1}_{G(\sigma)}) \cap M^1([0,1];\ \tilde{S}^{1-\chi,1}_{G(\sigma)}) \cap G^{(\sigma)}(I;\tilde{S}^{0,1}_{G(\sigma)})$ and $\alpha_r(t,\xi) \geq \alpha_0 > 0$. Moreover

$$\alpha(t)-\alpha_r(t,\xi) \in M^0([0,T];\ \tilde{S}^{-\chi,1}_{G(\sigma)}) \cap G^{(\sigma)}(I;\tilde{S}^{-\chi,1}_{G(\sigma)}), \tag{4.4}$$

therefore we can factorize $P(t,x,D_t,D_x)$ in (1.1) as follows:

$$P(t,x,D_t,D_x) = (D_t-\alpha_r(t,D_x)\mu'(x,D_x)+b(t,x))(D_t+\alpha_r(t,D_x)\mu'(x,D_x)) + \\ + R(t,x,D_x) \tag{4.5}$$

with $R(t,x,\xi) \in M^\circ([0,T];\ \tilde{S}^{2-\chi,1}_{G(\sigma)}) \cap G^{(\sigma)}(I;\tilde{S}^{2-\chi,1}_{G(\sigma)})$.

From factorization (4.5), following [13], we reduce problem (1.9) to a system form

$$\begin{cases} L_1(t,x,D_t,D_x)U = 0 & (t,x) \in [0,T] \times R^n \\ U(0,x) = G_1(x) & x \in R^n \end{cases} \tag{4.6}$$

for

$$L_1 = D_t-\alpha_r(t,D_x)\begin{bmatrix} \mu'(x,D_x) & 0 \\ 0 & -\mu'(x,D_x) \end{bmatrix} + (b'_{j,k}(t,x;D_x))_{j,k=1,2}, \tag{4.7}$$

$b'_{j,k} \in M^\circ([0,T_0];\ \tilde{S}^{1-\chi,1}_{G(\sigma)}) \cap G^{(\sigma)}(I;\tilde{S}^{1-\chi,1}_{G(\sigma)})$,

and finally to the problem (4.1) with L given by (4.2), by using the properties (4.4).

Next we want to construct a parametrix for the Cauchy problem (4.1); Concerning the diagonal part of L, from the results of [10], we note there exist $e^j(t,s) \in M^1([0,T_0]^2;\ S^{\infty,\sigma}_{G(\sigma)}) \cap G^{(\sigma)}(I^2;S^{\infty,\sigma}_{G(\sigma)})$, $j = 1,2$, such that

$(D_t-\lambda_j'(t,x,D_x)+b_{j,j}(t,x,D_x))e^j_{\phi_{2j}(t,s)}(t,s;x;D_x)$ is a σ-regularizing operator with kernel in $M^\circ([0,T_0]^2;\ G^{(\sigma)}(R^n\times R^n))$ and $e^j(s,s;x,\xi)=1$. On the other hand from commutation laws (3.7) and (3.8) we have for $p(t,t_1,s)\in M^\circ(\Delta;S^{\infty,\mu}_{G^{(\sigma)}})$, $\Delta=\{(t,t_1,s)\in[0,T_0]^3;\ s\gtrless t_1\gtrless t\}$:

$$\int_s^t p_{\Phi_{k,\ell}(t,t_1,s)}(t,t_1,s;x,D_x)u(x)dt_1 = \int_s^t (\Theta\, p)_{\Phi_{\ell,k}(t,t_1,s)}(t,t_1,s;x,D_x)u(x)dt_1 \tag{4.8}$$

where $u\in G^{(\sigma)'}(R^n)$ and $\Theta: M^m(\Delta;S^{\infty,\mu}_{G^{(\sigma)}})\to M^m(\Delta;S^{\infty,\mu}_{G^{(\sigma)}})$ for $m=0,1$ is defined by

$$\Theta p(t,t_1,s)=p(t,\theta(t,t_1,s),s)\alpha(t_1)\alpha(\theta(t,t_1,s))^{-1}. \tag{4.9}$$

These remarks lead us to look for an $E(t,s)$ of the type

$$E(t,s)=\begin{bmatrix} e^1_{\phi_1(t,s)}(t,s,x,D_x) & 0\\ 0 & e^2_{\phi_2(t,s)}(t,s,x,D_x)\end{bmatrix}+ \left(\int_s^t p^{k,\ell}_{\Phi_{k,j(k)}(t,t_1,s)}(t,t_1,s;x,D_x)dt_1\right)_{k,\ell=1,2}, \tag{4.10}$$

where $j(k)$ is the permutation $j(1)=2$, $j(2)=1$, to have $LE(t,s)$ equal to a matrix of σ-regularizing operators with kernel in $M^\circ([0,T_0]^2,\ G^{(\sigma)}(R^n\times R^n))$.

Take in (4.10) the amplitudes $p^{k,\ell}(t,t_1,s)$ as asymptotic sums $p^{k,\ell}(t,t_1,s)\sim\sum_{j\geq 0}p_j^{k,\ell}(t,t_1,s)$ in $M^1(\Delta;S^{\infty,\sigma}_{G^{(\sigma)}})$, $p_j^{k,\ell}(t,t_1,s;x,\eta)=$ $=\tilde p_j^{k,\ell}(t,t_1,s;y_k(t,t_1,s;x,\eta),\eta)$ with $y_k(t,t_1,s;x,\eta)=\nabla_\xi\Phi_{k,j(k)}(t,t_1,s;x,\eta)$.

From Theorem 2.10, Proposition 3.5, commutation laws (3.7), (3.8) and Remark (4.8) we have

$$LE(t,s) = -i \begin{bmatrix} p^{1,1}_{\phi_2(t,s)}(t,t,s) & \tilde{e}^2_{\phi_2(t,s)}(t,s)+p^{1,2}_{\phi_2(t,s)}(t,t,s) \\ \tilde{e}^1_{\phi_1(t,s)}(t,s)+p^{2,1}_{\phi_1(t,s)}(t,t,s) & p^{2,2}_{\phi_1(t,s)}(t,t,s) \end{bmatrix} +$$

$+ \left(\int_s^t q^{k,}_{\Phi_{k,j(k)}(t,t_1,s)}(t,t_1,s)dt_1\right)_{k,\ell=1,2}$ + matrix of σ-regularizing operators

where $-i\tilde{e}^k_{\phi(t,s)}(t,s) = b_{j(k),k}(t,x,D_x)e^k_{\phi(t,s)}(t,s;x,D_x)$, mod. σ-regularizing operators, and $q^{k,\ell}(t,t_1,s;x,\eta) = \tilde{q}^{k,\ell}(t,t_1,s;y_k(t,t_1,s;x,\eta),\eta) \sim \sum_{j\geq 0} \tilde{q}^{k,\ell}_j(t,t_1,s;\ y_k(t,t_1,s;x,\eta),\eta)$ with

$$(\tilde{q}^{k,\ell}_0(t,t_1,s;y,\eta))_{k,\ell=1,2} = \tag{4.11}$$

$$= (D_t - \begin{bmatrix} \tilde{\alpha}_{1,1}(t,t_1,s) & \tilde{\alpha}_{1,2}(t,t_1,s)\Theta \\ \tilde{\alpha}_{2,1}(t,t_1,s)\Theta & \tilde{\alpha}_{2,2}(t,t_1,s) \end{bmatrix})(\tilde{p}^{k,\ell}_0(t,t_1,s))_{k,\ell=1,2},$$

and

$$(\tilde{q}^{k,\ell}_j(t,t_1,s;y,\eta))_{k,\ell=1,2} = \tag{4.12}$$

$$= (D_t - \begin{bmatrix} \tilde{\alpha}_{1,1}(t,t_1,s) & \tilde{\alpha}_{1,2}(t,t_1,s)\Theta \\ \tilde{\alpha}_{2,1}(t,t_1,s)\Theta & \tilde{\alpha}_{2,2}(t,t_1,s) \end{bmatrix})(\tilde{p}^{k,\ell}_j(t,t_1,s))_{k,\ell=1,2}$$

$$- (r^{k,\ell}_j(t,t_1,s))_{k,\ell=1,2}\ ,\ j \geq 1.$$

Here $\tilde{\alpha}_{k,\ell}(t,t_1,s;y,\eta) = \alpha_{k,\ell}(t,t_1,s;x_k(t,t_1,s;y,\eta),\eta)$,

$$\alpha_{k,k}(t,t_1,s;x,\eta) = 2^{-1}i \sum_{j,h=1}^{n} \partial_{\xi_j} \partial_{\xi_h} \lambda'_k(t,x,\nabla_x\Phi_{k,j(k)}(t,t_1,s))$$

$$\partial_{x_j}\partial_{x_h}\Phi_{k,j(k)}(t,t_1,s) + b_{k,k}(t,x,\nabla_x\Phi_{k,j(k)}(t,t_1,s)),$$

$\tilde{\alpha}_{k,j(k)}(t,t_1,s;y,\eta) = \alpha_{k,j(k)}(t,t_1;x_k(t,t_1,s;y,\eta),\eta)$,

$$\alpha_{k,j(k)}(t,t_1,s;x,\eta) = b_{k,j(k)}(t,x,\nabla_x\Phi_{k,j(k)}(t,t_1,s)),$$

$x_k(t,t_1,s;y,\eta)$ is the inverse function of $y = \nabla_\xi\Phi_{k,j(k)}(t,t_1,s;x,\eta)$, Θ is defined by (4.9) and the sequences $(\tilde{r}_j^{k,\ell}(t,t_1,s))_{j\geq 1}$, completely determined by the terms of L and $(\tilde{p}_h^{k,\ell})_{0\leq h<j}$, satisfy condition (2.4) uniformly with respect to $(t,t_1,s) \in \Delta$.

Thus solving inductively the following transport equations of Volterra type in the class $M^1([0,T_o]^3;\ S^{\infty,\sigma}_{G^{(\sigma)}})$

$$(\tilde{P}_o^{k,\ell}(t,t_1,s)_{k,\ell=1,2} = \begin{bmatrix} 0 & -\tilde{\tilde{e}}^2(t_1,s) \\ -\tilde{\tilde{e}}^1(t_1,s) & 0 \end{bmatrix} + \tag{4.13$_o$}$$

$$+\ i\int_{t_1}^{t} \begin{bmatrix} \tilde{\alpha}_{1,1}(\tau,t_1,s) & \tilde{\alpha}_{1,2}(\tau,t_1,s) \\ \tilde{\alpha}_{2,1}(\tau,t_1,s) & \tilde{\alpha}_{2,2}(\tau,t_1,s) \end{bmatrix} (\tilde{P}_o^{k,\ell}(\tau,t_1,s))_{k,\ell=1,2}\, d\tau,$$

$$(\tilde{P}_j^{k,\ell}(t,t_1,s))_{k,\ell=1,2} = \tag{4.13$_j$}$$

$$= i\int_{t_1}^{t} \begin{bmatrix} \tilde{\alpha}_{1,1}(\tau,t_1,s) & \tilde{\alpha}_{1,2}(\tau,t_1,s)\Theta \\ \tilde{\alpha}_{2,1}(\tau,t_1,s)\Theta & \tilde{\alpha}_{2,2}(\tau,t_1,s) \end{bmatrix} (\tilde{P}_j^{k,\ell}(\tau,t_1,s))_{k,\ell=1,2}\, d\tau\ +$$

$$+\ i\int_{t_1}^{t} (\tilde{r}_j^{k,\ell}(\tau,t_1,s))_{k,\ell=1,2}\, d\tau,\quad j \geq 1,$$

$\tilde{\tilde{e}}^k(t_1,s;y,\eta) = \tilde{e}^k(t_1,s;x_k(t_1,t_1,s;y;\eta),\eta)$, $k = 1,2$, we can prove:

<u>THEOREM 4.1</u>: There exists $E(t,s)$ of the form (4.10), defined on $[0,T_o]^2$ with a smaller T_o if necessary, such that $LE(t,s)$ is equal to matrix of σ-regularizing operators with kernels in $M^o([0,T_o]^2;\ G^{(\sigma)}(R^n \times R^n)$ and $E(s,s) = I$.

5. PROPAGATION OF SINGULARITIES

Since the Cauchy problems (1.9) and (4.1) are well-posed in the $G^{(\sigma)}$ class, the solution $U(t,x)$ of problem (4.1) is given by $U(t,x) = E(t,0)G(x) + V(t,x)$ with $E(t,s)$ the parametrix we have constructed in Section 4 and $V(t,x) \in M^0([0,T_0]; G^{(\sigma)}(R^n))$. Theorem 4.1 yields:

THEOREM 5.1: Let $U(t,x)$ be the solution of the Cauchy problem (4.1). Then for $0 < t_0 \leqq T_0$ we have

$$WF_{(\sigma)}(U(t_0,\cdot)) \subset \bigcup_{\nu=0}^{1} \Gamma_\nu(t,[0,t_0], WF_{(\sigma)}(G)). \tag{5.1}$$

To end the proof of Theorem 1.3 we have to replace $[0,t_0]$ in (5.1) with the closure of the set $F_0(t_0)$ defined by (3.9). From Theorem 2.7 this is possible if we prove that $p^{k,\ell}(t,t_1,0;x,\ ) \in G^{(\sigma)}_{t_1}([0,t]\setminus\bar{F}_0(t); S^{\infty,\sigma}_{G^{(\sigma)}})$. So we turn to transport equations $(4.13)_0$ and $(4.13)_j$, $j \geqq 1$, with $s = 0$.

We represent the solution of $(4.13)_0$ with $s = 0$ as the limit of the successive approximations

$$(\tilde{p}_0^{k,\ell}(t,t_1,0))^0_{k,\ell=1,2} = (\tilde{p}_0^{k,\ell}(t_1,t_1,0))_{k,\ell=1,2} \tag{5.2^0}$$

$$(\tilde{p}_0^{k,\ell}(t,t_1,0))^h_{k,\ell=1,2} = (\tilde{p}_0^{k,\ell}(t_1,t_1,0))_{k,\ell=1,2} + \tag{5.2^h}$$

$$+ i\int_{t_1}^{t} \begin{vmatrix} \tilde{\alpha}_{1,1}(\tau,t_1,0) & \tilde{\alpha}_{1,2}(\tau,t_1,0)\Theta \\ \tilde{\alpha}_{2,1}(\tau,t_1,0)\Theta & \tilde{\alpha}_{2,2}(\tau,t_1,0) \end{vmatrix} (\tilde{p}_0^{k,\ell}(\tau,t_1,0))^{h-1}_{k,\ell=1,2} d\tau, h \geqq 1.$$

From $\tilde{\tilde{e}}^k(t_1,0) \in G^{(\sigma)}(I;S^{\infty,\sigma}_{G^{(\sigma)}})$, $k = 1,2$, $\tilde{\alpha}_{k,\ell}(t,t_1,0) \in G^{(\sigma)}(I^2,\tilde{S}^{1-\chi,1}_{G^{(\sigma)}})$, $k,\ell \in \{1,2\}$, and considering the action of the operator Θ in $(5.2)^h$ we can prove inductively that the terms of $(\tilde{p}_0^{k,\ell}(t,t_1,0))^h_{k,\ell=1,2}$ belong to $M^1_{t_1}([0,t]\setminus\bar{F}_0(t); S^{\infty,\sigma}_{G^{(\sigma)}})$ uniformly with respect to $h \in Z_+$. Thus we have that each term of $(\tilde{P}_0^{k,\ell}(t,t_1,0))_{k,\ell=1,2}$ is in $M^1_{t_1}([0,t]\setminus\bar{F}_0(t); S^{\infty,\sigma}_{G^{(\sigma)}})$. Applying similar arguments to $(\partial^j_{t_1}\tilde{p}_0^{k,\ell}(t,t_1,0))_{k,\ell=1,2}$ we can prove by induction with respect to $j \in Z_+$ that $\tilde{p}_0^{k,\ell}(t,t_1,0) \in G^{(\sigma)}_{t_1}([0,T]\setminus\bar{F}_0(t), S^{\infty,\sigma}_{G^{(\sigma)}})$, $k,\ell \in \{1,2\}$.

By the same method we obtain $\tilde{p}_j^{k,\ell}(t,t_1,0) \in G_{t_1}^{(\sigma)}([0,t]\smallsetminus\bar{F}_0(t), S^{\infty,\sigma}_{G^{(\sigma)}})$ for $j \geq 1$ and $k,\ell \in \{1,2\}$ and from this $p^{k,\ell}(t,t_1,0) \in G_{t_1}^{(\sigma)}([0,T]\smallsetminus\bar{F}_0(t), S^{\infty,\sigma}_{G^{(\sigma)}})$, ending the proof of Theorem 1.3.

Bibliography

[1] F. Colombini, E. De Giorgi, S. Spagnolo, Sur les équations hyperboliques avec des coefficients qui ne dépendent que du temps. Ann. Scuola Normale Superiore Pisa, Classe di Scienze, Vol. VI, n. 3 (1979), 511-559.

[2] F. Colombini, E. Jannelli, S. Spagnolo. Well-posedness in the Gevrey classes of the Cauchy problem for a non-strictly hyperbolic equation with coefficients depending on time. Ann. Scuola Normale Superiore Pisa, Classe di Scienze, Vol. X, n. 2 (1983), 291-312.

[3] E. Jannelli, Gevrey well-posedness for a class of weakly hyperbolic equations. Journ. of Math. of Kyoto Univ., 24-4 (1984), 763-778.

[4] T. Nishitani. Sur les équations hyperboliques à coefficients qui sont hölderiens en t et de la classe de Gevrey en x. Bull. de Sc. Math. 107 (1983), 113-138.

[5] Y. Ohya, S. Tarama, Le problème de Cauchy a caractéristiques multiples dans la class de Gevrey-coefficients hölderiens en t. To appear.

[6] E. Jannelli, On the symmetrization of the principal symbol of hyperbolic equations. Proceedings of this workshop.

[7] S. Hashimoto, T. Matsuzawa, Y. Morimoto, Opérateurs pseudo-différentiels et classes de Gevrey. Comm. Partial Differential Equations 8 (1983), 1277-1289.

[8] V. Iftimie, Opérateurs hypoelliptiques dans des espaces de Gevrey. Bull. Soc. Sci. Math. R.S. Roumanie 27 (1983), 317-333.

[9] L. Zanghirati, Pseudo-differential operators of infinite order and Gevrey classes. Ann. Univ. Ferrara, Sez. VII, 31 (1985), 197-219.

[10] L. Cattabriga, L. Zanghirati, Fourier integral operators of infinite order on Gevrey spaces and applications to the Cauchy problem for certain hyperbolic operators. To appear.

[11] L. Cattabriga, D. Mari, Parametrix of infinite order on Gevrey spaces to the Cauchy problem for hyperbolic operators with one multiple characteristic. Ricerche Mat., to appear.

[12] S. Mizohata, Propagation de la régularité au sens de Gevrey pour les opérateurs différentiels a multiplicité constante. J. Vaillant, Sèminaire equations aux dériveés partielels hyperboliques et holomorphes, Hermann Paris, 1984, 106-133.

[13] K. Taniguchi, Fourier integral operators in Gevrey classes on R^n and the fundamental solution for a hyperbolic operator. Publ. RIMS, Kyoto Univ. 20 (1984), 491-542.

[14] Y. Morimoto, K. Taniguchi, Propagation of wave front sets of solutions of the Cauchy problems for hyperbolic equations in Gevrey Classes. Osaka Journ. of Math. Vol. 23, n. 4, 765-814.

[15] H. Kumano-Go, Pseudo-differential operators. MIT Press, 1981.

Massimo Cicognani

Dipartimento di Matematica,
Università di Bologna,
Piazza di Porta San Donato 5,
Bologna,
Italy.

P. GÉRARD

Regularization by averaging for solutions of partial differential equations

INTRODUCTION

Let $P = P(x,D_x,\omega)$ be a linear partial differential operator depending on a parameter ω lying in a probability space (Ω,μ). Let $u = u(x,\omega)$ be a solution of the equation $Pu = f$; we are interested in the regularity of the mean value $\tilde{u} = \int_\Omega u \, d\mu(\omega)$, knowing the regularity of u and f.

This problem occurs naturally for instance in Physical Kinetics (see [1]). A general answer has been given by Golse-Lions-Perthame-Sentis in [8] in the case of a "Transport equation" ($P = \omega.\partial_x$). The proof in [8] admits a straightforward extension to the case of any differential operator with constant coefficients.

In this talk we want to show how methods of microlocal analysis allow us to generalize this result to some classes of differential operators with variable coefficients.

Notations: (Ω,μ) is a measured space, where μ has finite mass. $P = p(x,\xi,\omega)$ is a bounded family of classical symbols on $T^*\mathbf{R}^n$, indexed by $\omega \in \Omega$; we may always assume that p is of order 1. P is a pseudodifferential operator of symbol p; we study the H^s regularity of u, knowing that

$$u \in L^2(\mathbf{R}^n \times \Omega, dxd\mu), \; Pu \quad L^2(\mathbf{R}^n \times \Omega, dxd\mu). \tag{1}$$

Thus we may assume that p is homogeneous.

1. THE CASE OF CONSTANT COEFFICIENTS

Let us first recall the result in [8]:

THEOREM 1 (Golse-Lions-Perthame-Sentis): Assume that p does not depend on x, i.e. $p = p(\xi,\omega)$. Then, given any $\delta \in]0,1[$, the following conditions are equivalent:

(i) the condition (1) ensures $\tilde{u} \in H^\delta(\mathbb{R}^n)$.

(ii) there exists $C > 0$ such that $\mu\{\omega/|p(\xi,\omega)| \leq \varepsilon\} \leq C\varepsilon^{2\delta}$ for all $\xi \in S^{n-1}$.

PROOF: (i) ⇒ (ii). By the Banach theorem we have the following inequality:

$$\left\| \int u d\mu \right\|_\delta \leq C(\|u\|_0 + \|Pu\|_0). \tag{2}$$

Then (ii) follows applying (2) to $u(x,\omega) = e^{ix.\xi/\varepsilon} \psi(x)\theta_\varepsilon(\omega)$, with $\psi \in C_0^\infty$ and $\theta_\varepsilon(\omega) = 1_{\|p(\xi,\omega\| \leq\varepsilon\}}$.

(ii) ⇒ (i). Taking the Fourier transform of $\tilde{u}$, we get

$$\int \hat{u}(\xi,\omega)d\mu(\omega) = J_1 + J_2, \text{ where } J_1 = \int_{|p(\xi,\omega)|\leq 1} \hat{u}d\mu,$$

$$J_2 = \int_{|p(\xi,\omega)|\geq 1} \hat{u}d\mu.$$

Using (ii) and the homogeneity of p, we obtain

$$|J_1|^2 \leq C|\xi|^{-2\delta} \int |\hat{u}|^2 d\mu.$$

By the Schwarz inequality, we have

$$|J_2|^2 \leq \int_{|p(\xi,\omega)| \geq 1} 1/|p(\xi,\omega)|^2 d\mu(\omega). \int |p(\xi,\omega)\hat{u}(\xi,\omega)|^2 d\mu(\omega)$$

$$\leq |\xi|^{-2} \int_{t\geq 1/|\xi|} d\nu(t)/t^2 \int |p(\xi,\omega)\hat{u}(\xi,\omega)|^2 d\mu(\omega),$$

where ν is the image of μ under the map $|p(\xi/|\xi|,.)|$. Integration by parts gives

$$\int_{t\geq 1/|\xi|} d\nu(t)/t^2 = - \nu\{s/s \leq 1/|\xi|\}|\xi|^2 + \int_{t\geq 1/|\xi|} 2\nu\{s/s\leq t\}/t^3 \leq C|\xi|^{2-2\delta},$$

since $\delta < 1$.

Summing up and integrating in ξ, we obtain (2).

REMARKS:

a) The above theorem admits straightforward local and microlocal formulations.

b) The mean value $\tilde{u}$ is as regular as the tubes around $\{p = 0\}$ are μ-small for a fixed ξ. When Ω is a manifold and μ is a density, the decay index of the μ-measure of those tubes is of course related to the singularities of the set $\{p(\xi,\cdot) = 0\}$. For instance, if this set is a smooth manifold, (ii) holds with $\delta = 1/2$. A few more details can be found in [4].

c) Example. Take $P = \partial_t - \omega.\partial_x$, then the equation $Pu = 0$, $u_{|t=0} = u_o(x)\varphi(\omega)$ is solved by $u(t,x,\omega) = u_o(x + t\omega)\varphi(\omega)$. If $\varphi \in L^2$ and is compactly supported, $\int \varphi \, d\omega = 1$, and if $u_o \in H^s$, then Theorem 1 asserts that $\int u(t,x,\omega)d\omega \in H^{s+1/2}_{loc}$, which gives the classical formula for a section of the trace operator in Sobolev spaces.

2. GENERALIZATIONS

In order to extend the previous proof to some differential operator P, we have to study the smoothness of u and $\tilde{u}$ using a transformation under which P acts essentially as a multiplier. For instance, if P has constant coefficients as above, Fourier transformation is of course the good choice.

A. We start with the dual case of 1. Here $p = a(x,\omega)|\xi|$, and the relevant transformation is the Littlewood-Paley decomposition, as described for instance in Coifman-Meyer [3].

THEOREM 2: Let $a = a(x,\omega) \in L^\infty(\Omega, \mathrm{Lip}(\mathbb{R}^n))$. The following conditions are equivalent (for a given $\delta \in]0,1[$):

(i) $(u \in L^2(\mathbb{R}^n \times \Omega, dxd\mu),\ au \in L^2(\Omega, H^1(\mathbb{R}^n)) \Rightarrow \int u d\mu \in H^\delta(\mathbb{R}^n)$.

(ii) there exists $C > 0$ such that $\mu\{\omega / |a(x,\omega)| \leq \varepsilon\} \leq C\varepsilon^{2\delta}$ for all $x \in \mathbb{R}^n$.

PROOF (i) $\Rightarrow$ (ii). Test inequality (1) on $u(y,\omega) = \varepsilon^{-n/2}\psi((y-x)/\varepsilon)\theta_\varepsilon(\omega)$, with $\psi \in C_o^\infty$ and $\theta_\varepsilon(\omega) = 1_{\{|a(x,\omega)| \leq \varepsilon\}}$.

(ii) $\Rightarrow$ (i). We use the Littlewood-Paley decomposition of u with respect to x:

$$u = S_o u + \sum_o^\infty \Delta_p u,$$

where $\Delta_p = \varphi(2^{-p}D_x)$, $\varphi = \varphi(\xi)$ is a C^∞ function supported in the ring

$C_0 = \{\xi/1/2 \leq |\xi| \leq 4\}$ such that $\sum_{-\infty}^{+\infty} \varphi(2^{-q}\xi) = 1$.

Then (see [3]) (1) is equivalent to

$$\int\int \sum_0^\infty |(1 + 2^p|a(x,\omega)|)\Delta_p u(x,\omega)|^2 \, dx d\mu(\omega) < \infty,$$

and we have to prove that

$$\int \Sigma \, 2^{2p\delta} |\Delta_p \tilde{u}(x)|^2 dx < \infty .$$

But $\Delta_p \tilde{u}(x) = \int \Delta_p u(x,\omega) d\mu(\omega) = J_1 + J_2$, where $J_1 = \int_{|a(x,\omega)| \leq 2^{-p}}$, $J_2 = \int_{|a(x,\omega)| \geq 2^{-p}}$; J_1 and J_2 can be estimated exactly as in §1. Inequality (2) follows after summing in p and integrating in x.

Application: Let $\alpha = \alpha(x)$ be a locally Lipschitz continuous function on $\mathbb{R}^n$, and let $v \in L^2_{loc}$ such that $\alpha v \in H^1_{loc}$. We are interested in the convolution product $v_* \varphi$ with $\varphi \in L^2$, compactly supported. After applying a cut-off to α and v, we are led to use Theorem 2 with $a(x,\omega) = \alpha(x-\omega)$, $u(x,\omega) = v(x-\omega)\varphi(\omega)$. Thus $v_*\varphi \in H^{\min(\delta,1)}_{loc}$, if $\mu\{\omega/|\omega| \leq R, \, |\alpha(\omega)| \leq \varepsilon\} \leq C_R \varepsilon^{2\delta}$ for all $R > 0$, (here μ stands for Lebesgue measure).

Classical examples are $\alpha(x) = x_1$, $\alpha(x) = |x|$. The latter shows that we may have $\delta > 1$. In order to take advantage of this, it is natural to introduce conormal distributions as in Bony [2]:

$$H^{0,k} = \{v \in L^2_{loc}, \; x^\beta v \in H^{|\beta|}_{loc} \text{ for } |\beta| \leq k\}.$$

Those spaces are described by the Littlewood-Paley decomposition as follows:

$$(v \in H^{0,k}) \iff \int \sum_0^\infty |(1+2^p|x|)^k \, \Delta_p v(x)|^2 dx < \infty$$

(for a compactly supported v, for instance).
Then we can modify the above proof as follows. With $u(x,\omega) = v(x-\omega)\varphi(\omega)$ we write

$$|J_2|^2 \leq \int_{|\omega| \leq R, |x-\omega| \geq 2^{-p}} 1/|x-\omega|^{2k} \, d\omega \, . \int |x-\omega|^{2k} \, |\Delta_p u(x,\omega)|^2 d\omega$$
$$\leq C_R \, 2^{-np} \int |2^{kp}|x-\omega|^k \, \Delta_p v(x-\omega)|^2 |\varphi(\omega)|^2 d\omega$$

as long as $k > n/2$. Summing up, we have proved

COROLLARY: If $v \in H^{0,k}(\mathbb{R}^n)$ with $k > n/2$, $\varphi \in L^2$ is compactly supported, the convolution product $v_*\varphi$ is locally in $H^{n/2}$.

B. Elliptic degenerate operators

As a second generalization of Theorem 1, we are going to handle the case of a semi-bounded symbol. The main tool will be an appropriate decomposition of the phase space following Hörmander [9] and Melin [10].

THEOREM 3: Suppose there exists a convex cone Γ in $\mathbb{C}$, with opening angle $< \pi$, such that $p(x,\xi,\omega) \in \Gamma$ for all $(x,\xi,\omega) \in T^*\mathbb{R}^n \times \Omega$. Given $\delta \in]0,1[$, the following conditions are equivalent:

(i) the condition (1) ensures that $\tilde{u} \in H^\delta(\mathbb{R}^n)$.

(ii) there exists $C > 0$ such that $\mu\{\omega / |p(x,\xi,\omega)| \leq \varepsilon\} \leq C\varepsilon^{2\delta}$ for all $(x,\xi) \in S^*\mathbb{R}^n$.

PROOF: First remark that the hypothesis of degenerate ellipticity ($p \in \Gamma$) can be written $|\mathrm{Im} p| \leq \gamma \mathrm{Re} p$, after applying a rotation to p if necessary. Then, using also the 1-homogeneity of p, we obtain the inequality

$$|\xi|^{-1}|\partial_x p|^2 + |\xi||\partial_\xi p|^2 \leq C \ \mathrm{Re} p \qquad (3)$$

which is the only information we will need about degenerate ellipticity.

(i) ⇒ (ii). Test inequality (2) on $u(y, \ ) = \varepsilon^{-n/4} e^{i(y-x),\xi/\varepsilon}\psi((y-x)/\varepsilon^{1/2})\,\theta_\varepsilon(\omega)$, with $\psi \in C_0^\infty$ and $\theta_\varepsilon(\omega) = 1_{\{|p(x,\xi,\omega\}|\leq\varepsilon\}}$ and use inequality (3).

(ii) ⇒ (i). We follow closely the proof of Lemma 22.4.13 in [9]. Introduce the (1/2,1/2) metric $g_h = (h|\xi|+1)|dx|^2 + h^2|d\xi|^2/(h|\xi|+1)$ where $h \in]0,1]$. The metric g_h is slowly varying and σ-temperate uniformly when $0 < h \leq 1$. Moreover we have $g_h = h^2 g_h^\sigma$.

Consider the covering of the phase space by balls of constant g_h-radius: $\mathbb{R}^n \times \mathbb{R}^n = \cup B_j$, where $B_j = \{(x,\xi)/h|\xi-\xi_j| \leq C(1+h|\xi_j|)^{1/2}, |x-x_j| \leq c(1+h|\xi_j|)^{-1/2}\}$, and introduce a bounded sequence $\varphi_j = \varphi_j(x,\xi)$ in $S(1,g_h)$ such that $\mathrm{supp}\,\varphi_j \subset B_j$, $\Sigma\varphi_j^2 = 1$. Set $\Phi_j = \varphi_j(x,D)$.

LEMMA 1: There exists $h \in]0,1[$ such that:

a) $\Sigma\|\Phi_j u\|_0^2 \sim \|u\|_0^2$; b) for any $\delta > 0$, $\Sigma\, |\xi_j|^{2\delta}\, \|\Phi_j u\|_0^2 \sim \|u\|_\sigma^2$ mod. $\|u\|_0^2$

SKETCH OF THE PROOF: a) φ is a symbol of $S(1,g_h)$ taking its values in l^2. Thus $\Sigma\, \|\Phi_j u\|_0^2 \leq C\, \|u\|_0^2$. Conversely, $\sigma(\Phi_j^* \Phi_j) = \varphi_j^2 + r_j$, where $\Sigma r_j \in S(h,g_h)$. Therefore $\|u\|_0^2 = (\Sigma\, \varphi_j^2 u,u) \leq \Sigma\, \|\Phi_j u\|_0^2$, and we get a) choosing h small enough. b) Fix h as above. Then $|\xi| \sim |\xi_j|$ on B_j. Apply a) to $\Lambda^\delta u$ and remark by Taylor's formula that $\varphi_j|\xi|^\delta = \varphi_j|\xi_j|^\delta + r_j$, where $r \in S(|\xi|^{\delta-1/2},g)$ is l^2-valued. Hence we obtain $|\Sigma|\xi_j|^{2\delta}\|\Phi_j u\|_0^2 - \|u\|^2_\delta| \leq C\|u\|^2_{\delta-\frac{1}{2}}$, which leads to b).

It remains to check that P acts essentially as a multiplier on the Φ_j decomposition.

LEMMA 2: If p satisfies inequality (3), we have

$$\Sigma|p(x_j,\xi_j)|^2 \quad \|\Phi_j u\|_0^2 \leq C(\, \|Pu\|_0^2 + \|u\|_0^2) \tag{4}$$

PROOF: By Lemma 1, a), $\Sigma\, \|\Phi_j Pu\|_0^2 \leq C\, \|Pu\|_0^2$, and by Taylor's formula joined to the classical asymptotic expansion

$$\varphi_j \# p = p(x_j,\xi_j)\varphi_j + \partial_x p(x_j,\xi_j)\alpha_j + \partial_\xi p(x_j,\xi_j)\beta_j + r_j, \tag{5}$$

where α, β, r are l^2-valued g-symbols of orders $|\xi|^{-1/2}$, $|\xi|^{1/2}$ and 1 respectively. Moreover B_j contains supp α_j and supp β_j. Thus, by the same arguments, we can write

$$\varphi_j \# p = p(x_j,\xi_j)\varphi_j + \alpha_j \# \partial_x p + \beta_j \# \partial_\xi p + r_j'.$$

Hence $\Sigma|p(x_j,\xi_j)|^2 \quad \|\Phi_j u\|_0^2 \leq C(\, \|Pu\|_0^2 + \|u\|_0^2 + \|\partial_x p(x,D)u\|_{-\frac{1}{2}}^2 +$
$+ \|\partial_\xi p(x,D)u\|_{\frac{1}{2}}^2)$.

In view of (3) and the sharp Garding inequality, the lemma is proved.

END OF THE PROOF OF THEOREM 3: In view of Lemma 1,b), we only have to estimate $\|\Phi_j u\|_0^2$.

$$\Phi_j u = \int \Phi_j u \, d\mu = J_1 + J_2, \text{ where } J_1 = \int_{|p(x_j,\xi_j,\omega)| \leq 1} \Phi_j u d\mu,$$

$$J_2 = \int_{|p(x_j,\xi_j,\omega)| \geq 1} \Phi_j u d\mu.$$

It remains to estimate $\|J_1\|_0$ and $\|J_2\|_0$ as in the proof of Theorem 1; for J_2, we obtain $\|J_2\|_0^2 \leq C|\xi_j|^{-2\delta} \int |p(x_j,\xi_j,\omega)|^2 \quad \|\Phi_j u\|_0^2 \, d\mu(\omega)$, and we conclude using Lemma 2.

REMARKS:

1. As in the section A, (ii) may occur with some $\delta > 1$, (see [4]). Again we introduce $H^{0,k}(P) = \{u \in L^2, P^i u \in L^2 \text{ for } 0 \leq i \leq k\}$. Iterating identity (5), Lemma 2 extends to

LEMMA 3: If p satisfies inequality (3), we have

$$\Sigma\, |p(x_j,\xi_j)|^{2k} \quad \|\Phi_j u\|_0^2 \leq C \sum_0^k \|P^i u\|_0^2 \tag{4'}$$

Then, modifying the above proof as in the end of Section A, it is easy to prove the

COROLLARY: Given any $\delta > 0$ such that (ii) holds, if $u \in L^2(\Omega, H^{0,k}(P))$, $k > \delta$, then $\tilde{u} \in H^\delta$.

2. Results on averaging solutions of real principal type equations and applications can be found in [4] and (with other methods) in [5] and [7], where P is allowed to contain derivatives in ω. Other aspects will be discussed in [6].

Bibliography

[1] C. Bardos, F. Golse, B. Perthame, R. Sentis: The Non-Accretive Radiative Transfer Equations; existence of solutions and Rosseland Approximation, C.E.A. Report N-2496; to appear in J. Funct. Anal.

[2] J.-M. Bony: Second microlocalization and propagation of singularities for semi linear hyperbolic equations. Preprint Universite de Paris-Sud, to appear.

[3] R. Coifman, Y. Meyer: Au-delà des opérateurs pseudo-différentiels. Astérisque, vol. 57, 1978.

[4] P. Gérard: Moyennes de solutions d'équations aux dérivées partielles. Séminaire Equations aux dérivées partielles 1986-1987, Exposé n° 11, Ecole Polytechnique, Paris (to appear).

[5] P. Gérard: Régularité de moyennes de solutions d'équations aux dérivées partielles, Journées "E.D.P." de Saint-Jean-de-Monts 1987, to appear.

[6] P. Gérard, F. Golse: In preparation.

[7] F. Golse: Résultats de moyennisation pour des équations aux dérivées partielles. To appear in Rendiconti del Seminario Matematico di Torino.

[8] F. Golse, P.-L. Lions, B. Perthame, R. Sentis: Regularity of the Moments of a solution of a Transport Equation. Preprint Ecole Normale Supérieure, to appear in J. Funct. Anal.

[9] L. Hörmander: The Analysis of Linear Partial Differential Operators, tome 3. Springer-Verlag, 1985.

[10] A. Melin: Lower bounds for pseudo-differential operators. Ark. Mat. 9, 117-140 (1971).

Patrick Gérard
Département de Mathématiques
Ecole Normale Supérieure
45, rue d'Ulm
F-75 230 Paris Cedex 05
France.

D. GOURDIN

Problèmes de Cauchy hyperboliques nonlineaires

ABSTRACT: Nonlinear hyperbolic Cauchy problems *

In 1969, Madame Choquet-Bruhat constructed asymptotic waves for quasi-linear systems at the neighourhood of a solution u_0 for which the linearized system admit double characteristic, involving the subcharacteristic polynomial connected with this linearized system and extending in such a way some results of J. Vaillant; she gave also applications to relativistic magneto hydrodynamics.

On the other hand, in 1985, N. Iwasaki has obtained C^∞ well posedness Cauchy problem results extending to non linear operators the effective hyperbolicity property and he has given applications to Euler system, to Monge-Ampère hyperbolic system and some non linear wave equations.

We describe here other classes of non linear weakly hyperbolic operators for which the C^∞ local Cauchy problem is well posed. These operators admit linearized operators which have characteristics with any constant multiplicities. For this purpose we use the method of N. Iwasaksi and the theory of Nash-Moser.

---

* Actes du Congrès "Half-year program on Evolution Equations - Pisa, January to June, 1987, 3 Hyperbolic Equations (9 marzo - 4 aprile)". Consiglio Nazionale delle ricerche - G.N.A.F.A. Universita di Pisa, Scuola Normale Superiore di Pisa.

Après la théorie de J. Leray sur l'hyperbolicité stricte [16] et sa résolution du problème de Cauchy local, avec des données sur des hypersurfaces spatiales, dans des espaces de Sobolev pour des équations (ou des systèmes à parties principales diagonales) quasi-linéaires strictement hyperboliques et la thèse de Ph. Dionne prolongeant cette théorie [6], J. Leray et Y. Ohya ont résolu en 1966 le problème de Cauchy local dans des espaces de Gevrey pour des équations et des systèmes diagonaux quasi-linéaires hyperboliques non stricts [17], [20].

Madame Choquet-Bruhat a complété ces résultats, sans conditions supplémentaires sur l'opérateur et dans des espaces de Gevrey en montrant d'abord que les systèmes quasi-linéaires, non diagonaux, étaient diagonalisables et elle les a appliqués à la magnétohydrodynamique relativiste [2], [4]; signalons aussi des travaux de Cl. Wagschal pour les équations non linéaires ([22]) et de K. Kajitani pour les systèmes quasi-linéaires ([14], [15]) dans ces espaces.

Il est bien connu, surtout pour les équations linéaires que, dès que l'on se place dans des espaces de Sobolev ou dans C^∞, le problème de Cauchy non caractéristique même local, n'est pas bien posé pour les opérateurs hyperboliques non stricts, sans hypothèse supplémentaire.

C'est ce qui apparait aussi, en non linéaire, dans les travaux suivants.

En utilisant des développements asymptotiques, inspirés de la méthode WKB pour les équations différentielles ordinaires, de la forme:

$u \sim \sum_{p=0}^{+\infty} \omega^{-p} u_p(x,\omega\varphi)$, Madame Choquet-Bruhat a construit en 1969 des ondes asymptotiques pour des systèmes quasi-linéaires d'équations au voisinage d'une solution u_0 de ce système pour une phase φ solution de l'équation caractéristique du système linéarisé en u_0, sous certaines hypothèses ([3]). En particulier, dans le cas faiblement hyperbolique à caractéristique double pour u_0, elle retrouve et prolonge au quasi-linéaire des résultats dus à J. Vaillant pour les systèmes linéaires, en imposant une condition de divisibilité par le facteur double du polynôme sous-caractéristique (attaché au système linearise au point u_0) dont la définition avait été introduite par J. Vaillant [21].

D'autre part, N. Iwasaki a obtenu récemment des résultats de résolubilité locale C^∞ en prolongeant aux opérateurs non linéaires la notion d'hyperbolicité effective [12] et les a appliqués aux équations d'Euler, aux équations hyperboliques de Monge-Ampère et à certaines équations des ondes non linéaires [13].

Nous allons décrire ici d'autres classes d'opérateurs non linéaires hyperboliques non stricts pour lesquelles le problème de Cauchy local C^∞ est bien posé, en nous inspirant de la méthode de N. Iwasaki basée sur le théorème de Nash-Moser [11], et, pour les systèmes, des résultats de Madame Choquet-Bruhat et J. Vaillant.

I. THEOREME DE NASH-MOSER; APPLICATION AU PROBLEME DE CAUCHY

1. Définitions et notations

Soit $p \in C^\infty_R(\Omega \times \mathbf{R}^N)$ une fonction C^∞ des variables X et Y, à valeurs réelles

$X = (x_0,x) \in \Omega$ ouvert de $\mathbf{R}^{r+1}$ contenant l'origine 0; $x_0 \in \mathbf{R}$ et
$x = (x_1,\ldots,x_r) \in \mathbf{R}^r$.

$Y = \{y_\alpha\}_{\alpha \in A} \in \mathbf{R}^N$ où $A = \{\alpha = (\alpha_0,\alpha_1,\ldots,\alpha_r) \in \mathbf{N}^{r+1};\ |\alpha| \leq m\}$ et

N = card A.

On note $\zeta = (\xi_0,\xi)$ la variable duale de la variable X où $\xi_0 \in \mathbf{R}$ et
$\xi = (\xi_1,\ldots,\xi_r) \in \mathbf{R}^r$ et U un ouvert de $\mathbf{R}^N$ contenant l'origine.

Considérons l'opérateur aux dérivées partielles non linéaire, d'ordre m, noté

$$\Phi : C^\infty_R(\Omega) \longrightarrow C^\infty_R(\Omega)$$

et défini par:

$$\Phi(y) = p(X,\{\partial^\alpha y\}_{\alpha \in A}) = p(X,\partial^\alpha y) \quad \text{(notation)} \qquad (1)$$

où $\quad y \in C^\infty_R(\Omega)$ et $\partial^\alpha y = \dfrac{\partial^{|\alpha|} y}{(\partial x_0)^{\alpha_0}(\partial x_1)^{\alpha_1},\ldots,(\partial x_r)^{\alpha_r}}$.

Soit $\Phi'(y) \in L(C^\infty_R(\Omega))$ la dérivée de Fréchet de Φ en y (ou opérateur linéarisé de Φ en $y \in C^\infty_R(\Omega)$) défini par:

$$\Phi'(y)z = \sum_{|\alpha| \leq m} \frac{\partial p}{\partial y_\alpha}(X,\partial^\alpha y)\partial^\alpha z \qquad (2)$$

pour tout $z \in C^\infty_R(\Omega)$; c'est-à-dire, en posant:

$$\psi(Y)z = \sum_{|\alpha|\leq m} \frac{\partial p}{\partial y_\alpha}(X,Y)\partial^\alpha z \tag{3}$$

on a

$$\Phi'(y) = \Psi(\{\partial^\alpha y\}_{\alpha\in A})$$

pour tout $y \in C^\infty_R(\Omega)$.

<u>DEFINITION 1</u>: On dira que Φ est hyperbolique strictement (resp. faiblement ou non strictement) relativement à ξ_0 et à l'ouvert U de $\mathbb{R}^N$ si l'une des conditions équivalentes suivantes est réalisée:

i) $\psi(Y)$ est hyperbolique strictement (resp. faiblement) relativement à ξ_0 pour tout $Y \in U$.

ii) $q_m(X,Y,\zeta) = \sum_{|\alpha|=m} \frac{\partial p}{\partial y_\alpha}(X,Y)\zeta^\alpha$ est hyperbolique strictement (resp. faiblement) relativement à ξ_0 pour tout $X \in \Omega$ et tout $Y \in U$.

iii) $q_m(X,Y,\zeta)$ admet m racines réelles en ξ_0 distinctes deux à deux pour tout $\xi \neq 0$ (resp. distinctes ou non) pour tout $X \in \Omega$ et tout $Y \in U$.

2. <u>Contre-exemple au théorème classique d'inversion locale</u> ([11])

Si E est un espace de Banach, on sait que $F : E \to E$ est inversible au voisinage de $e_0 \in E$ si F est dérivable au voisinage de e_0 et $F'(e_0)$ est inversible.

Ce résultat n'est plus vrai lorsque E est de Fréchet; par exemple, $F : C^\infty([-1,1]) \to C^\infty([-1,1])$ défini par $F(f) = f - xf\frac{df}{dx}$ n'est pas inversible au voisinage de $f_0 = 0$ bien que $F'(0) =$ Identite.

3. <u>Théorème de Nash-Moser et application au problème de Cauchy</u>

On dispose cependant du résultat d'inversion locale pour les espaces E et F gradués réguliers (définis à partir de chaînes d'espaces de Banach et d'opérateurs régularisants) et les fonctions $G : V \subseteq E \to F$ régulières admettant des dérivées $G'(e)$ inversibles ($\forall\, e \in V$) telles que $H:V \times F \to E$, définie par $H(e,f) = [G'(e)]^{-1}f$, soit régulière (V étant un ouvert de E); c'est le théorème de Nash-Moser (cf. [11]).

Soit $\tilde{\Omega} = \{X = (x_0,x) \in \mathbb{R}^{r+1}; -\varepsilon \leq x_0 \leq \varepsilon(1 - |x|^2)\} \subset \Omega$ pour $\varepsilon > 0$ et

$$C_+^\infty(\tilde{\Omega}) = \{y \in C_R^\infty(\tilde{\Omega});\ \text{supp } y \subset R^+ \times \mathbf{R}^r\}$$

où $R^+ = [0, +\infty[$.

Supposons

(Hyp 1) $\quad \text{Supp } p(X,0) \subset \Omega^+ = \Omega \cap (R^+ \times \mathbf{R}^r)$.

Alors par le théorème des accroissements finis avec reste intégral, dans les espaces de Fréchet, l'opérateur Φ défini par (1) vérifie:

$$\varphi = \Phi_{|C_+^\infty(\tilde{\Omega})} : C_+^\infty(\tilde{\Omega}) \to C_+^\infty(\tilde{\Omega}). \tag{4}$$

D'autre part, $C_+^\infty(\tilde{\Omega})$ est gradué à l'aide de la chaîne des espaces de Sobolev sur $\tilde{\Omega}$ (ou de façon équivalente à partir des classes Höldériennes sur $\tilde{\Omega}$) et régulier ([11]); de plus, φ est régulière ([11]).

Soit $\tilde{V}$ l'ouvert de $C_+^\infty(\tilde{\Omega})$ défini par:

$$\tilde{V} = \{y \in C_+^\infty(\tilde{\Omega});\ \{\sup_{x\in\tilde{\Omega}} |\partial^\alpha y|\}_{|\alpha| \leq m} \in U\}. \tag{5}$$

Alors si $\varphi'(y)$ est inversible pour tout $y \in \tilde{V}$ et si l'inverse $[\varphi'(y)]^{-1}$ définit un opérateur Θ:

$$\Theta : \tilde{V} \times C_+^\infty(\Omega) \to C_+^\infty(\tilde{\Omega}) \tag{6}$$

$$(y,z) \to [\varphi'(y)]^{-1} z$$

tel que Θ soit régulier, l'opérateur φ sera localement inversible dans $\tilde{V}$.

Nous allons donner au paragraphe II des conditions suffisantes d'inversibilite des $\varphi'(y)$.

La démonstration de la régularité des fonctions Θ correspondantes (définies par (6)) se fait en construisant des estimations, dans les espaces de Sobolev $H_s(\tilde{\Omega})$, du type suivant:

$$(E_s) \qquad \|z\|_s \leq C_s(\|\varphi'(y)z\|_{s+\ell} + \|a\|_{s+\ell}\|\varphi'(y)z\|_\ell)$$

où ℓ est independante de s, a représente les coefficients de $\varphi'(y)$, C_s est une constante positive indépendante de a, y, z (pourvu que a appartienne à un ensemble borné de $C_+^\infty(\tilde{\Omega})$); on suit pour cela une méthode inspirée de celle de N. Iwasaki ([12]).

On obtiendra alors un theoreme de resolubilite localé du probleme de Cauchy C^∞.

II. QUELQUES CLASSES D'OPERATEURS FAIBLEMENT HYPERBOLIQUES NON LINEAIRES POUR LESQUELLES LE PROBLEME DE CAUCHY C^∞ LOCAL EST BIEN POSE ([9],[10])

1. Hypothèses pour les opérateurs scalaires ([9] - [10])

Le polynôme caractéristique de $\Phi'(y)$ défini par

$$q_m(X,Y,\zeta) = \sum_{|\alpha|=m} \frac{\partial p}{\partial y_\alpha}(X,Y)\zeta^\alpha$$

vérifie les hypothèses suivantes:

(Hyp 2) $q_m(X,0,1,0,\dots,0) = \frac{\partial p}{\partial y_{1,0,\dots,0}}(X,0) \neq 0$

et $\inf_{X\in\Omega} |q_m(X,0,1,0,\dots,0| > 0.$

(Hyp 3) q_m admet la décomposition en facteurs irréductibles dans $\mathbb{R}[\zeta]$ suivante (pour X fixé dans Ω et Y fixé dans U)

$$q_m(X,Y,\zeta) = [H_0(X,Y,\zeta)]^2 \sum_{s=1}^{\sigma} H_s(X,Y,\zeta)$$

où H_s est C^∞ en X et Y et polynomial en ζ irréductible dans $\mathbb{R}[\zeta]$ $(s=0,\dots,\sigma)$.

(Hyp 4) Le polynôme $\prod_{s=0}^{\sigma} H_s(X,Y,\zeta)$ est strictement hyperbolique par rapport à ξ_0 ($\forall X \in \Omega$, $\forall Y \in U$).

(Hyp 5) $H_0(X,Y,\zeta) = H_0'(X,Y)H_0''(X,\{y_\alpha\}_{|\alpha|\leq m-1},\zeta)$

avec

$H_0' \in C_R^\infty(\Omega \times U)$

H_0'' de classe C^∞ en X et $\{y_\alpha\}_{|\alpha|\leq m-1}$ et polynomial irréductible dans $R[\zeta]$.

Dans ces conditions, Φ est faiblement hyperbolique par rapport à ξ_0 et à U.

2. Enonces des résultats pour les opérateurs scalaires ([9] - [10])

PROPOSITION 1 ([9] - [10]): ($\varphi'(y)$ est inversible pour tout $y \in \tilde{V}$ si il existe $K(X,Y,\zeta)$ de classe C^∞ en X et Y dans $\Omega \times U$ et polynomial dans $\mathbb{R}[\zeta]$ tel que:

$$q'_{m-1} = \sum_{|\alpha|=m-1} \frac{\partial p}{\partial y_\alpha}(X,Y)\zeta^\alpha - \frac{1}{2}\sum_{\lambda=0}^{r}\sum_{|\alpha|=m}\left[\frac{\partial^2 p}{\partial x_\lambda \partial y_\alpha}(X,Y) + \sum_{|\beta|\leq m-1}\frac{\partial^2 p}{\partial y_\alpha \partial y_\beta}(X,Y)y_{\beta+1_\lambda}\right]\alpha_\lambda \zeta^{\alpha-1_\lambda} \tag{7}$$

vérifie:

(Hyp 6) $q'_{m-1} = H_0(X,Y,\zeta)K(X,Y,\zeta)$

pour tout $X \in \Omega$, $Y \in U$ et $\zeta \in \mathbb{R}^{r+1}$, où $1_\lambda = (0,\ldots,0,1,0,\ldots,0)$ (1 se trouvant à la λ ème colonne).

PREUVE: D'après les résultats bien connus sur les opérateurs hyperboliques faiblement, linéaires, à caractéristiques de multiplicité constante au plus deux, $\varphi'(y)$ est inversible si et seulement si le polynôme sous-caractéristique de $\varphi'(y)$ est divisible par H_0: c'est la condition de Levi, ([18], [19]).

Ce polynôme sous-caractéristique s'écrit:

$$q''_{m-1} = \sum_{|\alpha|=m-1} \frac{\partial p}{\partial y_\alpha}(X,Y)\zeta^\alpha - \frac{1}{2}\sum_{\lambda=0}^{r}\sum_{|\alpha|=m}\left[\frac{\partial^2 p}{\partial x_\alpha \partial y_\alpha}(X,Y) + \sum_{|\beta|\leq m}\frac{\partial^2 p}{\partial x_\alpha \partial y_\beta}(X,Y)y_{\beta+1_\lambda}\right]\alpha_\lambda \zeta^{\alpha-1_\lambda} \tag{8}$$

dans lequel on remplace y_γ par $\partial^\gamma y$ et Y par $\{\partial^\alpha y\}_{\alpha \in A}$.

Or, cette expression est linéaire par rapport à chaque paramètre y_γ tel que $|\gamma| = m+1$.

Donc H_0 divisera q''_{m-1} si et seulement si

$$\left.\begin{array}{l} H_0 \text{ divise } \displaystyle\sum_{\substack{\lambda=0 \\ \beta+1_\lambda=\gamma}}^{r}\sum_{|\beta|=m}\sum_{|\alpha|=m}\alpha_\lambda \frac{\partial^2 p}{\partial y_\alpha \partial y_\beta}\zeta^{\alpha-1} \\ \text{pour chaque } \gamma = (\gamma_0,\ldots,\gamma_r) \text{ tel que } |\gamma| = m+1 \end{array}\right\} \tag{9}$$

et

H_0 divise q'_{m-1}. (10)

D'après (Hyp 3), (9) équivaut à:

$$H_0 \text{ divise } \frac{\partial H_0}{\partial y_\beta}\frac{\partial H_o}{\partial \xi_\lambda} \quad (\forall\beta \quad \text{tel que } |\beta| = m)\ (\forall\lambda \quad = 0,\dots,r). \tag{11}$$

En particulier (11) est satisfaite lorsque:

$$H_0(X,Y,\zeta) = H'_0(X,Y)H''_0(X,\{y_\alpha\}_{|\alpha|\leq m-1},\zeta) \tag{12}$$

où $H'_0 \in C^\infty_R(\Omega \times U)$,

H''_0 est de classe C^∞ en X et $\{y_\alpha\}_{|\alpha|\leq m-1}$ et polynomial irréductible en ζ ((11) $\Leftrightarrow$ (12) lorsque $d^0_\zeta H_0 = 1$).
D'où le résultat puisque (9) $\Leftrightarrow$ (hyp 6) et (12) $\Leftrightarrow$ (Hyp 5).

DEFINITION 2 ([9]-[10]: On désigne par $H_{m,2}(\Omega)$ la classe des opérateurs aux dérivées partielles faiblement hyperboliques d'ordre m non linéaires Φ définis par (1) tels qu'il existe U voisinage ouvert de $0 \in \mathbf{R}^N$ pour lequel les hypothèses (Hyp 1) - (Hyp 6) sont vérifiées.

DEFINITION 3 ([9]-[10]): On désigne par $C_{m,2}(\Omega)$ la classe des opérateurs Φ aux dérivées partielles d'ordre m non linéaires définis par:

$$\Phi(y) = p(X,\partial^\alpha y) = g(X,k\ h^2 y, e\ h\ y, \{\partial^\alpha y\}_{|\alpha|\leq m-2}) \tag{13}$$

tels que:

$$h(X,\partial^\alpha),\ k(X,\partial^\beta),\ e(X,\partial^\gamma) \tag{14}$$

sont des operateurs aux dérivées partielles linéaires d'ordres respectifs ρ, τ et $m-1-\rho$ (avec $m = 2\rho + \tau$) à coefficients $C^\infty_R(\Omega)$ et $x_0 = 0$ n'est pas caractéristique pour h et k.

h k est strictement hyperbolique par rapport à ξ_0 dans Ω, (15)

$$g(X,u,v,\{\omega_\alpha\}_{|\alpha|\leq m-2}) \in C^\infty_R(\Omega \times \mathbf{R} \times \mathbf{R} \times \mathbf{R}^{N'}), \tag{16}$$

$$\text{Supp } g(X,0,0,0) \subset \Omega+ \tag{17}$$

$$\frac{\partial g}{\partial u}(X,0,0,0) \neq 0 \text{ et } \inf_{X\in\Omega} \left|\frac{\partial g}{\partial u}(X,0,0,0)\right| > 0 \tag{18}$$

PROPOSITION 2 [9]-[10]: $\mathcal{C}_{m,2}(\Omega) \subset \mathcal{H}_{m,2}(\Omega)$.

PREUVE: Les linéarisés sont bien décomposables au sens de J.C. De Paris ([5]) donc vérifient la condition de Levi; les hypothèses (Hyp 1) - (Hyp 6) sont satisfaites.

On obtient alors le théorème:

THÉORÈME 1([9] - [10]): Soit $\Phi \in \mathcal{H}_{m,2}(\Omega)$; alors pour tout $f \in C_+^\infty(\Omega)$, il existe $\tilde{\Omega} \subset \Omega$ et $y \in C_+^\infty(\tilde{\Omega})$ unique tel que

$$\Phi(y) = p(X,\partial^\alpha y) = f \text{ dans } \tilde{\Omega}.$$

De plus, dans l'estimation (E_s), on a $\ell = -(m-2)$.

EXEMPLE 1:

$$\Omega = \mathbf{R}^2,\ m = 2$$

$$\Phi(y) = p(X,\partial^\alpha y) = \partial_0^2 y + y^2\partial_x^2 y - 2y\,\partial_0\partial_x\, y - (\partial_0 y)(\partial_x y)$$
$$+ y(\partial_x y)^2 + b(X,y)$$

alors $\Phi \in \mathcal{H}_{2,2}(\mathbf{R}^2)$ si supp $b(X,0) \subset R^+ \times R$.

EXEMPLE 2:

$$g\Big(X,\big(\partial_0 - \sum_{k=1}^{r} a_k(X)\partial_k\big)^2 y, b(X)\big(\partial_0 - \sum_{k=1}^{r} a_k(X)\partial_k\big)y, c(X)y\Big) \in \mathcal{C}_{2,2}(\mathbf{R}^{r+1})$$

si g vérifie (16), (17), (18).

Dans cet exemple 2, l'équation $\Phi(y) = f$ se ramène à l'intégration d'une équation différentielle ordinaire du second ordre le long des courbes intégrales de l'opérateur

$$\partial_0 - \sum_{k=1}^{r} a_k(X)\partial_k \text{ où } \partial_k = \frac{\partial}{\partial x_k} \ (k = 0,\ldots,r).$$

Le théorème 1 reste vrai pour les opérateurs plus généraux du type suivant.

<u>DEFINITION 4</u> ([9]-[10]): On appelle $C_{m,d}(\Omega)$ la classe des opérateurs Φ aux dérivées partielles d'ordre m non linéaires définis par:

$$\Phi(y) = p(X,\partial^{\alpha}y) = g(X,(h_1)^{d_1}(h_2)^{d_2} \ldots (h_{\mu})^{d} y, \tag{19}$$

$$e_1(h_1)^{d_1-1}(h_2)^{d_2-1} \ldots (h_{\mu})^{d_{\mu}-1} y,$$

$$\ldots\ldots\ldots\ldots\ldots\ldots\ldots\ldots\ldots\ldots\ldots\ldots$$

$$e_j(h_1)^{(d_1-j)_+}(h_2)^{(d_2-j)_+} \ldots (h_{\mu})^{(d_{\mu}-j)_+} y,$$

$$\ldots\ldots\ldots\ldots\ldots\ldots\ldots\ldots\ldots\ldots\ldots\ldots$$

$$e_{d-1}(h_1)^{(d_1-d+1)_+}(h_2)^{(d_2-d+1)_+} \ldots (h_{\mu})^{(d_{\mu}-d+1)_+} y,$$

$$\{\partial^{\alpha}y\}_{|\alpha|\leqq m-d})$$

où

(i) $d = \sup\{d_k \ ; \ 1 \leqq k \leqq \mu\}$

$(d_k-j)_+ = \sup(d_k - j,0)$;

(ii) les h_j (resp. e_j) sont des opérateurs aux dérivees partielles linéaires, à coefficients $C^{\infty}_R(\Omega)$, d'ordre

$$\rho_j \text{ (resp. } \theta_j = m-j - \sum_{k=1}^{\mu} \rho_k(d_k-j)_+) \text{ et}$$

$x_0 = 0$ n'est pas caractéristique pour les h_j.

(iii) $\sum_{k=1}^{\mu} d_k\rho_k = m$

(iv) $g(X,v_m,v_{m-1},\ldots,v_{m-d+1}, \{\omega_{\alpha}\}_{|\alpha|\leqq m-d}) \in C^{\infty}_R(\Omega \times R^{N'})$

(v) $\text{Supp } g(X,0,0,\ldots,0,\{0\}) \subset \Omega^+$

(vi) $\frac{\partial g}{\partial v_m}(X,0,0,\dots,0,\{0\}) \neq 0$ et $\inf_{X\in\Omega} \left|\frac{\partial g}{\partial v_m}(X,0,\dots,\{0\})\right| > 0$.

(vii) $h_1 h_2 \dots h_\mu$ est strictement hyperbolique par rapport à ξ_0 dans Ω.

PROPOSITION 3 ([9]-[10]): Si $\Phi \in C_{m,d}(\Omega)$ alors il existe U voisinage ouvert de 0 dans R^N tel que Φ soit faiblement hyperbolique relativement à ξ_0 et à U et tel que $\Phi'(y)$ soit inversible pour tout $y \in \tilde{V}$ (quelque soit $\tilde{\Omega} \subset \Omega$ fixé).

PREUVE: Le polynome caractéristique de $\Phi'(y)$ s'écrit:

$$q_m(X,Y,\zeta) = \frac{\partial g}{\partial v_m}(X,v_m,\dots,v_{m-d+1},\{\omega_\alpha\}_{|\alpha|\leq m-d}) \times (H_1(X,\zeta))^{d_1}\dots(H_\mu(X,\zeta))^{d_\mu}.$$

Ce polynôme est donc faiblement hyperbolique relativement à ξ_0, $X \in \Omega$, $Y \in U$ pour un voisinage ouvert U de 0 dans $\mathbb{R}^N$ d'après les hypothèses (on a noté $H_j(X,\zeta)$ le polynôme caractéristique de h_j).

D'autre part, $\Phi'(y)$ est bien decomposable au sens de J.C. De Paris ([5]) pour tout $y \in \tilde{V}$; donc $\Phi'(y)$ est inversible pour tout $y \in \tilde{V}$.

On obtient alors:

THÉORÈME 2 ([9]-[10]): Si $\Phi \in C_{m,d}(\Omega)$, pour tout $f \in C_+^\infty(\Omega)$, il existe $\tilde{\Omega} \subset \Omega$ et $y \in C_+^\infty(\tilde{\Omega})$ unique tel que

$$\Phi(y) = p(X,\partial^\alpha y) = f \text{ dans } \tilde{\Omega}.$$

De plus, dans l'estimation (E_s), on a $\ell = -(m-d)$.

3. Cas des systèmes semi-linéaires du premier ordre ([9]-[10]):

Considérons

$$Ly = \sum_{\lambda=0}^{r} A^\lambda(X)\frac{\partial y}{\partial x_\lambda} + B(X,y) \tag{20}$$

où $A^\lambda(X) = (a_j^{i\lambda}(X))_{\substack{1\leq i\leq N\\1\leq j\leq N}}$ est une matrice de fonctions $C_R^\infty(\Omega)$,

$B(X,y) = (b^i(X,y))_{1\leq i\leq N}$ est une matrice colonne de fonctions de classe $C_R^\infty(\Omega \times \mathbb{R}^N)$,

$$y = {}^t(y_1,\dots,y_N) \in [C^\infty_R(\Omega)]^N.$$

La matrice caractéristique s'écrit $H(X,\xi) = \sum_{\lambda=0}^{r} A^\lambda(X)\xi_\lambda$; $K(X,\zeta)$ désigne la matrice des cofacteurs de $H(X,\zeta)$.

On suppose que:

$$\inf_{X\in\Omega} |\det A^0(X)| > 0. \tag{21}$$

$$\operatorname{Supp} B(X,0) \subset \Omega^+. \tag{22}$$

$$\det H(X,\zeta) = [H_0(X,\zeta)]^2 \prod_{s=1}^{\sigma} H_s(X,\zeta) \text{ est une décomposition} \tag{23}$$

en facteurs irréductibles H_s dans $\mathbb{R}[\zeta]$, de classe $C^\infty_R(\Omega)$ en X.

$$\prod_{s=0}^{\sigma} H_s(X,\zeta) \text{ est strictement hyperbolique par rapport à } \xi_0 \tag{24}$$

$(\forall X \in \Omega)$.

$$\operatorname{rang} H(X,\zeta)\big|_{H_0(X,\zeta)=0} = N-1 \text{ et } \inf_{X\in\Omega, |\xi|=1} |K_1^1(X,\xi)| > 0. \tag{25}$$

Les opérateurs linéarisés s'écrivent:

$$(L'y)z = \sum_{\lambda=0}^{r} A^\lambda(X) \frac{\partial z}{\partial x_\lambda} + \frac{\partial B}{\partial y}(X,y)z$$

avec

$$\frac{\partial B}{\partial y} = \begin{vmatrix} \frac{\partial b^1}{\partial y_1}, & \dots, & \frac{\partial b^1}{\partial y_N} \\ \vdots & & \vdots \\ \frac{\partial b^N}{\partial y_1}, & \dots, & \frac{\partial b^N}{\partial y_N} \end{vmatrix}$$

et $z = {}^t(z_1,\dots,z_N) \in [C^\infty_R(\Omega)]^N$.

Le polynôme sous caractéristique de L'y s'écrit ([21])

$$K = \sum_{i,j} \left[\frac{\partial b^i}{\partial y_j}(X,y) - \frac{1}{2}\sum_{\lambda=0}^{r} \frac{a_j^{i,\lambda}}{\partial x_\lambda}\right] K_i^1(X,\xi) K_1^j(X,\xi)$$

$$+ \frac{1}{2}\left(\sum_{\lambda=0}^{r} a_j^{i,\lambda}(X)\xi_\lambda\right) \sum_{\lambda=0}^{r} \left(\frac{\partial K_1^j}{\partial \xi_\lambda}\frac{\partial K_i^1}{\partial x_\lambda} - \frac{\partial K_1^j}{\partial x_\lambda}\frac{\partial K_i^1}{\partial \xi_\lambda}\right).$$

D'après la généralisation des conditions de Levi aux systèmes faiblement hyperboliques (cf. [1], [7], [8], [21]), on sait que L'(y) est inversible si et seulement si $K(X,y,\zeta)$ est divisible par $H_o(X,\zeta)$.

On en déduit le résultat suivant:

THÉORÈME 3 ([9]-[10]): Sous les hypothèses (29-(25), si il existe un ouvert U de $\mathbb{R}^N$ contenant 0 et une fonction $L(X,Y,\zeta)$ polynomiale en ζ, de classe $C_R^\infty(\Omega \times U)$ en (X,Y) telle que, pour $\forall X \in \Omega$, $\forall Y \in U$ on ait:

$$K(X,Y,\zeta) = H_o(X,\zeta)L(X,Y,\zeta),$$

alors quel que soit $f = {}^t(f_1,\ldots,f_N) \in [C_+^\infty(\Omega)]^N$, il existe $\tilde{\Omega}$ et $y = {}^t(y_1,\ldots,y_N) \in [C_+^\infty(\tilde{\Omega})]^N$ unique vérifiant

$$Ly = f \text{ dans } \tilde{\Omega}.$$

REMARQUES ([9]-[10]): Dans le cas matriciel quasi-linéaire du ler ordre, on retrouve la condition obtenue par Y. Choquet-Bruhat (cf. [3]). L'extension au cas matriciel non linéaire peut se faire en généralisant le schéma d'hypothèses exprimé dans 1), 2), 3) ([10]).

Bibliographie

[1] R. Berzin, J. Vaillant, Systèmes hyperboliques à caractéristiques multiples. Journal de Math. Pures et Appl., 58, 1979, p. 165-216.

[2] Y. Choquet-Bruhat, Diagonalisation des systèmes quasi-linéaires et hyperbolicité non stricte. Journal de Math. Pures et Appl., t. 45, 1966, p. 371-386.

[3] Y. Choquet-Bruhat, Ondes asymptotiques et approchées pour des systèmes d'équations aux dérivées partielles non linéaires (pages 137 et 142). Journal de Math. Pures et Appl., t. 48, 1969, p. 117-158.

[4] Y. Choquet-Bruhat, C. DeWitt-Morette and M. Dillard-Bleick, Analysis, Manifold and Physics. (North Holland, Amsterdam, 1977).

[5] J.Cl. De Paris, Problème de Cauchy oscillatoire pour un opérateur différentiel à caractéristiques multiples; lien avec l'hyperbolicité. Journal de Math. Pures et Appl., 51, 1972, p. 231-256.

[6] Ph. Dionne, Sur les problèmes de Cauchy hyperboliques bien posés. J. Analyse Math., 10 (1962), p. 1-90.

[7] D. Gourdin, Systèmes faiblement hyperboliques à caractéristiques multiples. Comptes Rendus Acad. Sci. Paris, Série A, t. 278 (1974), p. 269-272.

[8] D. Gourdin, Les opérateurs faiblement hyperboliques matriciels à caractéristiques de multiplicités constantes, bien décomposables et le problème de Cauchy associé. J. Maths. Kyoto Univ. (JMKYAZ), 17-3 (1977), p. 539-566.

[9] D. Gourdin, Problèmes de Cauchy hyperboliques non linéaires. Proposition de Note aux Comptes Rendus Acad. Sc. Paris (1987).

[10] D. Gourdin, Problèmes faiblement hyperboliques non linéaires. A paraître.

[11] R.S. Hamilton, Nash-Moser Inverse Function Theorem. Bulletin of the American Mathematical Society, Vol. 7, n° 1, 1982, p. 65-222.

[12] N. Iwasaki, Effectively hyperbolic equations. J. Maths of Kyoto Univ. 25-4 (1985), p. 727-743.

[13] N. Iwasaki, The strongly hyperbolic equation and its applications. RIMS (520) Kyoto Univ. (1985), p. 1-38.

[14] K. Kajitani, Local solution of Cauchy problem for non linear hyperbolic systems in Gevrey classes. Hokkaido Math. J., t. 12, 1983, n° 3, p. 434-460.

[15] K. Kajitani, The Cauchy problem for non linear systems. Bulletin des Sciences Math. 2ème série, 110 (1986), p. 3-48.

[16] J. Leray, Hyperbolic differential equations. Cours de Princeton (1952).

[17] J. Leray et Y. Ohya, Equations et systèmes non linéaires hyperboliques non strictes. Math. Ann., 170 (1967), p. 167-205.

[18] E.E. Levi, Caratteristiche multiple e problema di Cauchy. Ann. di Mat., 16 (1909), p. 161-201.

[19] S. Mizohata et Y. Ohya, Sur la condition de E.E. Levi concernant des équations hyperboliques. Publ. RIMS, Kyoto Univ. Ser. A, 4, 511-526 (1968).

[20] U. Murthy, A survey of some recent developpments in the theory of Cauchy problem. Bolletino U.M.I. (4), 4, (1971), 473-562.

[21] J. Vaillant, J. Maths Pures et Appl. (9), 47, (1968), p. 1-40.

[22] Cl. Wagschal, Séminaire J. Vaillant 1979-80 (Paris VI - I.H.P.).

Daniel Gourdin
Université des Sciences et Techniques
de Lille Flandres Artois,
U.F.R. de Mathématiques Pures et
Appliquées,
59655 Villeneuve d'Ascq Cedex
France.

Y. HAMADA

Les singularités des solutions du problème de Cauchy à données holomorphes

ABSTRACT: The singularities of the solutions of the Cauchy problem with holomorphic data

J. Leray [6], and L. Gårding, T. Kotake and J. Leray [2] have studied the singularities of the solution of the linear Cauchy problem with holomorphic data, when the initial surface has the characteristic points. They have proved that the solution ramifies around the analytic variety K of codimension 1 tangent to the initial surface S and it can be explicitly uniformized, except in the exceptional cases. In [6] and [2], they have supposed that the operators have simple characteristics.

Y. Choquet-Bruhat [1] has extended certain results of [6] and [2] to the case of a system of nonlinear differential equations.

In this paper we study the case of the operators with multiple characteristics. Then the solution has the essential singularities along the analytic variety K except in the exceptional cases. Thus, by following the reasonings of [6] and [2], we give a complement of certain results of [6] and [2].

J. Leray [6], et L. Gårding, T. Kotake et J. Leray [2] ont étudié les singularités de la solution du problème de Cauchy linéaire à données holomorphes, lorsque la surface initiale a des points caractéristiques. Ils ont démontré que la solution se ramifie autour d'un ensemble analytique K de codimension 1 tangent à la surface initiale S et peut être explicitement uniformisée, sauf dans des cas exceptionnels. Dans [6] et [2], ils ont supposé que les opérateurs ont les caractéristiques simples.

Y. Choquet-Bruhat [1] a étendu certains résultats de [6] et [2] au cas d'un système d'équations différentielles non linéaires.

Dans cet article nous étudions le cas des opérateurs à caractéristiques multiples. La solution a alors des singularités essentielles le long d'un ensemble analytique K sauf dans des cas exceptionnels. Ainsi, en suivant les raisonnements de [6] et [2], nous complétons les résultats de [6] et [2].

1. HYPOTHÈSES ET RÉSULTATS

Soit X un voisinage de l'origine 0 de $\mathbb{C}^{n+1}$. $x = (x_0,x')$, $x' = (x_1,\dots,x_n)$.

On considère $a(x,D)$ un opérateur différentiel à coefficients holomorphes sur X d'ordre m

$$a(x,D) = \sum_{|\alpha|\leq m} a_\alpha(x)D^\alpha,$$

où $D = (D_0,D_1,\dots,D_n)$, $D_i = \partial/\partial x_i$, $\alpha = (\alpha_0,\dots,\alpha_n) \in \mathbb{N}^{n+1}$,

$$|\alpha| = \sum_{i=0}^{n} \alpha_i, \quad D^\alpha = D_0^{\alpha_0} \dots D_n^{\alpha_n}.$$

Son polynôme caractéristique est noté $g(x,\xi)$. $\xi = (\xi_0,\dots,\xi_n)$.

Soit S : $\varphi(x) = 0$ ($\varphi(0) = 0$, $D_x\varphi(0) \neq (0)$) une surface régulière dans X.

Nous faisons l'

Hypothèse 1.1: $g(x,\xi)$ est de la forme $g(x,\xi) = g_1(x,\xi)^\mu \, g_2(x,\xi)$, où $g_i(x,\xi)$ $(i = 1,2)$ sont des polynômes homogènes en ξ de degrés m_i et μ est un entier ≥ 1.

(i) La surface S est non caractéristique relatif à $g_1(x,\xi)$ sauf sur l'ensemble analytique T de codimension 1 par rapport à S, où $T = \{x \in S;\ g_1(x,D_x\varphi(x)) = \alpha(x)=0\}$. $\alpha(x)$ est alors une fonction holomorphe au voisinage de 0, $\alpha(0) = 0$ et $\alpha(x) \not\equiv 0$ sur S.

(ii) La surface S n'est pas caractéristique relatif à $g_2(x,\xi)$. Donc on a $g_2(x,D_x\varphi(x)) \neq 0$ au voisinage de 0.

Nous étudions le problème de Cauchy

$$a(x,D)u(x) = v(x), \quad u(x) - w(x) \text{ s'annule m fois sur S}, \tag{1.1}$$

où v(x) et w(x) sont des fonctions holomorphes au voisinage de 0.

Nous avons dit qu'une fonction s'annule m fois sur S si ses dérivées d'ordre $\leqq m-1$ sont nulles sur S.

D'après le théorème de Cauchy - Kowalewski, le problème (1.1) admet une unique solution holomorphe au voisinage de $S \setminus T$.

Nous allons étudier le prolongement analytique de cette solution au voisinage de l'origine.

Pour cela faire, nous faisons en plus une hypothèse.

Soit $\Phi(\tau,x)$ la solution de l'équation

$$D_\tau\Phi(\tau,x) + g_1(x,D_x\Phi) = 0, \; \Phi(0,x) = \varphi(x). \tag{1.2}$$

Nous supposons que $\Phi(\tau,x)$ vérifie la condition suivante.

Hypothèse 1.2: $\Phi(\tau,0) \not\equiv 0$ pour τ suffisamment petit.

D'après l'hypothèse 1.2 et le théorème de préparation de Weierstrass, on a $\Phi(\tau,x) = d(\tau,x)D(\tau,x)$, où $d(\tau,x)$ est une fonction holomorphe non nulle au voisinage de (0,0). $D(\tau,x)$ est un pseudo-polynôme distingué en τ au point (0,0). $D(\tau,x)$ est irréductible, car $\Phi(0,x) = \varphi(x)$, $\varphi(0) = 0$ et $D_x\varphi(0) \neq (0)$.

Soit $\Delta(x)$ le discriminant de $D(\tau,x)$ en τ et $K = \{x;\ \Delta(x) = 0\}$. K est un ensemble analytique de codimension 1 tangent à S sur T et n'est pas régulier en général. On a aussi $K = \{x;\ \Phi(\tau,x) = D_\tau\Phi(\tau,x) = 0\}$. K est caractéristique relatif à $g(x,\xi)$, plus précisément relatif à $g_1(x,\xi)$.

Nous notons $\tau(x)$ la fonction algébroïde vérifiant $\Phi(\tau(x),x) = 0$. La fonction $\tau(x)$ se ramifie autour K.

Nous avons alors le

<u>THÉORÈME 1</u>: Sous les hypothèses 1.1 et 1.2, le problème de Cauchy (1.1)

admet une unique solution holomorphe $u(x)$ sur un revêtement fini $\mathfrak{R}_{fini}(V\setminus K)$ de $V\setminus K$, V étant un voisinage de 0.

Plus précisément, $u(x)$ est de la forme $u(x) = U(\tau(x),x)$; Pour $\mu > 1$, $U(\tau,x)$ est une fonction holomorphe sauf sur la surface $\{(\tau,x);\ D_\tau\Phi(\tau,\quad 0\}$, ayant des singularités essentielles sur cette surface. Donc $u(x)$ a des singularités essentielles sur K en général.

Pour $\mu = 1$, $U(\tau,x)$ est une fonction holomorphe sur un voisinage de $(0,0)$. $u(x)$ est donc une fonction algébroïde et elle est continue sur K. Ceci a été démontré dans [6] et [2].

Comme dans [2], nous introduisons une variable complexe σ et considérons le problème de Cauchy

$$\begin{cases} a(x,D)u(\sigma,x) = v(x), \\ u(\sigma,x)-w(x) \text{ s'annule } m \text{ fois sur } S(\sigma):\ \varphi(x) - \sigma = 0. \end{cases} \tag{1.3}$$

Nous avons alors le

<u>THÉORÈME 2</u>: Sous les hypothèses 1.1 et 1.2, le problème de Cauchy (1.3) admet une unique solution $u(\sigma,x)$ telle que $u(\Phi(\tau,x),x) = U(\tau,x)$ soit pour $\mu = 1$ une fonction holomorphe au voisinage de $(0,0)$, et soit pour $\mu > 1$ une fonction holomorphe sauf sur la surface $\{(\tau,x);\ D_\tau\Phi(\tau,x) = 0\}$, ayant des singularités essentielles sur cette surface.

Nous notons que le théorème 2 résulte du théorème 1 en choisissant $\sigma = 0$.

Nous donnons un exemple.

Considérons le problème de Cauchy

$$(2x_1D_0-D_1)^2u(x) = D_0u(x),\ u(0,x_1) = \gamma_1x_1^3,\ D_0u(0,u_1) = \gamma_2x_1,$$

où $$\gamma_1 = \sum_{n=0}^{\infty} (-1)^n \frac{\Gamma(n-\frac{3}{2})}{(2n)!},\quad \gamma_2 = \sum_{n=0}^{\infty} (-1)^{n+1} \frac{\Gamma(n-\frac{1}{2})}{(2n)!}\ .$$

La solution est la fonction

$$u(x) = \sum_{n=0}^{\infty} (-1)^n \frac{\Gamma(n-\frac{3}{2})}{(2n)!} (x_0 + x_1^2)^{\frac{3}{2}-n}\ x_1^{2n}\ ,$$

ayant une singularité essentielle sur $K = \{(x_0,x_1);\ x_0 + x_1^2 = 0\}$.

2. MÉTHODE DE RÉSOLUTION DU PROBLÈME DE CAUCHY (1.3)

En remplaçant $u(\sigma,x)$ par $u(\sigma,x) - w(x)$, nous pouvons supposer $w(x) \equiv 0$ dans (1.3). Nous posons

$$a(x,D) = h(x,D) - b(x,D),$$

où $h(x,D) = g_1(x,D)^{\mu}\, g_2(x,D)$; donc $b(x,D)$ est un opérateur différentiel d'ordre $m - 1$.

Pour étudier le problème (1.3), nous considérons le problème

$$\begin{cases} h(x,D)u_0(\sigma,x) = v(x), \\ u_0(\sigma,x) \text{ s'annule } m \text{ fois sur } S(\sigma), \\ h(x,D)u_k(\sigma,x) = b(x,D)u_{k-1}(\sigma,x), \\ u_k(\sigma,x) \text{ s'annule } m \text{ fois sur } S(\sigma), \text{ pour } k \geq 1. \end{cases} \tag{2.1}$$

On a alors $u(\sigma,x) = \sum_{k=0}^{\infty} u_k(\sigma,x)$.

Nous posons pour $k \geq 0$,

$$\begin{cases} u_k(\sigma,x) = D_\sigma^{(k+1)(\mu-1)}\hat{u}_k(\sigma,x), \\ \hat{u}_k(\sigma,x) \text{ s'annule } m + (k+1)(\mu - 1) \text{ fois sur } S(\sigma). \end{cases} \tag{2.2}$$

On a le

LEMME 2.1: Le problème (2.1) équivaut au problème de Cauchy suivant

$$\begin{cases} h(x,D)D_\sigma^{\mu-1}\hat{u}_0(\sigma,x) = v(x), \\ \hat{u}_0(\sigma,x) \text{ s'annule } m + \mu - 1 \text{ fois sur } S(\sigma), \\ h(x,D)D_\sigma^{\mu-1}\hat{u}_k(\sigma,x) = b(x,D)\hat{u}_{k-1}(\sigma,x), \\ \hat{u}_k(\sigma,x) \text{ s'annule } m + \mu - 1 \text{ fois sur } S(\sigma), \text{ pour } k \geq 1. \end{cases} \tag{2.3}$$

Note: Les solutions $\hat{u}_k$ du problème (2.3) s'annulent donc nécessairement $m + (k + 1)(\mu - 1)$ fois sur $S(\sigma)$.

PREUVE: Pour une fonction $f(\sigma,x)$, on pose

$$D_\sigma^{-1}f(\sigma,x) = \int_{\varphi(x)}^{\sigma} f(s,x)ds.$$

On a $D_\sigma^{\ell}[D_\sigma^{-k}f(\sigma,x)] = D_\sigma^{\ell-k}f(\sigma,x)$, si ℓ, $k \geqq 0$.

Si une fonction $f(\sigma,x)$ s'annule ℓ fois sur $S(\sigma)$, on a $D_\sigma^{-\ell}[D_\sigma^{\ell}f(\sigma,x)] = f(\sigma,x)$.

Soient $\hat{u}_k$ les solutions du problème (2.1) avec (2.2). On a

$$D_\sigma^{k(\mu-1)}h(x,D)D_\sigma^{\mu-1}\hat{u}_k = D_\sigma^{k(\mu-1)}b(x,D)\hat{u}_{k-1}.$$

En notant que, vu (2.2), $h(x,D)D_\sigma^{\mu-1}\hat{u}_k$ et $b(x,D)\hat{u}_{k-1}$ s'annulent $k(\mu-1)$ fois sur $S(\sigma)$ et en opérant $D_\sigma^{-k(\mu-1)}$ à cette équation, on obtient $h(x,D)D_\sigma^{\mu-1}u_k = b(x,D)\hat{u}_{k-1}$. Donc les $\hat{u}_k$ vérifient (2.3). Alors l'équivalence de (2.1) avec (2.2) et (2.3) résulte de l'unicité des solutions $\hat{u}_k$ des problèmes (2.1) avec (2.2) et (2.3). C.Q.F.D.

Pour étudier le problème (2.3), nous suivons les raisonnements de [2]. Nous considérons une fonction composée

$$F(\tau,x) = f(\Phi(\tau,x),x) = f(\sigma,x) \circ \Phi(\tau,x).$$

On a $D_\tau(f\circ\Phi) = D_\tau\Phi(D_\sigma f\circ\Phi)$, $D_i(f\circ\Phi) = D_i\Phi(D_\sigma f\circ\Phi) + D_i f\circ\Phi$.

Posons $f_j(\sigma,x) = (-D_\sigma)^j f(\sigma,x)$. On a une formule de dérivation de fonction composée.

LEMME 2.2 ([2]): Soit $P(x,D_x)$ un opérateur différentiel homogène en D_x d'ordre r. On a

$$[P(x,D_x)f]\circ\Phi = \sum_{j=0}^{r} P_j(\tau,x,D_x)(f_{r-j} \circ \Phi),$$

où $P_j(\tau,x,D_x)$ $(0 \leqq j \leqq r)$ sont des opérateurs d'ordre j ne dépendant que de P et Φ. En particulier, on a

$$P_0 = P(x, D_x\Phi),$$

$$P_1 = \sum_{i=0}^{n} P^{(i)}(x, D_x\Phi)D_i + c(P,\Phi),$$

$$c(P,\Phi) = \frac{1}{2}\sum_{i,j=0}^{n} P^{(i,j)}(x, D_x\Phi)D_iD_j\Phi.$$

On a noté $P^{(i)}(x,\xi) = D_{\xi_i} P(x,\xi)$, $P^{(i,j)}(x,\xi) = D_{\xi_i}D_{\xi_j}P(x,\xi)$.

Ce lemme s'écrit sous la forme suivante.

<u>LEMME 2.3</u>: Posons $P(x,\xi) = \sum_{|\alpha| = r} b_\alpha(x)\xi^\alpha$. On a

$$[P(x,D_x)f]\circ\Phi = P(x, D_x\Phi\mathfrak{D}_\tau + D_x)F(\tau,x)$$

$$= \sum_{|\alpha| = r} b_\alpha(x)(D_x\Phi\mathfrak{D}_\tau + D_x)^\alpha F(\tau,x)$$

$$= \sum_{j=0}^{r} P_j(\tau,x,D_x)\mathfrak{D}_\tau^{r-j}F(\tau,x),$$

où $\mathfrak{D} = -(1/D_\tau\Phi)D_\tau$.

Enfin on a

$$\mathfrak{D}_\tau(D_i\Phi\mathfrak{D}_\tau + D_i)F = (D_i\Phi\mathfrak{D}_\tau + D_i)\mathfrak{D}_\tau F,$$

et donc $\mathfrak{D}_\tau P(x,D_x\Phi\mathfrak{D}_\tau + D_x)F = P(x,D_x\Phi\mathfrak{D}_\tau + D_x)\mathfrak{D}_\tau F$.

<u>PREUVE</u>: Il suffit de démontrer le dernier. En effet, il résulte de ce que $[-D_\sigma D_i f]\circ\Phi = [D_i(-D_\sigma)f]\circ\Phi$. C.Q.F.D.

En employant ce lemme, on obtient

$$[g_1(x,D)u(\sigma,x)] \circ \Phi(\tau,x) = g_1(x, D_x\Phi\mathfrak{D}_\tau + D_x)U(\tau,x)$$

$$= \{g_1(x,D_x\Phi)\mathfrak{D}_\tau^{m_1} + \sum_{j=1}^{m_1} G_j(\tau,x,D_x)\mathfrak{D}_\tau^{m_1-j}\}U(\tau,x)$$

$$= \{[D_\tau + G_1(\tau,x,D_x)]\mathfrak{D}_\tau^{m_1-1} + \sum_{j=2}^{m_1} G_j(\tau,x,D_x)\mathfrak{D}_\tau^{m_1-j}\}U(\tau,x),$$

où $U(\tau,x) = u(\Phi(\tau,x),x)$ et $G_j(\tau,x,D_x)$ sont des opérateurs différentiels d'ordre j.

Posons $\hat{U}_k(\tau,x) = \hat{u}_k(\Phi(\tau,x),x)$ dans (2.2).

En utilisant cette formule et le lemme 2.3, le problème (2.3) se transforme en le problème suivant

$$\begin{cases} L(\tau,x,D_\tau,D_x,\mathfrak{D}_\tau)\hat{U}_0(\tau,x) = v(x), \\ \hat{U}_0(\tau,x) \text{ s'annule } m+\mu-1 \text{ fois sur } \tau = 0, \\ L(\tau,x,D_\tau,D_x,\mathfrak{D}_\tau)\hat{U}_k(\tau,x) = M(\tau,x,D_x,\mathfrak{D}_\tau)\hat{U}_{k-1}(\tau,x), \\ \hat{U}_k(\tau,x) \text{ s'annule } m+\mu-1 \text{ fois sur } \tau = 0, \text{ pour } k \geqq 1, \end{cases} \tag{2.4}$$

où $L(\tau,x,D_\tau,D_x,\mathfrak{D}_\tau) = L_\mu(\tau,x,D_\tau,D_x)\mathfrak{D}_\tau^{m-1} + \sum_{j=1}^{m-\mu} L_{\mu+j}(\tau,x,D_\tau,D_x)\mathfrak{D}_\tau^{m-j-1}$ et $M(\tau,x,D_x,\mathfrak{D}_\tau) = \sum_{j=0}^{m-1} M_j(\tau,x,D_x)\mathfrak{D}_\tau^{m-j-1}$, L_j, M_j étant des opérateurs différentiels holomorphes d'ordre j en D_τ, D_x. En particulier, $L_\mu = D_\tau^\mu - L'_\mu(\tau,x,D_\tau,D_x)$, L'_μ est un opérateur d'ordre μ en D_τ, D_x. L'_μ et $L_j(j \geqq \mu+1)$ sont d'ordre $\mu-1$ en D_τ.

Nous allons démontrer que le problème (2.4) admet une unique solution holomorphe $(\hat{U}_k(\tau,x))$ et estimer $\hat{U}_k(\tau,x)$ $(k \geqq 0)$.

Nous posons $\mathfrak{D}_\tau^j\hat{U}_k(\tau,x) = \hat{U}_{k,j}(\tau,x)$, $k \geqq 0$, $0 \leqq j \leqq m-1$. Alors nous pouvons écrire le problème (2.4) comme suit;

$$\begin{cases} D_\tau^\mu\hat{U}_{0,m-1} = L'_\mu\hat{U}_{0,m-1} + \sum_{j=1}^{m-\mu} L_{\mu+j}\hat{U}_{0,m-j-1} + v(x), \\ D_\tau^\mu\hat{U}_{k,m-1} = L'_\mu\hat{U}_{k,m-1} + \sum_{j=1}^{m-\mu} L_{\mu+j}\hat{U}_{k,m-j-1} + \sum_{j=0}^{m-1} M_j\hat{U}_{k-1,m-j-1}, \text{ pour } k \geqq 1, \\ D_\tau\hat{U}_{k,j} = -D_\tau\Phi\hat{U}_{k,j+1}, \text{ pour } k \geqq 0,\ 0 \leqq j \leqq m-2, \\ D_\tau^\ell\hat{U}_{k,m-1}(0,x) = 0, \text{ pour } k \geqq 0,\ 0 \leqq \ell \leqq \mu-1, \\ \hat{U}_{k,j}(0,x) = 0, \text{ pour } k \geqq 0,\ 0 \leqq j \leqq m-2. \end{cases} \tag{2.5}$$

Le problème (2.5) admet une unique solution holomorphe $(\hat{U}_{k,j}(\tau,x))$ $(k \geqq 0,\ 0 \leqq j \leqq m-1)$. ([2] et [8]). Dans le paragraphe suivant nous estimons $\hat{U}_{k,j}(\tau,x)$.

3. UNE ESTIMATION DE $\hat{U}_{k,j}$ ET PREUVE DU THÉORÈME 2

Pour estimer $\hat{U}_{k,j}$, nous utilisons la méthode des fonctions majorantes. Nous rappelons les résultats de [7] et [3].

PROPOSITION 3.1: Soit $\theta(z) = \sum_{n=0}^{\infty} c_n z^n$ $(c_n \geq 0)$ une série formelle vérifiant

$$(r - z)\ \theta(z) \gg 0, \text{ où } r > 0. \tag{3.1}$$

On a

(i) $\frac{1}{R-z}\ \theta(z) \ll \frac{1}{R-r}\ \theta(z)$, pour $R > r > 0$,

(ii) $\theta(z) \ll r^k D_z^k \theta(z)$, pour $k \geq 0$,

(iii) $(r-z) D_z^k \theta(z) \gg 0$, pour $k \geq 0$.

COROLLAIRE 3.1: Soit $A(\tau,x,D_\tau,D_x)$ un opérateur différentiel holomorphe sur $\{(\tau,x);|\tau| \leq R,\ |x_j| \leq R,\ 0 \leq j \leq n\}$, d'ordre m en D_τ, D_x et d'ordre p en D_τ.

Il existe alors une constante $B > 0$ ne dépendant que de A, R, r, p telle que pour tout $\theta(z)$ vérifiant (3.1), $z = \rho\tau + \sum_{i=0}^{n} x_i$, $\rho \geq 1$, si $u(\tau,x) \ll \theta(z)$, on a

$$A(\tau,x,D_\tau,D_x)u(\tau,x) \ll B\rho^p D_z^m \theta(z).$$

Posons pour p un entier ≥ 0

$$\theta_p(z) = \frac{p!}{(r - z)^{p+1}}, \quad \theta_{-p}(z) = \frac{z^p}{p!(r - z)}.$$

Alors $\theta_p(z)$ $(p \gtreqless 0)$ vérifie (3.1).

PROPOSITION 3.2: Pour tout p entier et $\ell \geq 0$, on a

$$\theta_p(z) \ll D_z^\ell \theta_{p-\ell}(z).$$

PROPOSITION 3.3: Pour tout $r > r' > 0$, il existe une constante C ne depéndant que de r, r', telle que pour tout p, $\ell \geqq 0$, on ait

$$|D_z^{\ell}\theta_{-p}(z)| \leqq C^{p+\ell+1}\ell!/p! \text{ pour } |z| \leqq r'.$$

En utilisant ces propositions, nous obtenons une estimation de la solution $(\hat{U}_{k,j}(\tau,x))$ du problème (2.5).

Dans le problème (2.5), nous pouvons supposer que v(x) est holomorphe sur $\{x;\ |x_i| \leqq r,\ 0 \leqq i \leqq n\}$ et que les opérateurs sont holomorphes sur $\{(\tau,x);\ \ |\tau| \leqq R,\ |x_i| \leqq R,\ 0 \leqq i \leqq n\}$, $(R > r > 0)$.

Nous avons alors la

PROPOSITION 3.4: Dans le problème (2.5), il existe une constante $\rho(\geqq 1)$ telle que pour tout $k \geqq 0$, $0 \leqq j \leqq m-1$, on ait

$$\hat{U}_{k,j}(\tau,x) << \psi_{k,j}(z) = \|v\|\, D_z^{k(m-1)+j}\theta_{-(k+1)(m+\mu-1)} \qquad (3.2)$$
$$z = \rho\tau + \sum_{i=0}^{n} x_i,$$

où $\|v\| = \max_{|x_i| \leqq r} |v(x)|$.

PREUVE: Soit $V(z) = \|v\|\ \theta_0(z) >> v(x)$. Dans (2.5), $\mathcal{L}'_\mu, \mathcal{L}_{\mu+j}, \mathcal{L}_0, \mathfrak{M}_j$ sont des opérateurs remplacés les coefficients de L'_μ, $L_{\mu+j}$, $-D_\tau\Phi$, M_j par leurs fonctions majorantes. Pour démontrer (3.2), il suffit de vérifier que $\psi_{k,j}$ satisfont aux équations majorantes;

$$\begin{cases} D_\tau^\mu\psi_{0,m-1} >> \mathcal{L}'_\mu\psi_{0,m-1} + \sum_{j=1}^{m-\mu} \mathcal{L}_{\mu+j}\psi_{0,m-j-1} + V(z), \\ D_\tau^\mu\psi_{k,m-1} >> \mathcal{L}'_\mu\psi_{k,m-1} + \sum_{j=1}^{m-\mu} \mathcal{L}_{\mu+j}\psi_{k,m-j-1} + \sum_{j=0}^{m-1} \mathfrak{M}_j\psi_{k-1,m-j-1}, \quad \text{pour } k \geqq 1, \\ D_\tau\psi_{k,j} >> \mathcal{L}_0\psi_{k,j+1}, \text{ pour } k \geqq 0,\ 0 \leqq j \leqq m-2. \end{cases} \qquad (3.3)$$

D'après le corollaire 3.1 et la proposition 3.2, on a avec une constante B (> 0)

$$\mathcal{L}'_{\mu}\psi_{0,m-1} + \sum_{j=1}^{m-\mu} \mathcal{L}_{\mu+j}\, \psi_{0,m-j-1} + V(z)$$

$$<< \|v\|\, \rho^{\mu-1}[BD_z^{m+\mu-1}\theta_{-(m+\mu-1)} + \sum_{j=1}^{m-\mu} BD_z^{m+\mu-1}\theta_{-(m+\mu-1)}] + \|v\|\,\theta_0$$

$$<< \|v\|\, (\rho^{\mu-1}mB + 1)D_z^{m+\mu-1}\theta_{-(m+\mu-1)}.$$

D'autre part, on a

$$D_\tau^{\mu}\psi_{0,m-1} = \|v\|\, \rho^{\mu}D_z^{m+\mu-1}\theta_{-(m+\mu-1)}.$$

Pour satisfaire à la première équation majorante dans (3.3) il suffit de choisir $\rho \geqq mB + 1$.

Ensuite, on a de même

$$\mathcal{L}'_{\mu}\psi_{k,m-1} + \sum_{j=1}^{m-\mu} \mathcal{L}_{\mu+j}\psi_{k,m-j-1} + \sum_{j=0}^{m-1} \mathfrak{m}_j\psi_{k-1,m-j-1}$$

$$<< \|v\|\, [\rho^{\mu-1}mBD_z^{k(m-1)+m+\mu-1}\theta_{-(k+1)(m+\mu-1)} + mBD_z^{k(m-1)}\,\theta_{-k(m+\mu-1)}]$$

$$<< \|v\|\, (\rho^{\mu-1}mB + mB)D_z^{k(m-1)+m+\mu-1}\theta_{-(k+1)(m+\mu-1)}$$

D'autre part, on a

$$D_\tau^{\mu}\psi_{k,m-1} = \|v\|\, \rho^{\mu}\, D_z^{k(m-1)+m+\mu-1}\theta_{-(k+1)(m+\mu-1)}.$$

Pour satisfaire à la deuxième équation majorante dans (3.3), il suffit de choisir $\rho \geqq 2mB$.

Enfin, pour satisfaire à la troisième équation majorante dans (3.3), il suffit de choisir $\rho \geqq B$.

Donc, si on choisit $\rho \geqq \max(mB + 1, 2mB)$, on obtient la proposition 3.4.

C.Q.F.D.

D'après les proposition 3.3 et 3.4, nous avons la

<u>PROPOSITION 3.5</u>: Pour $k \geqq 0$ et $0 \leqq j \leqq m-1$, on a

$$|\hat{U}_{k,j}(\tau,x)| \leqq \|v\|\, A^{k+j+1}j!/(k\mu)!, \quad \rho|\tau| + \sum_{i=0}^{n} |x_i| \leqq r' < r,$$

en particulier,

$$|\hat{U}_k(\tau,x)| = |\hat{U}_{k,0}(\tau,x)| \leq \|v\| A^{k+1}/(k\mu)!,$$

où A est une constante > 0.

Nous notons $K(\sigma) = \{(\sigma,x);\Phi(\tau,x) - \sigma = D_\tau\Phi(\tau,x) = 0\}$.

Soit $\tau(\sigma,x)$ la fonction algébroïde vérifiant $\Phi(\tau,x) - \sigma = 0$; $\tau(\sigma,x)$ se ramifie autour $K(\sigma)$.

Par définition on a

$$\hat{u}_k(\sigma,x) = \hat{U}_{k,0}(\tau(\sigma,x),x),$$

$$u_k(\sigma,x) = D_\sigma^{(k+1)(\mu-1)}\hat{u}_k(\sigma,x)$$

$$= (-\mathcal{D}_\tau)^{(k+1)(\mu-1)}\hat{U}_{k,0}(\tau,x)|_{\tau=\tau(\sigma,x)}.$$

Nous avons alors la

PROPOSITION 3.6: $D_x^\alpha D_\sigma^p \hat{u}_k(\sigma,x)$ $(k \geq 0)$ sont fonctions algébroïdes ramifiées sur $K(\sigma)$ pour $|\alpha| + p \leq m - 1$; elles sont contnues sur $K(\sigma)$.

Les fonctions $u_k(\sigma,x)$ $(k \geq 0)$ sont holomorphes sur le revêtement fini $\mathcal{R}_{fini}(V_1 \setminus K(\sigma))$ de $V_1 \setminus K(\sigma)$, V_1 étant un voisinage de (0,0). Les $u_k(\sigma,x)$ ont des singularités polaires sur $K(\sigma)$.

Sur tout ensemble compact Γ dans $\mathcal{R}_{fini}(V_1 \setminus K(\sigma))$, on a

$$|\hat{u}_k(\sigma,x)| \leq \|v\| A^{k+1}/(k\mu)!,$$

$$|u_k(\sigma,x)| \leq \|v\| C^{k+1}/k!,$$

où $C(> 0)$ est une constante dépendant de Γ.

Cette proposition 3.6 démontre le théorème 2. En effet, la série $u(\sigma,x) = \sum_{k=0}^{\infty} u_k(\sigma,x)$ converge sur Γ. La solution $u(\sigma,x)$ du problème (1.3) est donc holomorphe sur $\mathcal{R}_{fini}(V_1 \setminus K(\sigma))$ et elle a des singularités essentielles sur $K(\sigma)$.

On a aussi

$$u(\Phi(\tau,x),x) = U(\tau,x) = \sum_{k=0}^{\infty} (-\mathcal{D}_\tau)^{(k+1)(\mu-1)} \hat{U}_k(\tau,x).$$

Donc si $\mu > 1$, $U(\tau,x)$ est une fonction holomorphe sur

$$\tilde{V}_1 \setminus \{(\tau,x);\ D_\tau\Phi(\tau,x) = 0\}$$

et elle a en général des singularités essentielles sur

$$\{(\tau,x);\ D_\tau\Phi(\tau,x) = 0\},$$

$\tilde{V}_1$ étant un voisinage de (0,0).

Si $\mu = 1$, $U(\tau,x)$ est une fonction holomorphe sur $\tilde{V}_1$.

Références bibliographiques

[1] Y. Choquet-Bruhat, Uniformisation de la solution d'un problème de Cauchy non linéaire à données holomorphes, Bull. Soc. math. France, 94 (1966), 25-38.

[2] L. Gårding, T. Kotake et J. Leray, Uniformisation et développement asymptotique de la solution du problème de Cauchy linéaire à données holomorphes; analogies avec la théorie des ondes asymptotiques et approchées, Bull. Soc. math. France, 92 (1964), 263-361.

[3] Y. Hamada, J. Leray et C. Wagschal, Systèmes d'équations aux dérivées partielles à caractéristiques multiples: problème de Cauchy ramifié; hyperbolicité partielle. J. Math. pures et appl., 55 (1976), 297-352.

[4] Y. Hamada et A. Takeuchi, Sur le problème de Goursat holomorphe pour certains opérateurs à caractéristiques en involution, Japan. J. Math., 7 (1981), 307-318.

[5] Y. Hamada, J. Leray et A. Takeuchi, Prolongements analytiques de la solution du problème de Cauchy linéaire, J.Math. pures et appl., 64, (1985), 257-319.

[6] J. Leray, Uniformisation de la solution du problème linéaire analytique de Cauchy près de la variété qui porte les données de Cauchy, Bull. Soc. math. France, 85 (1957), 389-429.

[7] C. Wagschal, Problème de Cauchy analytique, à données méromorphes, J. Math. pures et appl., 51 (1972), 375-397.

[8] C. Wagschal, Une généralisation du problème de Goursat pour des systèmes d'équations intégro-différentielles holomorphes ou partiellement holomorphes, J. Math. pures et appl., 53 (1974), 99-132.

[9] C. Wagschal, Sur le problème de Cauchy ramifié, J. Math. pures et appl., 53 (1974), 147-164.

[10] Y. Hamada, Sur le problème de Cauchy holomorphe pour les opérateurs à caractéristiques multiples, lorsque la surface initiale a des points caractéristiques, à paraitre aux C.R. Acad. Sc. Paris, (1987).

Yûsaku Hamada
Département de Mathématiques
Université Technologique de Kyoto,
Matsugasaki, Sakyo-Ku, Kyoto, 606,
Japon.

E. HORST

The Vlasov-Maxwell system

0. INTRODUCTION

The relativistic Vlasov-Maxwell system (RVMS) describes the evolution of a plasma in Euclidean space $\underline{R}^3$. We treat the plasma as a continuum and assume that is so rarified that collisions between individual particles can be ignored. The *electric field* $E(t,x)$ and the *magnetic field* $B(t,x)$ ($t \in [0,\infty[$ denotes time and $x \in \underline{R}^3$ denotes position) are assumed to satisfy Maxwell's equations

$$\begin{aligned} \partial_t E &= \operatorname{curl}_x B - j(t,x) \qquad & \operatorname{div}_x E &= \rho(t,x) \\ \partial_t B &= - \operatorname{curl}_x E & \operatorname{div}_x B &= 0. \end{aligned} \tag{ME}$$

This is a very simple form of Maxwell's equations that is valid in the absence of dipoles (cf. [4, p. 35]), but the results can be generalized to other versions of Maxwell's equations (compare, e.g. [8]).

The plasma itself is described by Vlasov's equation. We use the relativistic version (cf. [9]). It makes more sense from the standpoint of physics and it avoids some complications connected with the speed of light (cf. [5, prop. 9]). The plasma at time t is given by a density $f(t,x,u)$ in phase space $\underline{R}^6 = \underline{R}^3 \times \underline{R}^3$. Vlasov's equation is the following first-order scalar linear partial differential equation

$$\partial_t f + \hat{u}\cdot\partial_x f + (E(t,x) + \hat{u} \wedge B(t,x))\cdot\partial_u f = 0. \tag{RVE}$$

u is the momentum, the velocity is given by $\hat{u} := (1 + u^2)^{-1/2}u$; it is obviously bounded by the speed of light, which for our equations is just unity. The wedge $\wedge$ denotes the cross product of R^3.

The relatively simple linear equations (ME) and (RVE) are coupled in a nonlinear way and it is this coupling that makes the problem difficult (and interesting): The *charge density* ρ and the *current density* j are computed with the coupling equations

$$\rho(t,x) = \int_{\underline{R}^3} f(t,x,u)du$$

$$j(t,x) = \int_{\underline{R}^3} \hat{u}f(t,x,u)du. \tag{CE}$$

In order to have a reasonable description of a plasma one should really introduce distribution functions f_+ for ions and f_- for electrons, but the mathematical investigation can be generalized in a straightforward manner and we prefer to adhere to the notationally simpler case of a single distribution f.

RVMS must, of course, be augmented with suitable initial-value and boundary-value conditions. We will always prescribe an initial value $f_0 = f(0,\cdot)$ for f, but will consider two different conditions for $K := (E,B)$: either an initial-value condition $K(0,\cdot) = K_0$ or the condition $\lim_{t\to\infty} \|K(t,\cdot)\|_2 = 0$. ($\|\cdot\|_2$ denotes the L^2-norm.)

1. SUITABLE FUNCTION SPACES AND LOCAL SOLUTIONS

What is the appropriate space for the solution? Of course, one first looks for conserved quantities. There is the *energy law*, which states that

$$2\int_{\underline{R}^6} \sqrt{1+u^2}f(t,x,u)dx\,du + \|K(t,\cdot)\|_2^2 = \text{const.}.$$

If f_0 is nonnegative, then $f(t,\cdot)$ is nonnegative for all times t and therefore both terms are uniformly bounded. Thus the L^2-norm of $K(t,\cdot)$ is a priori bounded.

Maxwell's equations constitute a symmetric hyperbolic system with constant coefficients and inhomogeneity j. They are easily solved by using the theory of strongly continuous semigroups in the Sobolev space $H^m(\underline{R}^3)$. On the other hand, the functions E and B are also part of the coefficients of Vlasov's equation. Here we want to make use of the theory of characteristics. This makes it necessary to assume that E and B have bounded derivatives (or are at least uniformly Lipschitz continuous). The obvious way to proceed is to use the Sobolev embedding theorem. For this one needs $m > 5/2$, say $m = 3$. The early existence proofs (cf. [10], [1] and [3]) all used this approach. (To be honest, these were existence proofs for the nonrelativistic equations, but the same ideas work in the relativistic case.)

It would be nice to have an a priori bound for K in the Sobolev space H^3; one could probably manage with less, but the above L^2-estimate seems to be nearly useless.

All these results are of a purely local nature in the time variable t. This is probably as far as one can get, using this approach. In order to estimate the H^3-norm of K, one must estimate the H^3-norm of j and this means – more or less – differentiating Vlasov's equation three times. This produces a rather complicated nonlinear term, and much information, that could possibly be used to prove global existence, is effectively lost.

2. ANOTHER APPROACH TO GLOBAL SOLVABILITY

In [5] R. Glassey and W. Strauss noticed the following: Given j and ρ there is an explicit formula for the solution K of Maxwell's equations. This formula can be derived from the well-known solution formula for the wave equation in three dimensions. There is the familiar loss of smoothness: In order to compute K one integrates an expression that contains the derivatives of j and ρ. However, if one makes use of Vlasov's equation, these derivatives can be eliminated by partial integration. This shows that K satisfies an identity of the following form

$$K(t,x) = \int_{|x-y|<t} a_1(x,y,w)f(t-|x-y|,y,w)|x-y|^{-2}\,dy\,dw$$

$$+ \int_{|x-y|<t} a_2(x,y,w)f(t-|x-y|,y,w)K(t-|x-y|,y)|x-y|^{-1}\,dy\,dw$$

$$+ \text{ a solution of the homogeneous Maxwell equations,}$$

where the integral kernel a_1 has values in $\underline{R}^6$ and a_2 has values in the space of 6 6-matrices. One must keep in mind that f is also a function of K, as it satisfies a partial differential equation with coefficients built from K. A similar identity can be written down for the derivatives of K. This equation consists of four integral operators.

In this manner one gets essentially a C^1-theory instead of the H^3-theory used earlier. At first, Glassey and Strauss did not succeed in finding global classical solutions, but they gave a very lucid necessary and sufficient condition for their existence: Assume that f_0 has compact support.

Then a global solution of RVMS exists, if and only if there exists an a priori bound for the mementa on every bounded time interval, i.e., if and only if for every $T \geqq 0$ there exists a $\beta \geqq 0$ such that $f(t,x,u) = 0$ for all $t \in [0,T], |u| > \beta$.

3. "SMALL" SOLUTIONS OF THE FIRST KIND

A problem closely related to RVMS is the Vlasov-Poisson System (VPS). This describes the same situation, but in the limit case of an infinite speed of light. (We won't bother to write down the equations). For this problem the magnetic field B vanishes identically. E is completely determined by f, in particular one cannot prescribe an initial value for E. A global existence proof for 'small' solutions of VPS was given by C. Bardos and P. Degond in [2]: If f_0 is small enough (in a suitable sense), there exists a unique global solution (f,E). This solution has the following asymptotic behaviour for $t \to \infty$

$$\begin{aligned} \|E(t,\cdot)\|_\infty &= O(t^{-2}) \\ \|\partial_x E(t,\cdot)\|_\infty &= O(t^{-3}\log t) \qquad\qquad (*) \\ \|E(t,\cdot)\|_2 &= o(1). \end{aligned}$$

We want to emphasize that (*) is not an a priori assumption in the case of VPS; a unique global solution just happens to exist and it satisfies these decay conditions. In the case of RVMS, we then suspected that there are also global solutions that satisfy (*) with E replaced by K.

We have shown the following

THEOREM 1 ([7]): Assume that $f_0 \in C^1(\underline{R}^6)$ has compact support, say $f_0(x,u) = 0$, if $x^2 + u^2 > R^2$. Then there exist $C, D \geqq 0$, depending only on R, such that $\|f_0\|, \|\nabla f_0\| \leqq C$ implies that RVMS has a unique solution (f,K) with

$$\begin{aligned} \|K(t,\cdot)\|_\infty &\leqq D(1+t)^{-2} \\ \|\partial_x K(t,\cdot)\|_\infty &\leqq D(1+t)^{-3}\log(2+t) \qquad\qquad (**) \\ \|K(t,\cdot)\|_2 &= o(1). \end{aligned}$$

The density f is a solution of Vlasov's equation in the sense of the theory of characteristics (i.e. an integral of the characteristic system of ordinary differential equations); the fields $K = (E,B)$ are mild solutions of Maxwell's equations in the sense of semigroup theory.

Sketch of the proof: In order to satisfy the condition $\|K(t,\cdot)\|_2 = o(1)$ one has to use a variant of the integral identity of Section 2: The integrations must be taken over the future instead of the past (i.e. one must replace $t - |x-y|$ by $t + |x-y|$). Moreover, the solution of the homogeneous Maxwell equations must vanish. This leads to an integral identity of the following form

$$K(t,x) = \int_{\underline{R}^6} b_1(x,y,w)f(t + |x-y|, y,w)|x-y|^{-2}\, dy\, dw$$

$$+ \int_{\underline{R}^6} b_2(x,y,w)f(t + |x-y|,y,w)K(t+ |x-y|,y)|x-y|^{-1}\, dy\, dw.$$

Again one should think of f as a function of K. This integral equation can be solved with the help of Banach's fixed point theorem. One finds a solution in a space of Lipschitz continuous functions that satisfy (**). The essential idea, due to Bardos and Degond, is to treat the problem as a small perturbation of the 'free' problem with $K = 0$. This yields that $\|j(t,\cdot)\|_\infty$ and $\|\rho(t,\cdot)\|_\infty$ decay of order t^{-3}, which is one of the main ingredients of the proof of (**).

4. "SMALL" SOLUTIONS OF THE SECOND KIND

Using the methods of [7], Glassey and Strauss then took a step into a different direction in [6]. They looked for small solutions of the initial-value problem with initial values for f *and* K. In this case one cannot expect the asymptotic behaviour given by (**), since for every small f_0 these exists *exactly one* K_0, such that the solution of the initial-value problem for RVMS satisfies (**) (namely $K(0,\cdot)$, (f,K) being the solution of Theorem 1 for the initial value f_0).

It turned out, that the estimates

$$\|K(t,\cdot)\|_\infty \leq D(1 + t)^{-2}$$

$$\|\partial_x K(t,\cdot)\|_\infty \leq D(1 + t)^{-3}\log(2 + t)$$

are more than is really needed. It is sufficient, to have this kind of decay well inside the light cone. The above estimates can be replaced by

$$|K(t,x)| \leqq D(t + |x| + 2k)^{-1}(t- |x| + 2k)^{-1}$$

$$|\partial_x K(t,x)| \leqq D(t + |x| + 2k)^{-1}(t- |x| + 2k)^{-2}\log(t+ |x| + 2k)$$

with a suitable constant k. Since all particles are moving with a velocity strictly below the speed of light, these estimates are as useful, as the stronger ones above. This makes it possible to prove the existence of global classical solutions of RVMS for small initial data f_0, K_0.

References

[1] K. Asano: On local solutions of the initial value problem for the Vlasov-Maxwell equation, Commun. Math. Phys. 106 (1986), 551-568.

[2] C. Bardos & P. Degond: Global existence for the Vlasov-Poisson equation in 3 space variables with small initial data, Ann. Inst. Henri Poincaré, Analyse non linéaire 2 (1985), 101-118.

[3] P. Degond: Local existence of solutions of the Vlasov-Maxwell equations and convergence to the Vlasov-Poisson Equations for infinite light velocity, Math. Meth. in the appl. Sci. 8 (1986), 533-558.

[4] G. Eder: Elektrodynamik, Mannheim 1967.

[5] R. Glassey & W. Strauss: Singularity formulation in a collisionless plasma could occur only at high velocities, Arch. Rat. Mech. Anal. 92 (1986), 59-90.

[6] R. Glassey & W. Strauss: Absence of shocks in an initially dilute collisionless plasma, preprint 1986.

[7] E. Horst: Global solutions of the relativistic Vlasov-Maxwell system of plasma physics, Habilitationsschrift 1986.

[8] R. Leis: Initial boundary value problems in mathematical physics, Stuttgart 1986.

[9] E.S. Weibel: L'équation de Vlasov dans la théorie spéciale de la relativité, Plasma Phys. 9 (1967), 665-670.

[10] S. Wollman: An existence and uniqueness theorem for the Vlasov-Maxwell system, Commun. Pure Appl. Math. 37 (1984), 457-462.

E. Horst
Abteilung Mathematik im FB IV,
Postfach 3825,
5500 Trier,
West Germany.

J-L. JOLY & J. RAUCH

Ondes oscillantes semi-linéaires à hautes fréquences

ABSTRACT: Semilinear oscillating waves

We want to study high frequently semilinear oscillating waves. The first result deals with simple waves, that is to say oscillating waves with just one phase function. In this case, even the multidimensional Cauchy problem can be handled: semilinear waves are showed to propagate and rigorous asymptotic developments are obtained.

As far as multiphase oscillations are concerned, (when nonlinear interactions may occur) we restrict ourselves to the one-space-dimension case. A resonance condition (which is a property of the web defined by the characteristic foliations of the hyperbolic system) plays a basic role in the study of the propagation and interaction of the semi linear oscillating waves; existence and regularity results are proved including a "sum law" for the oscillating energy.

Soit $L(x,\partial) = \sum_{k=0}^{n} A_k(x)\partial_k + B(x)$ un système $N \times N$ d'opérateurs différentiels d'ordre 1, a coefficients matriciels réguliers par rapport à $x = (x_0,x_1,\ldots,x_n)$ strictement hyperbolique par rapport à $(1,0,\ldots,0)$.

Les *ondes simples oscillantes linéaires* sont des fonctions u^{ε}, dépendant d'un paramètre $\varepsilon > 0$, solutions de $Lu^{\varepsilon} = 0$, admettant un développement asymptotique (d'ordre fini ou infini) de la forme

$$u^{\varepsilon}(x) \sim \sum_{j=0}^{\infty} \varepsilon^{j} a_j(x)e^{i\,\varphi(x)/\varepsilon}$$

où la phase réelle φ vérifie l'équation eikonale:

$$\det L_1(x,d\varphi(x)) = 0.$$

Une onde de ce type est donc pour ε voisin de 0 rapidement oscillantes dans les directions conormales au feuilletage caractéristique de codimension 1 défini par les lignes de niveau de φ. Une onde composée oscillante linéaire est une superposition linéaire d'ondes simples faisant intervenir plusieurs phases φ_k solutions de l'équation eikonale; on les rencontre naturellement dans le problème de Cauchy avec données initiales oscillantes étudié dans l'article fondamental de Lax [L].

Pour le problème semi-linéaire:

$$Lu + f(x,u) = b \qquad (1)$$

il est clair que le terme de couplage non linéaire défini par f, à cause des harmoniques qu'il introduit, oblige à modifier a priori la définition des ondes simples en rajoutant à la phase φ tous ses multiples entiers; on appelle donc *onde simple oscillante* toute solution de (1) admettant un développement asymptotique:

$$u^{\varepsilon}(x) \sim \sum_{j} \varepsilon^{j} U_j(x,\varphi(x)/\varepsilon)$$

les fonctions $U_j(x,\theta)$ étant 2π-périodiques en θ et devant satisfaire un système infini d'équations intégro-différentielles comme l'a montré Choquet-Bruhat [CB]. A ce stade, il restait toutefois à établir un résultat mathématique sur le fait que la propriété d'être une onde simple oscillante est effectivement propagée par un système hyperbolique semi-linéaire. C'est ce à quoi nous nous attachons dans la première partie, utilisant le résultat de Rauch-Reed [RR2] sur les ondes stratifiées bornées semi-linéaires.

Cette chose acquise, le passage des ondes oscillantes simples aux ondes composées pose, dans le cas semi-linéaire, une réelle difficulté par rapport au cas linéaire à cause des interactions entre les oscillations des divers modes. En effet, le couplage non linéaire fait appraître des termes de source oscillants de phase:

$$\psi = \sum_k n_k \varphi_k, \; n_k \in \mathbf{Z}$$

qui peuvent présenter les deux comportements extrêmes suivants:

a) interaction non résonnante si, localement pour tout x, $\det L_1(x,d\psi(x)) \neq 0$. Dans ce cas:

$$L^{-1}(ae^{i\psi/\varepsilon}) = \varepsilon b.e^{i\psi/\varepsilon} + O(\varepsilon^2)$$

avec $b = L_1(x,d\psi(x))^{-1}$ a. Il s'agit d'une inversion elliptique: l'amplitude résultante est de l'ordre de ε et locale, c'est-à-dire sans propagation;

b) interaction résonnante si, localement pour tout x, $\det L_1(x,d\psi(x)) = 0$. L'inversion se fait alors selon les lois de l'optique géométrique:

$$L^{-1}(ae^{i\psi/\varepsilon}) = be^{i\psi/\varepsilon} + O(\varepsilon)$$

l'amplitude de b étant transportée par des équations différentielles le long des rayons correspondant à ψ. Dans ce cas, il y a *création* d'une oscillation, avec *propagation* au-delà du support de a.

Evidemment, il peut y avoir des situations intermédiaires entre a) et b).

Dans le deuxième paragraphe, nous étudions le problème de l'interaction générale des oscillations dans le cas monodimensionnel et pour le problème de Cauchy. S'il y a résonnance, c'est-à-dire création d'oscillation, alors sur certains modes apparaissent effectivement des oscillations supplémentaires dont les phases sont différentes des phases associées aux conditions initiales. Par conséquent, les oscillations sont naturellement définies sous la forme:

$$U_k(x,\varphi_k^1(x)/\varepsilon,\ldots,\varphi_k^{r_k}(x)/\varepsilon).$$

qui fait intervenir plusieurs phases sur chaque mode. Les fonctions profils U_k sont déterminées par un problème de Cauchy pour des équations intégro-différentielles hyperboliques en x et intégrales par rapport aux variables oscillantes. De ces équations, on déduit les règles de propagation semi-linéaires des oscillations. Des développements asymptotiques de ce type, dits faiblement non linéaires, ont été utilisés pour trouver et décrire une variété de phénomènes intéressants (voir la bibliographie). Nous poursuivons cette étude en essayant de donner des énoncés de convergence rigoureux.

I. ONDES SIMPLES OSCILLANTES SEMI-LINÉAIRES DANS LE CAS MULTIDOMENSIONNEL

Tout ce qui suit est local, et se situe dans un voisinage $\Omega \subset \mathbb{R}^{1+n}$ d'un point $\bar{x} \in \{x_0 = 0\}$. On se donne φ une phase réelle définie sur l'ouvert Ω qu'on suppose lisse ($d\varphi \neq 0$) et satisfaisant det $L_1(d\varphi) = 0$ dans Ω. On note Γ le feuilletage défini sur Ω par les surfaces de niveau de φ. Dans la suite ω désignera un ouvert bien inclus dans Ω, contenant $\bar{x}$, et tel que ω soit contenu dans le domaine de détermination (relativement à L) de $\omega_- = \omega \cap \{x_0 < 0\}$.

Soit V l'algèbre de Lie des champs de vecteurs réguliers sur Ω tangents à Γ. Pour $s \in \mathbb{N}$, on note:

$$H^s_V(\omega) = \{u \in (L^2(\omega))^N, \ \forall\alpha \in \mathbb{N},\ \alpha \leq s,\ V_1 \dots V_\alpha\, u \in (L^2(\omega))^N,\ V_k \in V\}.$$

L'ensemble $H^s_V(\omega) \cap L^\infty(\omega)$ est l'algèbre des distributions *stratifiées* bornées de [RR2].

Etant donnée une fonction $U \in C^\infty(\bar{\omega} \times S^1;\ \mathbb{C}^N)$, la famille $(u^\varepsilon)_{\varepsilon>0}$ définie par $u^\varepsilon(x) = U(x,\varphi(x)/\varepsilon)$ est, de façon claire, bornée dans $H^s_V(\omega) \cap L^\infty(\omega)$.

DEFINITION: On dit qu'une famille $(u^\varepsilon)_{\varepsilon>0}$ est une *onde oscillante admettant un développement asymptotique* si, pour chaque m = 0,1,... il existe une fonction $U_m \in C^\infty(\omega \times S^1;\mathbb{C}^N)$ telles que, pour tout entier M:

$$u^\varepsilon(x) - \sum_{m=0}^{M} \varepsilon^m U_m(x,\varphi(x)/\varepsilon) = O(\varepsilon^{M+1})$$

dans $H^s_V(\omega) \cap L^\infty(\omega)$.

On ecrit alors:

$$u^{\varepsilon} \sim \sum_{m=0}^{\infty} \varepsilon^{m} U_m(\varphi/\varepsilon).$$

Le résultat suivant montre que la propriété d'être une onde oscillante simple est propagée par les systèmes hyperboliques semi-linéaires.

THEOREME 1: On suppose $s > \frac{n}{2} + \frac{3}{2}$. Soit $\mathcal{B}$ un borné de $(H^s_V(\Omega) \cap L^{\infty}(\Omega))^2$. Alors il existe $\omega \subset \Omega$ tel que:

a) Prolongement de la solution: pour tout $(b,v) \in \mathcal{B}$ vérifiant:

$$L(v) + f(x,v) = b \text{ dans } \omega_-$$

il existe un unique $u \in H^s_V(\omega) \cap L^{\infty}(\omega)$ tel que:

$$L(u) + f(x,u) = b \text{ dans } \omega$$

$$u - v_{|\omega_-} = 0.$$

b) Régularité: si $(b,v) \in (H^{s'}_V(\omega))^2 \cap \mathcal{B}$ avec $s' > s$ alors $u \in H^{s'}_V(\omega)$.

c) Développement asymptotique: si les données $(b^{\varepsilon},v^{\varepsilon})$ sont des ondes oscillantes de phase φ (définition précédente), les solutions u^{ε} correspondantes sont des ondes oscillantes, les profils U_m étant déterminés par un système infini d'équations.

COMMENTAIRES:

1. Les ondes oscillantes ne sont pas bornées dans les Sobolev standards. Le théorème d'existence habituel introduirait donc une famille d'ouvert ω_{ε} tendant vers $\emptyset$.

2. a) et b) forment le théorème de propagation pour les ondes stratifiées bornées de Rauch-Reed [RR2].

3. Les profils V_m et U_m qui interviennent dans les développements de v et u satisfont un système infini d'équations différentielles en x et intégrales en $\theta \in S^1$ qui sont non linéaires et pour lequel on démontre un résultat d'existence et d'unicité du prolongement des solutions.

4. Il y a une version "problème de Cauchy" du théorème 1 qui présente quelques difficultés techniques.

II. OSCILLATIONS MULTIPHASES EN 1.D; PROPAGATION ET INTERACTION SEMI-LINÉAIRE DES OSCILLATIONS

Considérons l'exemple le plus simple de résonance sur le système modèle:

$$X_1 u_1 = 0$$

$$X_2 u_2 = 0$$

$$X_3 u_3 = u_1 u_2$$

où X_1, X_2, X_3 sont trois champs de vecteurs sur $\mathbb{R}^{1+1}$ définissant un opérateur L strictement hyperbolique.

On se donne deux ondes simples entrantes de phases φ_1 et φ_2, c'est-à-dire:

$$u_k^\varepsilon = a_k \, e^{i\varphi_k/\varepsilon} \, , \; k = 1,2$$

avec $X_k \varphi_k = 0$, et les amplitudes a_k portées par deux tubes caractéristiques $\Gamma_1(K)$ et $\Gamma_2(K)$ se coupant selon un compact $K \subset \{x_0 > 0\}$. On suppose $u_3^\varepsilon = 0$ dans $\{x_0 \leqq 0\}$. La solution est complètement déterminée par l'intégrale oscillante:

$$u_3^\varepsilon = X_3^{-1} \left(a_1 a_2 e^{i(\varphi_1+\varphi_2)/\varepsilon}\right).$$

L'ensemble des points critiques de la phase est défini par:

$$C = \{x \in K; \; X_3(\varphi_1 + \varphi_2) = 0\}.$$

Si C est petit (i.e. ne contient pas de courbes intégrales de X_3) alors les phases étant supposées régulières, on a:

$$u_3^\varepsilon = o(1) \; L^\infty_{loc}$$

autrement dit, asymptotiquement il n'y a pas d'interaction des oscillations entrantes pour ε tendant vers 0.

A l'opposé, si C contient K, alors:

$$u_3^\varepsilon = e^{i(\varphi_1+\varphi_2)/\varepsilon}\ X_3^{-1}(a_1 a_2)$$

est une onde oscillante de phase $\varphi_1 + \varphi_2$, créée par interaction des deux ondes entrantes. Ce phénomène de *résonance* aura lieu dès qu'on pourra trouver des phases $(\varphi_1, \varphi_2, \varphi_3)$ solutions non triviales de:

$$X_k\ \varphi_k = 0,\ k = 1,2,3$$

$$\sum_{k=1}^{3} \varphi_k = 0$$

c'est le cas pour les exemples: $X_1 = \partial_0 + \partial_1$, $X_2 = \partial_0 - \partial_1$, $X_3 = \partial_0$ où $X_1 = \partial_0 + \partial_1$, $X_2 = \partial_0 + x_1\partial_1$, $X_3 = \partial_0$, mais pas pour $X_1 = \partial_0 + x_1\partial_1$, $X_2 = \partial_0 - x_1\partial_1$, $X_3 = \partial_0 + \partial_1$, ce qui montre la différence avec l'interaction des singularités. On démontre par exemple que l'existence d'une résonance pour les trois champs X_k est liée à la propriété classique de *fermeture hexagonale* des feuilletages Γ_k associés, qui se lit sur la figure suivante:

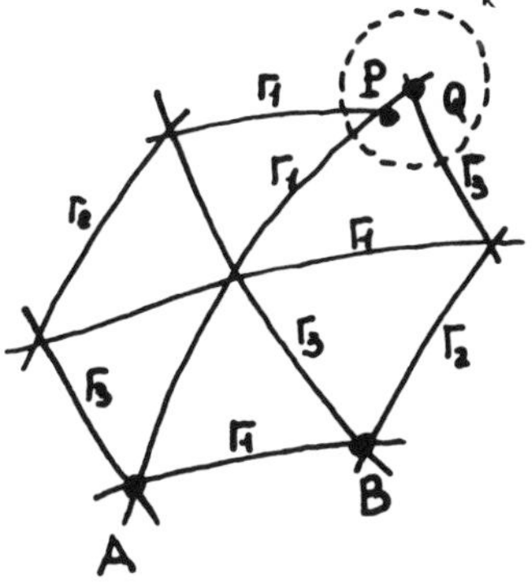

P = Q fermeture hexagonale

De façon générale, étant donnés N champs de vecteurs, C^∞ dans un ouvert $\Omega \subset \mathbb{R}^{1+1}$, associés à $L = \text{diag}(X_k)$ strictement hyperbolique, on appelle phases résonnantes toute classe (mod $\mathbb{R}^N$) de N-uple de fonctions réelles $(\varphi_1, \ldots, \varphi_N)$ vérifiant le système de N+2 équations:

$$X_k\ \varphi_k = 0,\ k = 1,\ldots,N$$

$$\sum_k d\varphi_k = 0.$$

On note S l'espace vectoriel de ces N-uples de phases résonnantes (mod $\mathbb{R}^N$). Si $N = 2$, $S = \{0\}$; si $N = 3$, $\dim S = 0$ ou 1, le second cas étant équivalent à la propriété de fermeture hexagonale. Génériquement, pour tout N, $S = \{0\}$.

Mais même si $S \neq \{0\}$, S ne peut être gros au moins dans le cas de Γ_k analytiques, puisqu'on démontre qu'alors

$$\dim S \leq \frac{(N-1)(N-2)}{2} \quad \text{(Poincaré, Blaschke-Bol [BB1]).}$$

Avant de définir les ondes oscillantes, on va préciser le choix des phases correspondant à chaque mode. On se donne d'abord un espace vectoriel Φ de fonctions réelles $C^\infty(\Omega)$, qu'on suppose *de dimension finie*. Notant $I_k = \{\varphi \in C^\infty(\Omega),\ X_k\varphi = 0 \text{ sur } \Omega\}$ pour $k = 1,\ldots,N$, on pose: $\Phi_k = \Phi \cap I_k$, $r_k = \dim \Phi_k$. L'entier $\ell = \Sigma \dim \Phi_k - \dim \Sigma\Phi_k$ est la dimension de l'ensemble des relations de résonance présentes dans l'espace Φ.

Soit $\varphi_k^1,\ldots,\varphi_k^{r_k}$ une base de Φ_k et notons $\vec{\varphi}_k = (\varphi_k^1,\ldots,\varphi_k^{r_k})$ pour $k = 1,\ldots,N$.

Les relations de résonance s'écrivent dans cette base au moyen de ℓ relations indépendantes, ce qui matriciellement donne:

$$R_1\,\vec{\varphi}_1 + \ldots + R_N\,\vec{\varphi}_N = 0$$

avec $R_j \in \text{Mat}\,(\ell,r_j)$ a coefficients réels.

Posons $\Theta_k = \{(\vec{\theta}_1,\ldots,\vec{\theta}_N) \in \mathbb{R}^{r_1+\ldots+r_N};\ \vec{\theta}_k = 0,\ \Sigma R_j\vec{\theta}_j = 0\}$.

Définissons E_{Θ_k} sur $C_{pp}(\mathbb{R}^{r_1+\ldots+r_N})$ par:

$$(E_{\Theta_k} f)(\theta_1) = \lim \frac{1}{T^k}\int_0^T \cdots \int_0^T f(\theta+t_1\vec{\eta}_1 + \ldots + t_r\vec{\eta}_r)dt_1 \ldots dt_r$$

où $(\vec{\eta}_1,\ldots,\vec{\eta}_r)$ est une base quelconque de Θ_k.

Si $f(\vec{\theta}_1,\ldots,\vec{\theta}_N) \sim \sum_\lambda a_\lambda e^{i(\vec{\lambda}_1\cdot\vec{\theta}_1+\ldots+\vec{\lambda}_N\cdot\vec{\theta}_N)}$ l'opérateur de moyenne conserve les termes du développement de Fourier de f dont l'indice λ est orthogonal à Θ_k et supprime les autres. Comme dans Θ_k on a $\theta_k = 0$, il conserve les termes d'indice λ tels que:

$$(\vec{\lambda}_1,\ldots,\vec{\lambda}_{k-1},\vec{\lambda}_{k+1},\vec{\lambda}_N)$$

soit dans l'orthogonal dans $\mathbb{R}^{\sum_{j\neq k} r_j}$ du noyau de la matrice:

$$M_k = (R_1,\ldots,R_{k-1},R_{k+1},\ldots,R_N)$$

c'est-à-dire une combinaison des lignes de cette matrice.

On en déduit que la moyenne de f se factorise sous la forme suivante:

$$E_{\Theta_k} f(\theta) = g(\theta_k, \sum_{j \neq k} \vec{R}_j \theta_j)$$

avec un unique $g \in C_{pp}(\mathbb{R}^{r_k} \times \mathbb{R}^{\ell})$.

Avant d'enoncer le théorème sur les solutions oscillantes, commençons par définir les fonctions oscillantes.

DEFINITION: On dit que la famille de fonctions complexes u^{ε} est oscillante de phases $(\varphi_1,\ldots,\varphi_r)$, sur un ouvert $\Omega \subset \mathbb{R}^2$ s'il existe $U(x,\vec{\theta}) \in C^{\circ}(\Omega;C_{pp}(\mathbb{R}^r))$ tel que:

$$u^{\varepsilon}(x) - U(x,\varphi_1(x)/\varepsilon,\ldots,\varphi_r(x)/\varepsilon)$$

tende vers 0 dans $L^{\infty}_{loc}(\Omega)$ quand ε tend vers 0. On note cette propriété:

$$u^{\varepsilon} \in \mathcal{O}_{\Omega}(\vec{\varphi}).$$

THÉORÈME: Supposons que l'espace des phases Φ satisfait la condition de non dégénérescence suivante:

$$k = 1,\ldots,N, \ \forall a \in C^{\circ}(\Omega), \ \forall \varphi \in \Phi \setminus I_k, \ X_k^{-1} a e^{i\varphi/\varepsilon} \underset{\varepsilon \to 0}{\longrightarrow} 0 \ L^{\infty}_{loc}(\Omega).$$

Soit v_k^{ε} satisfaisant $X_k v_k^{\varepsilon} = 0$ et u^{ε} la solution de:

$$\left.\begin{array}{l} X_k u_k^{\varepsilon} + F_k(x,u_k^{\varepsilon}) = 0 \\ u_k^{\varepsilon} - v_k^{\varepsilon}|_{\{x_o} = 0 \end{array}\right\} \quad k = 1,\ldots,N$$

Alors si $v_k^{\varepsilon} \in \mathcal{O}(\vec{\varphi}_k)$, les solutions u_k^{ε} sont ainsi des $\mathcal{O}_{\Omega}(\vec{\varphi}_k)$ et leur profil U_k uniques solutions du système intégro-différentiel suivant:

$$X_k U_k(x,\vec{\theta}_k) + E_\Theta F_k(U_1(x,\vec{\theta}_1),\ldots,U_N(x,\vec{\theta}_N))\Big|_{\sum_{j\neq k} R_j\vec{\theta}_j = -R_k\vec{\theta}_k}$$

$$U_k - V_k\big|_{\{X_o=0\}} = 0.$$

REMARQUE: 1) Les opérateurs intégraux E_k, c'est-à-dire les espaces Θ_k dépendent de Φ et de la structure des relations de résonance, c'est-à-dire de L.

2) Lorsque les X_k sont à coefficients constants et Φ formé de fonctions linéaires, on retrouve le résultat de [J].

3) D'autres exemples d'espaces Φ pour des champs à coefficients constants ou variables sont traités dans [JR3].

Terminons par un énoncé de propagation semi-linéaire de la répartition de l'énergie des oscillations. Cette propriété est formellement très semblable à celle que donnent Rauch et Reed [RR1] pour les singularités. Elle concerne la régularité par rapport aux variables oscillantes uniquement. On dit que u oscillante est de classe $s(s>\frac{1}{2})$ en $x \in \Omega$ si le profil U de u^ε vérifie $U(x,\cdot) \in H^s$ périodique. Supposons alors les données initiales v_k^ε *periodiques* et de *classe* $s_k(x)$. Construisons à partir des s_k les fonctions $x \to \sigma_k(x)$ comme dans [RR1], en utilisant *la loi de la somme*.

PROPOSITION: (loi de la somme pour les oscillations). Si v_k^ℓ est de classe s_k, alors u_k^ℓ est de classe σ_k.

References

[B-B] W. Blaschke and G. Boll, Geometrie der Gewebe, Springer (1938).

[CB] Y. Choquet-Bruhat, Ondes asymptotiques et approchées pour les systèmes non linéaires d'équations aux dérivées partielles, J. Math. Pures et Appliquées, 48 (1969).

[D] R. DiPerna, Convergence of the viscosity method for isentropic gas dynamics, Comm. Math. Phys. 91 (1983) 1-30.

[D-M] R. DiPerna and A. Majda, Nonlinear geometric optics, the validity of geometrical optics for weak solutions of conservation laws, Comm. Math. Physics, 98, (1985).

[F-F-McL] H. Flaschka, G. Forest, D. McLaughlin, Multiphase averaging and the inverse spectral solution of the Korteweg-de Vries equation, Comm. Pure Appl. Math. 33 (1980) 739-784.

[H] L. Hormander, The analysis of linear partial differential operators I, Springer-Verlag, Berlin Heidelberg, (1983).

[J] J.L. Joly, Sur la propagation des oscillations par un système hyperbolique semi-linéaire en dimension 1 d'espace, C.R. Acad. Sc. Paris, t.296 (25 Avril 1983).

[J-R1] J.L. Joly and J. Rauch, Ondes oscillantes semi-linéaires en dimension 1, Proceedings Journées E.D P. St. Jean de Monts, (1986).

[J-R2] J.L. Joly and J. Rauch, High frequency semilinear oscillations, Colloque pour le 60° anniversaire de P. Lax, Berkeley (1986).

[J-R3] J.L. Joly and J. Rauch, Ondes oscillantes semi-linéaires, en préparation.

[K-H] J. Keller and J. Hunter, Weakly nonlinear high frequency waves, Comm. Pure Appl. Math. XXXVI (1983) 547-569.

[H-M-R] J.K. Hunter, A. Majda and R. Rosales, Resonantly interacting, weakly nonlinear hyperbolic waves II: several spaces variables, Preprint.

[L1] P.D. Lax, Asymptotic solutions of oscillatory initial value problems, Duke Math. J. 24 (1957) 627-646.

[L2] P.D. Lax, Shock waves and entropy, in contributions to nonlinear functional analysis, ed. E.A. Zarantonello, Academic Press (1971).

[M-R] A. Majda and R. Rosales, Resonantly weakly nonlinear hyperbolic waves I, A single space variable.Studies in Applied Math., vol. 71 (1984).

[M-P-T] D. McLaughlin, G. Papanicolaou and L. Tartar, Weak limits of semilinear hyperbolic systems with oscillating data, Preprint.

[R] M. Rascle, Un résultat de "compacité par compensation à coefficients variables", application à l'élasticité non linéaire, C.R. Acad. Sc. Paris (1986).

[R-R1] J. Rauch and M. Reed, Nonlinear microlocal analysis of semilinear hyperbolic systems in one space dimension, Duke Math. J., vol. 49, (1986).

[R-R2] J. Rauch and M. Reed, Bounded stratified and striated solutions of hyperbolic systems, in Nonlinear partial differential equation and their application V, College France, Seminar H. Brezis et J.L. Lions eds. Pitman Publishers.

[T1] L. Tartar, Solutions oscillantes des équations de Carleman, Séminaire Goulaouic-Meyer-Schwartz, (1983).

[T2] L. Tartar, The compensated compactness method applied to systems of conservation laws, in Systems of Nonlinear Partial Differential Equations, J. Ball ed. Nato ASI series, C. Reidel Publ. (1983).

[W] G.B. Whitham, Linear and Nonlinear Waves, Wiley, New-York.

Jean-Luc Joly
Department of Mathematics
University of Bordeaux I,
351 Cours de la Libération,
33405, Talence,
France.

Jeffrey Rauch
Department of Mathematics
University of Michigan,
Angell Hall, Ann Arbor,
MI 48109,
U.S.A.

K. KAJITANI & S. WAKABAYASHI

Hypoelliptic operators in Gevrey classes

§1. INTRODUCTION

We studied in [14] the propagation of singularities of solutions of Micro-hyperbolic operators in Gevrey classes. In this paper we shall calculate the wave front sets of solutions of hypoelliptic operators in Gevrey classes by use of the method introduced in [13] and [14].

We denote by $\mathcal{E}^{\{\kappa\}}(R^n)$ the space of all f in $C^\infty(R^n)$ such that for any compact set K in R^n there are $C_K > 0$ and $A_K > 0$ satisfying $|D_x^\alpha f(x)| \leq C_K A_K^{|\alpha|} |\alpha| !^\kappa$ for x in K and α in Z_+^n. Then $\mathcal{E}^{\{\kappa\}}(R^n)$ is a locally convex space by a naturally induced topology. We denote $\mathcal{D}^{\{\kappa\}}(R^n) = \mathcal{E}^{\{\kappa\}}(R^n) \cap C_0^\infty(R^n)$ and $\mathcal{D}^{\{\kappa\}}{}'(R^n)$ the dual space of $\mathcal{D}^{\{\kappa\}}(R^n)$. Let $\kappa > 1$ and f in $\mathcal{D}^{\{\kappa\}}{}'(R^n)$. $WF_\kappa(f)$ is defined as the complement in $T^*(R^n)\setminus 0$ of the collection of all $(\hat{x},\hat{\xi})$ in $T^*(R^n)\setminus 0$ such that there are a neighbourhood U of $\hat{x}$ and a conic neighbourhood Γ of $\hat{\xi}$ such that for every ϕ in $\mathcal{E}^{\{\kappa\}}(R^n) \cap C_0^\infty(U)$ there are positive constants δ and C satisfying $|\widehat{\phi f}(\xi)| \leq \exp(-\delta|\xi|^{1/\kappa})$ for ξ in Γ.

Let $p(x,\xi)$ be a symbol in $S^m_{1,0}$ satisfying

$$|p^{(\alpha)}_{(\beta)}(x,\xi)| \leq C\, A^{|\alpha+\beta|} |\alpha+\beta|!^\kappa \langle\xi\rangle^{m-|\alpha|}, \tag{1.1}$$

for $(x,\xi) \in T^*(R^n) = R^{2n}$ and α,β in Z^n_+, $\langle\xi\rangle = (1 + |\xi|^2)^{1/2}$, $p^{(\alpha)}_{(\beta)}(x,\xi) = (i^{-1}\partial/\partial x)^\beta(\partial/\partial\xi)^\alpha p$ and $\kappa > 1$. Let $\Lambda(x,\xi)$ be a real analytic function satisfying

$$|\Lambda^{(\alpha)}_{(\beta)}(x,\xi)| \leq C_0 A_0^{|\alpha+\beta|} |\alpha+\beta|!\langle\xi\rangle^{1/\kappa_1-|\alpha|}, \tag{1.2}$$

for x,ξ in R^n and α,β in Z^n_+, where $\kappa_1 \geq \kappa$. We define

$$\lambda(\Lambda) = \inf_{L>0} \sup_{x\in R^n, |\xi|>L} |\mathrm{Re}\Lambda(x,\xi)|/|\xi|^{1/\kappa_1}.$$

We denote by $e^\Lambda(x,D)$ and $e^\Lambda(D,x)$ pseudo-differential operators of infinite order defined respectively,

$$e^{\Lambda}(x,D)u(x) = \int e^{ix\xi+\Lambda(x,\xi)}\hat{u}(\xi)d\!\!\!/\xi,$$

$$e^{\Lambda}(D,x)\ v(x) = \text{os-}\iint e^{i(x-x')\xi+\Lambda(x',\xi)}v(x')dx'd\!\!\!/\xi,$$

where os-$\iint$ means an oscillatory integral (c,f, [14]). Then it follows from Proposition 2,13 in [14] that there is a positive constant $\varepsilon_0 = \varepsilon_0(A_0,A)$ such that the product $p_\Lambda(x,D) = e^{\Lambda}(x,D)p(x,D)e^{-\Lambda}(D,x)$ is a pseudo-differential operator of order m for any Λ satisfying (1.2) with $\kappa_1 \geqq \kappa$ and $\lambda(\Lambda) \leqq \varepsilon_0$, and the symbol $p_\Lambda(x,\xi)$ belongs to $S^m_{1,0}$ and has an asymptotic expansion as follows:

$$|\{p_\Lambda(x,\xi)-\Sigma_{|\alpha+\beta|<N}\alpha!^{-1}\beta!^{-1}(p_{(\beta)}(x,\xi)\omega^\beta(\Lambda)\omega_\alpha(-\Lambda))^{(\alpha)}\}^{(\gamma)}_{(\delta)}| \qquad (1.3)$$

$$\leqq C_{N\gamma\delta}\ \langle\xi\rangle^{m-(1-1/\kappa_1)N-|\gamma|},$$

for x,ξ in R^n,γ,δ in Z^n_+ and N a nonnegative integer, where

$$\omega^\beta(\Lambda) = e^{-\Lambda(x,\xi)}(\partial/\partial\xi)^\beta e^{\Lambda(x,\xi)}, \qquad (1.4)$$

$$\omega_\alpha(-\Lambda) = e^{\Lambda(x,\xi)}(\partial/i\partial x)^\alpha e^{-\Lambda(x,\xi)}, \qquad (1.5)$$

and $C_{N\gamma\delta}$ depends only on N,γ,δ and ε_0.

We assume that for any compact set $K \subset R^n$ there are $C > 0$, $\theta \geqq 0$ and ℓ in R^1 such that for v in $C_0^\infty(K)$ and Λ satisfying (1.2) with $\lambda(\Lambda) \leqq \varepsilon_0$,

$$\|v\|_{m-\theta+\ell} \leqq C(\|p_\Lambda v\|_\ell + \|v\|_{m-\theta+\ell-1}), \qquad (1.6)$$

where $\|v\|_q$ stands for a norm of usual Sobolev space $H^q(R^n)$. Then we have

<u>THEOREM 1.1</u>: Let p(x,D) be a properly supported pseudo-differential operator of which symbol satisfies (1.1). Assume that (1.6) is valid. Then we have

$$WF_{\kappa_1}(u) = WF_{\kappa_1}(pu) \qquad (1.7)$$

for any u in $\mathcal{D}^{\{\kappa_1\}'}(R^n)$ and $\kappa_1 \geqq \kappa$.

We give a sufficient condition for (1.6).

THEOREM 1.2: Let $p(x,D)$ be a properly supported pseudo-differential operator of which symbol satisfies (1.1). Assume that for any compact set K in R^n there are $C > 0, \ell$ in R^1, $\theta \geq 0$ and σ in $(0,1)$ such that

$$\|u\|_{m-\theta+\ell} + \Sigma_{0<(1-\sigma)|\alpha+\beta|<\theta} \| \langle D\rangle^{\sigma|\alpha+\beta|} p^{(\alpha)}_{(\beta)} u\|_{\ell-|\beta|} \leq C\{\|pu\|_{\ell} + \|u\|_{m-\theta+\ell-1}\} \tag{1.8}$$

for any u in $C_0^\infty(K)$. Then (1.6) is valid for $\kappa_1 \geq \max\{\kappa, 1/\sigma\}$ and therefore $WF_{\kappa_1}(u) = WF_{\kappa_1}(pu)$ for any u in $\mathcal{D}^{\{\kappa_1\}\prime}(R^n)$.

REMARK: The condition (1.8) is a generalization of condition given by Durand in [3].

We describe some examples satisfying the condition (1.8).

EXAMPLE 1.3 (Hörmander [9]): Let $p(x,D) = \Sigma_{k=1}^r X_k^2 + X_0 + c(x)$, where $X_j = \Sigma_{i=1}^n a_j^i(x)\partial/\partial x_i$ $(j = 0,1,\ldots,n)$ and the coefficients $a_j^i(x)$ in $E^{\{\kappa\}}(R^n)$ are real valued and $c(x)$ in $E^{\{\kappa\}}(R^n)$. It follows from [9] that if the rank of Lie algebra generated by $X_0, X_1, \ldots, X_r$ equals to n, for any compact set in K in R^n there are $C = C(K) > 0$ and $\sigma = \sigma(K) > 0$ such that

$$\|u\|_\sigma + \Sigma_{|\alpha+\beta|=1} \| \langle D\rangle^{\sigma|\alpha+\beta|/2} p^{(\alpha)}_{(\beta)} u\|_{-|\beta|} \leq C \{\|pu\|_0 + \|u\|_0\},$$

for any u in $C_0^\infty(K)$. The number σ is given as follows. For $I = (i_1,\ldots,i_k)$ where i_ℓ in $\{0,1,\ldots,r\}$, we denote $X_I = \mathrm{ad}X_{i_1}\mathrm{ad}X_{i_2}\ldots\mathrm{ad}X_{i_{k-1}}(X_{i_k})$, where $\mathrm{ad}X(Y) = [X,Y]$ a commutator of X and Y. For any multi index I we associate a number $\sigma(I)$ defined by

$$I/\sigma(I) = \Sigma_{j=1}^k 1/\sigma_{i_j}, \text{ where } \sigma_{i_j} = \begin{cases} 1, & i_j = 1,\ldots,k \\ 1/2, & i_j = 0. \end{cases}$$

Let $\hat{x}$ be a point in K. If the vector fields $X_{I_1},\ldots,X_{I_n}$ are linearly independent at $\hat{x}$, they are so in small open neighbourhood $U(\hat{x})$ of $\hat{x}$. The family $\{U(\hat{x}); x \text{ in } K\}$ is an open covering of K. Let $\{U_1,\ldots,U_N\}$ be a finite covering of K and $\{X_{I_1^\ell},\ldots,X_{I_n^\ell}\}$ a independent family on $U_\ell (\ell = 1,\ldots,N)$.

We put $\sigma(K) = \inf_{j,\ell} \sigma(I_j^\ell)$. It follows from Theorems 1.1 and 1.2 that we have $WF_{\kappa_1}(u) = WF_{\kappa_1}(pu)$ for u in $\mathcal{D}^{\{\kappa_1\}'}(R^n)$ and $\kappa_1 \geqq \max(\kappa, 2/\sigma_{max})$, if $\sigma_{annuler} = \inf_{K\subset R^n} \{\sigma(K)\}$ is positive. Moreover when $X_0 = \Sigma_{k=1}^r c_k(x)X_k + \Sigma_{k=1}^r\Sigma_{j=1}^r b_{kj}(x)[X_k, X_j]$ where $c_j(x)$ and $b_{kj}(x)$ are in $E^{\{\kappa\}}(R^n)$, we have

$$\|u\|_{2\sigma} + \Sigma_{|\alpha+\beta|=1} \| \langle D\rangle^\sigma p_{(\beta)}^{(\alpha)} u\|_{-|\beta|} \leqq C\{\|pu\|_0 + \|u\|_0\},$$

for u in $C_0^\infty(K)$. Therefore in this case we have $WF_{\kappa_1}(u) = WF_{\kappa_1}(pu)$ for u in $\mathcal{D}^{\{\kappa_1\}'}(R^n)$ and $\kappa_1 \geqq \max(\kappa, 1/\sigma_{annuler})$.

REMARK: Derridj and Zuily proved in [2] that under the above condition $p(x,D) = \Sigma X_k^2 + X_0 + c$ (resp. $p = \Sigma X_k^2 + c$) is hypoelliptic in $E^{\{\kappa_1\}}(R^n)$ for $\kappa_1 \geqq \max\{\kappa, 2/\sigma_{max}\}$ (resp. $\kappa_1 > \max\{\kappa, 1/\sigma_{max}\}$). The optimality of Gevrey index κ_1 is studied by Métivier in [20].

EXAMPLE 1.4 (Hypoelliptic operator with loss of one derivative): Let $p(x,\xi)$ be a symbol in $S_{1,0}^m$ satisfying (1.1). Assume that there is $\gamma > 0$ such that the principal symbol $p_m(x,\xi)$ of $p(x,\xi)$ is contained in $\{z \text{ in } C^1; |\text{Im } z| \leqq \gamma \text{Re } z\}$ for (x,ξ) in R^{2n}. Then it follows from (22.4.30) and from 18.1.14 in [10] that for any compact set K in R^n there is $C > 0$ such that

$$\Sigma_{|\alpha+\beta|=1} \| \langle D\rangle^{|\alpha+\beta|/2} p_{(\beta)}^{(\alpha)} u\|_{-|\beta|} \leqq C\{\|pu\|_0 + \|u\|_0\},$$

for any u in $C_0^\infty(K)$. Moreover we assume

$$\|u\|_{m-1} \leqq C\{\|pu\|_0 + \|u\|_0\}, \tag{1.9}$$

for any u in $C_0^\infty(K)$. Then we have $WF_{\kappa_1}(u) = WF_{\kappa_1}(pu)$ for u in $\mathcal{D}^{\{\kappa_1\}'}(R^n)$ and $\kappa_1 \geqq \max(\kappa, 2)$. We know the sufficient condition in order that (1.9) holds (c.f. §22.4 in [10]).

REMARK: Iwasaki obtained in [12] the result of this example, verifying that $p(x,\xi)$ satisfies (1.10) and (1.12) ($\rho = \delta = 1/2$) below and constructing a parametrix of p(x,D) in Gevrey class.

<u>EXAMPLE 1.5</u> (Grusin type [5]): Let $(x,y) = (x,\tilde{y},\tilde{\tilde{y}})$ be in $R^n \times R^\ell \times R^{k-\ell}$ and (ξ,η) the dual variables of (x,y). Define

$$p(x,y,D_x,D_y) = \Sigma_M\, a_{\alpha\alpha'\gamma\gamma'}(x,y)x^\gamma\tilde{y}^{\gamma'}D_x^\alpha D_y^{\alpha'},$$

where $M = \{(\alpha,\alpha',\gamma,\gamma');\ \alpha,\gamma$ in Z_+^n, $\alpha' \in Z_+^k$, $\tilde{\gamma}' \in Z_+^\ell$, $(\alpha,M) + (\alpha',M') \leqq m$, $mh \geqq (\gamma,M) + (\tilde{\gamma}',\tilde{\sigma}') \geqq (\alpha,M) + (1+h)(\alpha',M')-m\}$, $M = (m/m_1,\dots,m/m_n)$, $M' = (m/m_1',\dots,m/m_k')$, $m = \max_{i,j}(m_i,m'_j)$, $\tilde{\sigma}' = (\sigma_1',\dots,\sigma_\ell')$ with $(1+h)m/m_j' > \sigma'_j \geqq 0$, $(j = 1,\dots,\ell)$, $h > 0$ and m_i,m'_j positive integers. We denote $(\alpha,M) = \Sigma_{j=1}^n\alpha_j(m/m_j)$, $(\alpha',M') = \Sigma_{j=1}^k\alpha'_j(m/m'_j)$, $(\gamma',\sigma') = \Sigma_{j=1}^\ell\gamma'_j\sigma'_j$, $M_0 = \{(\alpha,\alpha',\gamma,\gamma')$ in M; $(\gamma,M) + (\tilde{\gamma}',\tilde{\sigma}') = (\alpha,M) + (1+h)(\alpha',M')-m\}$, $M_0^0 = \{(\alpha,\alpha',\gamma,\gamma')$ in M_0; $(\alpha,M) + (\alpha',M') = m\}$ and

$$L(x,y,D_x,\eta) = \Sigma_{M_0}\, a_{\alpha\alpha'\gamma\gamma'}(0,0,\tilde{\tilde{y}})x^\gamma\tilde{y}^{\tilde{\gamma}'}\eta^{\alpha'}D_x^\alpha,$$

$$L_0(x,y,\xi,\eta) = \Sigma_{M_0^0}\, a_{\alpha\alpha'\gamma\gamma'}(0,0,\tilde{\tilde{y}})x^\gamma\tilde{y}^{\tilde{\gamma}'}\eta^{\alpha'}\xi^\alpha.$$

We assume that the coefficients $a_{\alpha\alpha'\gamma\gamma'}(x,y)$ are in $E^{\{\kappa\}}(R_x^n \times R_y^k)$, $L_0(x,y,\xi,\eta) \neq 0$ for $|x| + |\tilde{y}| \neq 0$ and $|\xi| + |\eta| \neq 0$, and the equation $L(x,y,D_x,\eta)v(x) = 0$ in R_x^n has no solution in $S(R_x^n)$ for any (y,η) in R^{2k} with $|\eta| = 1$. Then if $m_0(1+h) > m'^0$, $\bar{\sigma}(m'^0-m_0)/(mh) < m_0/m^0$ and $\sigma'^0 < (1+h)m/m'^0$ are valid, it follows from Grusin [5], Taniguti [27] and Parenti and Rodino [23] that we can estimate

$$\Sigma_{(\alpha,M)\leq m}\|\langle D\rangle_y^{m_0'(m-(\alpha,M)/(1+h)m}D_x^\alpha u\|_0 + \Sigma_{|\alpha+\beta|\leq N}\|\langle D_x, D_y\rangle^{\rho|\alpha|-\delta|\beta|}p_{(\beta)}^{(\alpha)}u\|_0$$

$$\leqq C_N\ \{\|pu\|_0 + \|u\|_0\},$$

for u in $C_0^\infty(K)$, where $\langle D_x,D_y\rangle = (1+|D_x|^2 + |D_y|^2)^{1/2}$, $p_{(\beta)}^{(\alpha)} = D_x^{\alpha(1)}D_y^{\alpha(2)}\partial_\xi^{\beta(1)}\partial_\eta^{\beta(2)}p$, $(\alpha = (\alpha^{(1)},\alpha^{(2)})$ and $\beta = (\beta^{(1)},\beta^{(2)}))$, $m_0 = \min\{m_j\}$, $m^0 = \max\{m_j\}$, $m'_0 = \min\{m'_j\}$, $m'^0 = \max\{m'_j\}$, $\sigma'^0 = \max\{\sigma'_j\}$, $\bar{\sigma} = \max\{m/m_0,\sigma'^0\}$, $\rho = \min\{m'_0/(1+h)m^0, m'_0/m'^0, m_0/m^0\}$ and $\delta = \max\{1/(1+h), \sigma'^0m'^0/(1+h)m, \bar{\sigma}(m'^0-m_0)/(mh)\}$. Then we have $WF_{\kappa_1}(u) = WF_{\kappa_1}(pu)$ for any u

in $\mathcal{D}^{\{\kappa_1\}\prime}(R^{n+k})$, if $\kappa_1 \geqq \max\{\kappa, 1/\rho, 1/(1-\delta)\}$. For example, in the case of $p = D_x + ix^h D_y^m$, we take $m_1 = 1$, $m'_1 = m$, $h > 0$ and $\sigma'_1 = 0$. Then if $(1+h) > m$, we can take $\rho = m/(1+h)$ and $\delta = \max\{1/(1+h), (m-1)/h\}$.

REMARK: Rodino in [24] proved without assumption $m_0(1+h) > m'^0$ that pu in $\mathcal{E}^{\{\kappa_1\}}(R^{n+k})$ and u in $C^\infty(R^{n+k})$ imply u in $\mathcal{E}^{\{\kappa_1\}}(R^{n+k})$ for $\kappa_1 \geqq \max\{\kappa, m/m_0, m/m'_0\}$. Moreover Parenti and Rodino in [32] indicated an example such that the condition $m_0(1+h) > m'^0$ is necessary in order that $WF(u) = WF(pu)$ holds for any distribution u. In [26] Sakurai treats a case of $m_j = m_i' = m$ ($j=1,\ldots,n$ and $i=1,\ldots,k$) and $\sigma'_i = 0$ ($i = 1,\ldots,k$).

EXAMPLE 1.6 (Hörmander [8]): Let $0 < \delta \leqq \rho < 1$ and $p(x,\xi)$ be a symbol in $S^m_{1,0}$ which satisfies (1.1) and for any compact set K in R^n there are $C > 0$, $R > 0$ and $\theta \geqq 0$ such that

$$|p(x,\xi)| \geqq C|\xi|^{m-\theta}, \tag{1.10}$$

$$|p^{(\alpha)}_{(\beta)}(x,\xi)| \leqq C_{\alpha\beta}|\xi|^{-\rho|\alpha|+\delta|\beta|}|p(x,\xi)|, \tag{1.11}$$

for x in K, ξ in R^n with $|\xi| \geqq R$ and α, β in Z^n_+, where $C_{\alpha\beta} > 0$ depends only on α, β and K. Then if $\delta < \rho$, it follows from Hörmander [8] that we have (1.6) for $\sigma \leqq \min\{\rho, 1-\delta\}$. Therefore we have $WF_{\kappa_1}(u) = WF_{\kappa_1}(pu)$ for u in $\mathcal{D}^{\{\kappa_1\}\prime}(R^n)$ and $\kappa_1 \geqq \max\{\kappa, 1/\rho, 1/(1-\delta)\}$, if $p(x,D)$ is properly supported.

REMARK: The hypoellipticity in Gevrey class of this example was studied by many authors (c.f. [1], [6], [11], [15], [16], [17], [18], [22], [25], [31]) under the more restircted assumption on the Gevrey index κ_1. Iwasaki in [12] and Taniguti in [28] treated this example and constructed a parametrix of p(x,D) in Gevrey class under the assumptions (1.10) and

$$|p^{(\alpha)}_{(\beta)}(x,\xi)| \leqq M^{|\alpha+\beta|}\{|\alpha|!^\kappa|\xi|^{-|\alpha|}+|\alpha|!^{\kappa\rho}|\xi|^{-\rho|\alpha|}\} \times \{|\beta|!^\kappa+|\beta|!^{\kappa(1-\delta)}|\xi|^{\delta|\beta|}\}|p(x,\xi)|\}. \tag{1.12}$$

for x in K, ξ in R^n with $|\xi| \geqq R$ and α, β in Z^n_+. In [14] Kajitani and Wakabayashi constructed a parametrix of p(x,D) in Gevrey class under the conditions (1.1), (1.10) and (1.11) instead of (1.12).

REMARK: When $\rho < 1$ and $\delta > 0$, we can prove by mean of interpolation theorem (c.f. (2.6) in Wakabayashi [30]) that (1.11) is valid for any α,β in Z_+^n, if (1.11) holds for $|\alpha+\beta| = 1$. Hence it suffices to verify (1.11) for $|\alpha+\beta|=1$. Moreover we note that (1.1), (1.10) and (1.11) imply (1.12). In fact, we have from (1.1) and (1.10)

$$|p_{(\beta)}^{(\alpha)}(x,\xi)| \leq CA^{|\alpha+\beta|}|\alpha+\beta|!^{\kappa}|\xi|^{\theta-|\alpha|}|p(x,\xi)|.$$

Furthermore noting that $|\alpha+\beta|!^{\kappa} \leq 2^{\kappa|\alpha+\beta|}|\alpha|!^{\kappa}|\beta|!^{\kappa}$ and $|\beta|!^{\kappa}|\xi|^{\theta} \leq C_1^{|\beta|+1}\{|\beta|!^{\kappa} + |\beta|!^{(1-\delta)\kappa}\langle\xi\rangle^{\delta|\beta|}\}$, for $\delta|\beta| \geq 2\theta$, we obtain (1.12) from the above estimate.

To compare the conditions (1.10) and (1.11) with Hörmander's conditions in the case of constant coefficients (C.F. [10]), we shall give a characterization equivalent to the conditions (1.10) and (1.11). To do so, we extend a symbol $p(x,\xi)$ to an almost analytic function in C^{2n} as follows:

$$p(x+iy,\xi+i\eta) = \Sigma_{\alpha,\beta}\, \alpha!^{-1}\beta!^{-1}(\partial/\partial x)^{\beta}(\partial/\partial\xi)^{\alpha}p(x,\xi)(iy)^{\beta}(i\eta)^{\alpha}\chi(b_{|\beta|}y, b_{|\alpha|}\eta/\langle\xi\rangle),$$

where $b_0 = 1$ and b_j tends to infinity rapidly for $j \to \infty$, and $\chi(x,\eta)$ in $C_0^{\infty}(R^{2n})$ with $\chi = 1$ in a neighbourhood of the origin in R^{2n} (c.f. [19]). When $\delta > 0$ and $\rho < 1$, we denote

$$p_{\theta,\rho,\delta}(x,\xi;y,\eta) = \Sigma_{(1-\rho)|\alpha|+\delta|\beta|\leq\theta}\alpha!^{-1}\beta!^{-1}(\partial/\partial x)^{\beta}(\partial/\partial\xi)^{\alpha}p(x,\xi)y^{\beta}\eta^{\alpha}. \tag{1.13}$$

Then we have

THEOREM 1.7: Let $p(x,\xi)$ be in $S^m_{1,0}$. Then the following conditions are equivalent.

(i) (1.10) and (1.11) are valid for some $\theta \geq 0$, $0 < \rho < 1$ and $\delta > 0$.

(ii) For any compact set K there are $C_K > 0$, $\lambda_K > 0$ and $R_K > 0$ such that

$$|p(x+iy,\xi+i\eta)| \geq C_K|\xi|^{m-\theta}, \tag{1.14}$$

for x in K, ξ in R^n with $|\xi| \geq R_K$ and (y,η) in R^{2n} satisfying

$$|y||\xi|^{\delta} + |\eta||\xi|^{-\rho} \leq \lambda_K. \tag{1.15}$$

(iii) $p_{\theta,\rho,\delta}(x,\xi;y,\eta)$ satisfies (1.14) for x in K, ξ in R^n with $|\xi| \geq R_K$ and (y,η) in C^{2n} satisfying (1.15).

When $p(x,\xi)$ is real analytic and its principal symbol $p_m(x,\xi)$ is positively homogeneous in ξ, we can characterize the conditions (1.10) and (1.11) as the property of complex zeros of the symbol p. We define for x,ξ in R^n,

$$\Sigma(x,\xi) = \{(y,\eta) \text{ in } C^{2n};\ \ p(x+y,\xi+\eta) = 0\},$$

$$d(x,\xi) = \inf_{(y,\eta) \text{ in } \Sigma(x,\xi)} \{|y||\xi|^{\delta} + |\eta||\xi|^{-\rho}).$$

Then we have

THEOREM 1.8: Let $p(x,\xi)$ in $S^m_{1,0}$ be real analytic in $R^n \times\{|\xi| \geq R\}$. Assume that there are $\varepsilon > 0$ and $p_m(x,\xi)$ positively homogeneous in ξ of degree m such that $p(x,\xi) - p_m(x,\xi)$ in $S^{m-\varepsilon}_{1,0}$. Then the following conditions are equivalent.

(i) For any compact set K in R^n there are $\theta \geq 0$, $\delta \geq 0$ and $0 \leq \rho \leq 1$ such that (1.10) and (1.11) are valid.

(ii) For any compact set K in R^n there are $\lambda_K > 0$ and $R_K > 0$ such that $d(x,\xi) \geq \lambda_K$ for x in K and ξ in R^n with $|\xi| \geq R_K$.

REMARK: In the case of constant coefficients Theorem 1.8 was derived by Hörmander in [10].

THEOREM 1.9: Let p(x,D) be a properly supported pseudo-differential operator of which symbol satisfies (1.1). When $0 < \delta < \rho < 1$, if one of the conditions of Theorem 1.7 is valid, we have $WF_{\kappa_1}(u) = WF_{\kappa_1}(pu)$ for u in $\mathcal{D}^{\{\kappa_1\}'}(R^n)$ and $\kappa_1 \geq \max\{\kappa,1/\rho,1/(1-\delta)\}$. Moreover when $0 < \delta = \rho < 1$, if for any compact set K in R^n and for any $\lambda > 0$ there are $C_{K\lambda} > 0$ and $R_{K\lambda} > 0$ such that $|p_{\theta,\rho,\delta}(x,\xi;y,\eta)| \geq C_{K\lambda}|\xi|^{m-\theta}$ for x in K, $|\xi| \geq R_K$ and (y,η) in C^{2n} with $|y||\xi|^{\delta} + |\eta||\xi|^{-\rho} \leq \lambda$, we have $WF_{\kappa_1}(u) = WF_{\kappa_1}(pu)$ for u in $\mathcal{D}^{\{\kappa_1\}'}(R^n)$ and $\kappa_1 \geq \max\{\kappa,1/\rho,1/(1-\rho)\}$.

REMARK: Assume that the condition of Theorem 1.8 is valid. Then if $d(x,\xi) \to \infty$ for $|\xi| \to \infty$, we obtain the conclusion of Theorem 1.9 in the case of $\delta = \rho$.

Finally we shall give a sufficient condition to be hypoelliptic in C^∞ sense. Let $p(x,\xi)$ be a symbol in $S^m_{1,0}$. We take $\Lambda(x,\xi)$ as follows:

$$\Lambda(x,\xi) = \phi(x,\xi)\log\langle\xi\rangle,$$

where $\phi(x,\xi)$ is in $S^0_{1,0}$. We assume that for any compact set K in R^n and for any $M > 0$ there are $\theta = \theta(K)$, $\ell = \ell(K)$ and $C = C(K,M) > 0$ such that

$$\|v\|_{m-\theta+\ell} \leqq C(K,M) \ \{\|p_\Lambda v\|_\ell + \|v\|_{m-\theta+\ell-1}\}, \tag{1.16}$$

for v in $C_0^\infty(K)$ and $\Lambda(x,\xi) = \phi(x,\xi)\log\langle\xi\rangle$ with $|\mathrm{Re}\ \phi(x,\xi)| \leqq M$, (x,ξ) in R^{2n}.

Then we have

THEOREM 1.10: Let $p(x,D)$ be a properly supported pseudo-differential operator of which symbol is in $S^m_{1,0}$. Assume that (1.16) is valid. Then we have $WF(u) = WF(pu)$ for u in $\mathcal{D}'(R^n)$.

We can give a sufficient condition for (1.16) similarly to (1.8) as follows: for any compact set K in R^n there are $\theta \geqq 0$, ℓ in R^1 and $C > 0$ such that

$$\|v\|_{m-\theta+\ell} \leqq C \ \{\|pv\|_\ell + \|v\|_{m-\theta+\ell-1}\}, \tag{1.17}$$

and for any $\varepsilon > 0$ there is $C_\varepsilon > 0$ such that

$$\Sigma_{0<|\alpha+\beta|\leqq[\theta]}\|(\log\langle D\rangle)^{|\alpha+\beta|}p^{(\alpha)}_{(\beta)}v\|_{\ell-|\beta|} \leqq \varepsilon\|pv\|_\ell + C_\varepsilon\|v\|_{m-\theta+\ell-1}, \tag{1.18}$$

for v in $C_0^\infty(K)$, where $[\theta]$ stands for the largest integer which does not exceed θ.

It is easy to derive (1.16) from (1.17) and (1.18). In fact, since $e^\Lambda(x,D)$ and $e^{-\Lambda}(D,x)$ are pseudo-differential operators of finite order, by mean of symbolic calculus of pseudo-differential operators we have for any positive integer N

$$r_N(x,D) = p_\Lambda(x,D) - \Sigma_{|\alpha+\beta|\leqq N} \omega_\alpha^\beta(x,D)p_{(\beta)}^{(\alpha)}(x,D),$$

where $r_N(x,\xi)$ is in $S_{1,0}^{m-N}$ and $\omega_\alpha^\beta(x,\xi)$ satisfies

$$|\partial_\xi^\gamma D_x^\delta \omega_\alpha^\beta(x,\xi)| \leqq C_{\alpha\beta\gamma\delta} (\log \langle\xi\rangle)^{|\alpha+\beta|}\langle\xi\rangle^{-|\beta|-|\gamma|},$$

for x, ξ in R^n. Hence (1.17) and (1.18) imply (1.16).

<u>THEOREM 1.11</u>: Let $p(x,D)$ be a properly supported pseudo-differential operator of which symbol $p(x,\xi)$ is in $S_{1,0}^m$. If (1.17) and (1.18) are valid, then we have $WF(u) = WF(pu)$ for u in $\mathcal{D}'(R^n)$.

Instead of (1.18) we assume

$$\Sigma_{0<|\alpha+\beta|\leqq[\theta]} \|(\log\langle D\rangle)^{|\alpha|}p_{(\beta)}^{(\alpha)}v\|_{\ell-|\beta|} \leqq \varepsilon\|pv\|_\ell + C_\varepsilon \|v\|_{m-\theta+\ell-1}, \tag{1.19}$$

for v in $C_0^\infty(\Omega)$ and any $\varepsilon > 0$. Then we have

<u>THEOREM 1.12</u>: Let $p(x,D)$ be a properly supported pseudo-differential operator of which symbol is in $S_{1,0}^m$ and Ω an open set in R^n. Assume that (1.17) and (1.19) are valid. Then if u in $\mathcal{D}'(R^n)$ and $p(x,D)u$ in $C^\infty(\Omega)$, u is in $C^\infty(\Omega)$.

<u>REMARK</u>: The conditions (1.17) and (1.18) was introduced essentially by Morimoto in [21]. Morimoto obtained the result of Theorem 1.11 under the following condition,

$$\|(\log\langle D\rangle)^m u\|_0 + \Sigma_{0<|\alpha+\beta|<m} \|(\log\langle D\rangle)^{|\alpha+\beta|}p_{(\beta)}^{(\alpha)}u\|_{-|\beta|}$$
$$\leqq \varepsilon\|pu\|_0 + C_\varepsilon\|u\|_0, \tag{1.20}$$

for u in $C_0^\infty(K)$ and for any $\varepsilon > 0$. This condition (1.20) is a generalization of one of Fedii [4] and Trèves [29]. Hoshiro in [7] gave a different proof of Theorem 1.11 under the assumption (1.20).

In Section 2 we shall prove Theorems 1.1 and 1.2. In Section 3 we explain Theorems 1.7, 1.8 and 1.9, and finally we shall give a proof of Theorems 1.10

and 1.11 in Section 4.

§2. PROOF OF THEOREMS 1.1 AND 1.2

It is sufficient to prove that $WF_{\kappa_1}(u) \subset WF_{\kappa_1}(pu)$. Let $(\hat{x},\hat{\xi})$ be not in $WF_{\kappa_1}(pu)$. We define for $\varepsilon > 0$,

$$\Lambda_\varepsilon(x,\xi) = \phi(x,\xi)\langle\xi\rangle^{1/\kappa_1-\varepsilon} - \varepsilon_2\langle\xi\rangle^{1/\kappa_1},$$

where $\phi(x,\xi) = \varepsilon_1(\varepsilon_0-|x-\hat{x}|^2-|\xi/\langle\xi\rangle-\hat{\xi}/\langle\hat{\xi}\rangle|^2)/(1+|x-\hat{x}|^2)$ and $\varepsilon_2 = \varepsilon_0\varepsilon_1/2(2+\varepsilon_0),(\varepsilon_0,\varepsilon_1 > 0)$. Then

$$\Lambda_0(x,\xi) \geqq \varepsilon_2\langle\xi\rangle^{1/\kappa_1}, \tag{2.1}$$

for $|x-\hat{x}|^2 + |\xi/\langle\xi\rangle-\hat{\xi}/\langle\xi\rangle|^2 \leqq \varepsilon_0/2$. Moreover for any $\varepsilon > 0$ there is $C_\varepsilon > 0$ such that

$$\Lambda_\varepsilon(x,\xi) \leqq -(\varepsilon_2/2)\langle\xi\rangle^{1/\kappa_1}, \tag{2.2}$$

for $|\xi| \geqq C_\varepsilon$. We note that $\Lambda_\varepsilon(x,\xi)$ satisfies (1.2) with C_0 and A_0 independent of ε and for $\varepsilon \geqq 0$

$$\lambda(\Lambda_\varepsilon) \leqq \varepsilon_1(\varepsilon_0 + 3). \tag{2.3}$$

Then it follows from Lemma 2.14 in [14] that if ε_1 is small, we have

$$e^{-\Lambda_\varepsilon}(D,x)e^{\Lambda_\varepsilon}(x,D) = I + q_\varepsilon(x,D) + r_\varepsilon(x,D), \tag{2.4}$$

where $q_\varepsilon(x,\xi)$ and $r_\varepsilon(x,\xi)$ satisfy

$$|q_{\varepsilon(\beta)}^{(\alpha)}(x,\xi)| \leqq \varepsilon_1 A_1^{|\alpha+\beta|+1}|\alpha+\beta|!^{\kappa_1}\langle\xi\rangle^{1/\kappa_1-1-|\alpha|}, \tag{2.5}$$

$$|r_{\varepsilon(\beta)}^{(\alpha)}(x,\xi)| \leqq A_\alpha^{1+|\beta|}|\beta|!^{\kappa_1} e^{-\gamma_0\langle\xi\rangle^{1/\kappa_1}}, \tag{2.6}$$

for x, ξ in R^n and $\varepsilon \geqq 0$, where A_1, A_α and $\gamma_0 > 0$ are independent of ε and ε_1. Hence if we choose $\varepsilon_1 > 0$ sufficiently small, $I + q_\varepsilon(x,D)$ has an inverse

$(I+q_\varepsilon)^{-1}$. Then from (2.5) it follows that we have $\hat{\delta} > 0$ independent of ε such that

$$\|e^{\delta\langle D\rangle^{1/\kappa_1}}(I+q_\varepsilon)^{-1}u\|_\ell \leqq C_\ell \|e^{\delta\langle D\rangle^{1/\kappa_1}}u\|_\ell, \tag{2.7}$$

for any δ with $|\delta| \leqq \hat{\delta}$ and $\ell \in R^1$. In fact we have $e^{\delta\langle D\rangle^{1/\kappa_1}}(I+q_\varepsilon)^{-1} = (I+\tilde{q}_\varepsilon)^{-1}e^{\delta\langle D\rangle^{1/\kappa_1}}$, where $\tilde{q}_\varepsilon(x,D) = e^{\delta\langle D\rangle^{1/\kappa_1}}q_\varepsilon(x,D)e^{-\delta\langle D\rangle^{1/\kappa_1}}$ is a pseudo-differential operator of which symbol $\tilde{q}_\varepsilon(x,\xi)$ is in $S^0_{1,0}$.

Let $\chi(x,\xi)$ be a symbol satisfying (1.1) with $m = 0$ such that $\chi = 1$ in $B_{2\varepsilon_0}(\hat{x},\hat{\xi}) = \{(x,\xi) \text{ in } R^{2n};\ |x-\hat{x}|^2 + |\xi/\langle\xi\rangle-\hat{\xi}/\langle\hat{\xi}\rangle|^2 \leqq 2\varepsilon_0\}$ and $\chi = 0$ for $(x,\xi) \notin B_{3\varepsilon_0}(\hat{x},\hat{\xi})$.

<u>LEMMA 2.1</u>: Let $\Lambda_\varepsilon(x,\xi)$ and $\chi(x,\xi)$ be given as above. If u in $\mathcal{D}^{\{\kappa_1\}\prime}(R^n)$ satisfies

$$\|e^{\Lambda_\varepsilon}(I+q_\varepsilon)^{-1}\chi u\|_\ell \leqq C, \tag{2.8}$$

for any $\varepsilon > 0$, where ℓ in R^1 and $C > 0$ are independent of ε, then $(\hat{x},\hat{\xi}) \notin WF_{\kappa_1}(u)$.

<u>PROOF</u>: (2.8) implies that $\langle D\rangle^\ell e^{\Lambda_\varepsilon}(I+q_\varepsilon)^{-1}\chi u$ converges weakly in $L^2(R^n)$ for $\varepsilon \to 0$. Hence $e^{\Lambda_0}(I+q_0)^{-1}\chi u(= g)$ is in $H^\ell(R^n)$. From (2.4) we have

$$\chi u = e^{-\Lambda_0}(D,x)g - r_0(I+q_0)^{-1}\chi u.$$

We choose $\chi_0(x,\xi)$ satisfying (1.1) with $m = 0$ and $\mathrm{supp}\chi_0 \subset B_{\varepsilon_0/2}(\hat{x},\hat{\xi})$. Then

$$e^{\delta\langle D\rangle^{1/\kappa_1}}\chi_0\chi u = e^{\delta\langle D\rangle^{1/\kappa_1}}\chi_0 e^{-\Lambda_0}(D,x)g - e^{\delta\langle D\rangle^{1/\kappa_1}}\chi_0 r_0(I+q_0)^{-1}\chi u.$$

If $0 < \delta < \min\{\varepsilon_2/2, \gamma_0/2\}$, ($\gamma_0$ given in (2.6)), it follows from (2.1), (2.6) and (2.7) that the right side of the above relation belongs to $L^2(R^n)$. Hence we have $(\hat{x},\hat{\xi}) \notin WF_{\kappa_1}(u)$. Q.E.D.

We put $pu = f$. Assume $(\hat{x},\hat{\xi}) \notin WF_{\kappa_1}(pu)$. Then we have

$$e^{\Lambda_\varepsilon} p\chi u = e^{\Lambda_\varepsilon}[p,\chi]u + e^{\Lambda_\varepsilon}\chi f = f_\varepsilon$$

and moreover from (2.4)

$$p_{\Lambda_\varepsilon} e^{\Lambda_\varepsilon}(I+q_\varepsilon)^{-1}\chi u = f_\varepsilon - e^{\Lambda_\varepsilon} pr_\varepsilon (I+q_\varepsilon)^{-1}\chi u. \tag{2.9}$$

We choose $\chi_1(x)$ in $\mathcal{D}^{\{\kappa\}}(R^n)$ such that $\chi_1(x) = 1$ for $|x-\hat{x}| \leqq 4\varepsilon_0$. Then we note

$$\text{supp } (1-\chi_1) \cap \text{supp}\chi = \phi. \tag{2.10}$$

We put $v_\varepsilon = \chi_1 e^{\Lambda_\varepsilon}(I+q_\varepsilon)^{-1}\chi u$. From (2.9)

$$p_{\Lambda_\varepsilon} v_\varepsilon = f_\varepsilon - e^{\Lambda_\varepsilon} pr_\varepsilon (I+q_\varepsilon)^{-1}\chi u - p_{\Lambda_\varepsilon}(\chi_1 - 1)e^{\Lambda_\varepsilon}(I+q_\varepsilon)^{-1}\chi u. \tag{2.11}$$

It follows from (2.2) that v_ε is in $C_0^\infty(R^n)$ for $\varepsilon > 0$. Hence we obtain by virtue of (1.6)

$$\|v_\varepsilon\|_{m-\theta+\ell} \leqq C\{\|p_{\Lambda_\varepsilon} v\|_\ell + \|v_\varepsilon\|_{m-\theta+\ell-1}\}, \tag{2.12}$$

where C is independent of ε. We prove that there is $C > 0$ independent of ε such that

$$\|p_{\Lambda_\varepsilon} v\|_\ell \leqq C, \tag{2.13}$$

for any $\varepsilon > 0$. We note that it follows from (2.10) and Proposition 3.4 in [14] that there is $\delta > 0$ independent of ε such that

$$\|(\chi_1 - 1)e^{\Lambda_\varepsilon}(I+q_\varepsilon)^{-1}\chi u\|_\ell \leqq C\,\|e^{-\delta\langle D\rangle^{1/\kappa_1}}\chi u\|_0, \tag{2.14}$$

for any $\varepsilon > 0$ and ℓ in R^1, if we take ε_1 sufficiently small. Moreover noting that for any $\rho > 0$ and $\gamma > 0$ there is $C_{\rho\gamma} > 0$ such that

$$\|v\|_{m-\theta+\ell-1} \leqq \rho\,\|v\|_{m-\theta+\ell} + C_{\rho\gamma}\|e^{-\gamma\langle D\rangle^{1/\kappa_1}} v\|_0,$$

for any v in $H^{m-\theta+\ell}$, we obtain from (2.12), (2.13) and (2.14)

$$\|e^{\Lambda_\varepsilon}(I+q_\varepsilon)^{-1}\chi u\|_{m-\theta+\ell} \leqq C$$

for any $\varepsilon > 0$. Hence we have $(\hat{x},\hat{\xi}) \notin WF_{\kappa 1}(u)$ by virtue of Lemma 2.1. Now we prove (2.13). Since (2.6) and (2.14) imply that the second term and the third one of the right side of (2.11) are uniformly bounded in $\varepsilon > 0$ respectively, it suffices to prove

$$\|f_\varepsilon\|_\ell \leqq \| e^{\Lambda_\varepsilon}[p,\chi]u\|_\ell + \|e^{\Lambda_\varepsilon}\chi f\|_\ell \leqq C,$$

for $\varepsilon > 0$. Since $(\hat{x},\hat{\xi}) \notin WF_{\kappa_1}(f)$, we have $\|e^{\Lambda_\varepsilon}\chi f\|_\ell \leqq C$ for $\varepsilon > 0$, if we take ε_0 small. $\Lambda_\varepsilon(x,\xi) \leqq -\varepsilon_1\varepsilon_0 \langle\xi\rangle^{1/\kappa_1}$ for (x,ξ) in supp $\chi^{(\alpha)}_{(\beta)}(|\alpha+\beta| \neq 0)$ and for $\varepsilon \geqq 0$ implies $\|e^{\Lambda_\varepsilon}[p,\chi]u\|_\ell \leqq C$ for $\varepsilon > 0$. Thus we obtain (2.13) and therefore completed the proof of Theorem 1.1.

<u>PROOF OF THEOREM 1.2</u>: Since $\omega^\beta(\Lambda)$ and $\omega_\alpha(-\Lambda)$ are given by (1.4) and (1.5), we can estimate

$$|\omega^{\beta(\gamma)}_{\ (\delta)}(\Lambda)| \leqq C_{\beta\gamma\delta}\lambda(\Lambda)\langle\xi\rangle^{-(1-1/\kappa_1)|\beta|-|\gamma|}$$

$$|\omega^{(\gamma)}_{\alpha(\delta)}(-\Lambda)| \leqq C_{\alpha\gamma\delta}\lambda(\Lambda)\langle\xi\rangle^{|\alpha|/\kappa_1-|\gamma|},$$

for $\lambda(\Lambda) \leqq 1$. Hence by mean of symbolic calculus of pseudo-differential operators we can expand from (1.3)

$$p_\Lambda(x,D) = p(x,D) + \Sigma_{(1-\sigma)|\alpha+\beta|<\theta}\omega^\beta_\alpha(x,D)p^{(\alpha)}_{(\beta)}(x,D) + r(x,D), \tag{2.15}$$

where $\omega^\beta_\alpha(x,\xi)$ and $r(x,\xi)$ satisfy

$$|\omega^{\beta(\gamma)}_{\alpha(\delta)}(x,\xi)| \leqq C_{\alpha\beta\gamma\delta}\lambda(\Lambda)\langle\xi\rangle^{\sigma|\alpha+\beta|-|\gamma|},$$

$$|r^{(\gamma)}_{(\delta)}(x,\xi)| \leqq C_{\gamma\delta}\lambda(\Lambda)\langle\xi\rangle^{m-\theta-|\gamma|},$$

if $\kappa_1 \geqq \max\{\kappa,1/\sigma\}$. Therefore if we take $\varepsilon_0 > 0$ sufficiently small, we obtain (1.6) for $\lambda(\Lambda) \leqq \varepsilon_0$ from (1.8) and (2.15).

§3. PROOF OF THEOREMS 1.7 AND 1.8

It is trivial that (i) → (ii) and (ii) → (iii). Therefore we prove that (iii) → (i). We note that Cauchy's integral formula implies

$$p^{(\alpha)}_{(\beta)}(x,\xi) = (\partial/\partial y)^{\beta}(\partial/\partial\eta)^{\alpha}p_{\theta,\rho,\delta}(x,\xi;0,0)$$

$$= \alpha!\beta!\ (2\pi i)^{-n} \iint_{\Gamma} p_{\theta,\rho,\delta}(x,\xi;y,\eta)/\Pi^{n}_{j=1}y_j^{\ \beta_j+1}\eta_j^{\alpha_j+1}dyd\eta,$$

for $(1-\rho)|\alpha|+\delta|\beta| \leq \theta$, where $\Gamma = \{(y,\eta)$ in C^{2n}; $|y_j| = \varepsilon|\xi|^{-\delta}$, $|\eta_j| = \varepsilon|\xi|^{\rho}$, $j = 1,\dots,n\}$ and $\varepsilon > 0$. Since $p_{\theta,\rho,\delta}(x,\xi;y,\eta)$ is a polynomial in (y,η), it follows from (11.1.3) in Hörmander [10] that

$$|p_{\theta,\rho,\delta}(x,\xi;y,\eta)| \leq 2^{\ell}|p_{\theta,\rho,\delta}(x,\xi;0,0)| = 2^{\ell}|p(x,\xi)|,$$

for (y,η) in Γ, where $\ell = \max_{(1-\rho)|\alpha|+\delta|\beta|\leq\theta}\{|\alpha+\beta|\}$. Hence we obtain (1.11) for $(1-\rho)|\alpha|+\gamma|\beta| \leq \theta$. We have (1.11) trivially for $(1-\rho)|\alpha|+\delta|\beta| > \theta$ from (1.10). Thus we have proved Theorem 1.7.

PROOF OF THEOREM 1.8: Taylor's expansion of $p(x+y,\xi+\eta)$ with respect to (y,η) implies (i) → (ii). We prove (ii) → (i). Let $(\hat{x},\hat{\xi})$ be fixed, where $|\hat{\xi}| = 1$. Choose $(\hat{y},\hat{\eta}) \in R^{2n}$ so that $|\hat{y}| = 1$, $|\hat{\eta}| = 1$ and $p_m(\hat{x}+t\hat{y}, \hat{\xi}+t\hat{\eta}) \not\equiv 0$ in t. We put

$$f(t,x,\xi,r) = r^m p(x+t\hat{y}, (\xi+t\hat{\eta})r^{-1}).$$

Since $p(x,\xi)-p_m(x,\xi)$ is in $S^{m-\varepsilon}_{1,0}$ $(\varepsilon > 0)$ and $p_m(x,\xi)$ is homogeneous in ξ of degree m, $f(t,x,\xi,r)$ is continuous in (x,ξ,r) and $f(t,x,\xi,0) = p_m(x+t\hat{y},\xi+t\hat{\eta})$. Since f is analytic in t and continuous in (x,ξ,r), Weierstrass preparation theorem implies that there exist $\delta = \delta(\hat{x},\hat{\xi}) > 0$, $\ell = \ell(\hat{x},\hat{\xi})$ and continuous functions $g(t,x,\xi,r)$, $t_j(x,\xi,r)$ $(j = 1,\dots,\ell)$ such that

$$f(t,x,\xi,r) = g(t,x,\xi,r)\Pi^{\ell}_{j=1}\ (t-t_j\ (x,\xi,r))$$

$$g(t,x,\xi,r) \neq 0$$

for $|t| + |x-\hat{x}| + |\xi-\hat{\xi}| + |r| < \delta$. The assumption $d_K(x,\xi) \geq \lambda_K$ implies that

$|t_j(x,\xi,r)| \geqq \lambda_K(|\hat{y}|r^\delta + |\hat{\eta}|r^{1-\rho}) = \lambda_K(r^\delta + r^{1-\rho})$. Hence we obtain for $\xi = \tilde{\xi}r^{-1}$, $r = |\xi|^{-1}$, $\tilde{\xi} = \xi/|\xi|$,

$$|p(x,\xi)| = r^{-m}r^m|p(x,\tilde{\xi}r^{-1}) = r^{-m}|f(0,x,\tilde{\xi},r)| \geqq \lambda_K r^{-m+\theta(\hat{x},\hat{\xi})},$$

for $|x-\hat{x}| + |\tilde{\xi} - \hat{\xi}| + r < \delta(\hat{x},\hat{\xi})$, where $\theta(\hat{x},\hat{\xi}) = \ell(\hat{x},\hat{\xi})\max\{1-\rho,\delta\}$. Since $(\hat{x},\hat{\xi})$ varies in a compact set $K \times \{|\xi| = 1\}$, there is a positive integer N such that $K \times \{|\xi| = 1\}$ is contained in an open covering $\cup_{j=1}^N \{(x,\tilde{\xi});$ $|x-\hat{x}^{(j)}| + |\tilde{\xi}-\hat{\xi}^{(j)}| < \delta(\hat{x}^{(j)},\hat{\xi}^{(j)})\}$. Then we can take $\theta = \max_{1\leqq j\leqq N} \theta(\hat{x}^{(j)},\hat{\xi}^{(j)})$. Thus we obtained (1.10). We can derive (1.11) analogously to the proof of Theorem 1.7 by use of $p_{\theta,\rho,\delta}(x,\xi;y,\eta)$.

PROOF OF THEOREM 1.9: If the assumptions in Theorem 1.9 are valid, we can construct a parametrix of p(x,D). In fact, in the case of $0 < \delta < \rho < 1$, it follows from Hörmander [8]. In the case of $0 < \delta = \rho < 1$, we have for any $\lambda > 0$,

$$|p^{(\alpha)}_{(\beta)}(x,\xi)| \leqq C\lambda^{-|\alpha+\beta|}|\xi|^{\rho(|\beta|-|\alpha|)}|p(x,\xi)|, \text{ if } x \text{ in } K \text{ and } |\xi| \geqq R_{K\lambda},$$

where C is independent of λ. Therefore we have $q(x,\xi) = p(x,\xi)^{-1}$ for $|\xi| \geqq R_{K\lambda}$ and x in K such that $q(x,D)p(x,D) = I+r(x,D)$, where $\|r(x,D)u\|_\ell \leqq C_\ell\lambda^{-1}\|u\|_\ell$. If we take $\lambda > 0$ sufficiently large, we obtain a parametrix of p(x,D) and hence the inequality (1.8). Thus Theorem 1.2 implies Theorem 1.8.

§4. PROOF OF THEOREM 1.10

We can give an analogous proof of Theorem 1.10 to one of Theorem 1.1. It suffices to prove that $(\hat{x},\hat{\xi}) \notin WF(pu)$ and u in $\mathcal{D}'(R^n)$ imply $(\hat{x},\hat{\xi}) \notin WF(u)$. Let $\chi(x,\xi)$ be in $S^0_{1,0}$ such that $\chi = 1$ on $B_{3\varepsilon_1}(\hat{x},\hat{\xi}) = \{(x,\xi)$ in R^{2n}; $|x-\hat{x}|^2 + |\xi/\langle\xi\rangle-\hat{\xi}/\langle\hat{\xi}\rangle|^2 \leqq 3\varepsilon_1\}$ and χ in $C_0^\infty(B_{4\varepsilon_1})$, and let u in $\mathcal{D}'(R^n)$. Then there is s_1 in R^1 such that $\chi(x,D)u$ belongs to Sobolev space $H^{s_1}(R^n)$. Choose $\varepsilon_1 > 0$ sufficiently small such that χpu is in $H^s(R^n)$ for any s in R^1. We prove that χu is in $H^s(R^n)$ for any $s \geqq s_1$. We take $\Lambda_\delta(x,\xi)$ as follows:

$$\Lambda_\delta(x,\xi) = \phi(x,\xi)\log\langle\xi\rangle - N\log(1 + \delta\langle\xi\rangle),$$

where $\phi(x,\xi)$ is in $S^0_{1,0}$ and $\phi = s$ for (x,ξ) in $B_{\varepsilon_1}(\hat{x},\hat{\xi})$ and $\phi = s_1-\ell-m$ for $(x,\xi) \notin B_{2\varepsilon_1}(\hat{x},\hat{\xi})$ and $\phi \leqq s$ and $N = s + m + \ell - s_1 + 1$, and $\delta \geqq 0$. Then noting that $e^{\Lambda_\delta(x,\xi)} \leqq \delta^{-N}\langle\xi\rangle^{s_1-m-\ell-1}$ for $\delta > 0$ and $e^{\Lambda_\delta(x,\xi)}$ is in $S^{s_1-m-\ell}_{1,0}$ for $\delta > 0$, we can see that $e^{\Lambda_\delta}(x,D)\chi u$ is in $H^{m+\ell}(R^n)$ for $\delta > 0$ and $e^{-\Lambda_\delta}(D,x)e^{\Lambda_\delta}(x,D) = I+q_\delta(x,D)$ is in $S^0_{1,0}$ and $q_\delta(x,\xi)$ satisfies

$$|q^{(\alpha)}_{\delta(\beta)}(x,\xi)| \leqq C_{\alpha\beta}\langle\xi\rangle^{-1-|\alpha|}(\log\langle\xi\rangle)^{|\alpha+\beta|},$$

where $C_{\alpha\beta}$ is independent of $0 \leqq \delta \leqq 1$. Hence we have an elliptic pseudo-differential operator $Q_\delta(x,D)$ such that $(I+q_\delta(x,D))Q_\delta(x,D)-I = R_\delta(x,D)$ is a smoothing operator satisfying

$$\|R_\delta u\|_q \leqq C_{q\ell}\|u\|_\ell, \tag{3.1}$$

for any q, ℓ in R^1, where $C_{q\ell}$ is independent of $0 \leqq \delta \leqq 1$. Let $\chi_0(x)$ be in $C^\infty_0(|x-\hat{x}| \leqq 6\varepsilon_1)$ and $\chi_0 = 1$ for $|x-\hat{x}| \leqq 5\varepsilon_1$. Then we have

$$p_{\Lambda_\delta}(\chi_0 e^{\Lambda_\delta}(x,D)Q_\delta u) = p_{\Lambda_\delta}(\chi_0-1)e^{\Lambda_\delta}Q_\delta\chi u+e^{\Lambda_\delta}pR_\delta\chi u+e^{\Lambda_\delta}[p,\chi]u+e^{\Lambda_\delta}\chi pu. \tag{3.2}$$

Since supp $(\chi_0-1) \cap$ supp $\chi = \phi$, we have

$$\|(\chi_0-1)e^{\Lambda_\delta}Q_\delta\chi u\|_q \leqq C_q\|u\|_{s_1}, \tag{3.3}$$

for any q in R^1. Moreover noting that (3.1) and $e^{\Lambda_\delta(x,\xi)} \leqq \langle\xi\rangle^{s_1-\ell-m}$ on supp $\chi^{(\alpha)}_{(\beta)}$ $(|\alpha+\beta| \neq 0)$, we can see that $e^{\Lambda_\delta}pR_\delta\chi u$ and $e^{\Lambda_\delta}[p,\chi]u$ are uniformly bounded in $H^\ell(R^n)$ for $\delta \to 0$. Since (1.16) is valid for v in $H^{m+\ell}(R^n)$ with suppv in K, if $(\hat{x},\hat{\xi})$ is not in WF (pu), (1.16) and (3.2) imply

$$\|\chi_0 e^{\Lambda_\delta}Q_\delta\chi u\|_{m-\theta+\ell} \leqq C, \tag{3.4}$$

for $0 < \delta < 1$, where C is independent of δ. Therefore noting that $e^{\Lambda_0(x,\xi)} \geqq \langle\xi\rangle^s$ for (x,ξ) in $B_{\varepsilon_1}(\hat{x},\hat{\xi})$, we have $(\hat{x},\hat{\xi}) \notin WF(u)$ from (3.3) and (3.4). Thus we completed the proof of Theorem 1.10.

To prove Theorem 1.12, we take $\Lambda_\delta(x,\xi) = \phi(x)\log\langle\xi\rangle-N\log(1+\delta\langle\xi\rangle)$, where

$\phi = s$ for $|x-\hat{x}|^2 \leq \varepsilon_1$ and $\phi = s_1-\ell-m$ for $|x-\hat{x}|^2 \geq 2\varepsilon_1$ and $N = s + m-\ell-s_1 + 1$. Then repeating the above argument, we obtain Theorem 1.12.

References

[1] P. Bolley et J. Camus: Régularité Gevrey et itérés pour d'opérateurs hypoelliptiques, Comm. in P.D.E. 6 (1981), 1057-110.

[2] M. Derridj et C. Zuily: Sur la régularité Gevrey des opérateurs de Hörmander, J. Math. pures et appl. 52 (1973), 309-336.

[3] M. Durand; Régularité Gevrey d'une classe d'opérateurs hypoelliptiques, J. Math. pures et appl. 57 (1978), 323-360.

[4] V.S. Fedii: On a criterion for hypoellipticity, Math. Sobornik 85 (1971), 15-45.

[5] V.V. Grusin: On a class of hypoelliptic operators, Math. Sobornik 83 (1970), 458-476.

[6] S. Hashimoto, T. Matsuzawa et Y. Morimoto: Opérateurs pseudo-differentiels et classe de Gevrey, Comm. in P.D.E. 8 (1983), 1277-1289.

[7] T. Hoshiro, Microlocal energy method of Mizohata and hypoellipticity, J. Math. Kyoto Univ. 28 (1988), 1-12.

[8] L. Hörmander, Pseudo-differential operators and hypoelliptic equations, Proc. Symp. Pure Math. 10 (1966), 138-183.

[9] L. Hörmander: Hypoelliptic second order differential equations, Acta Math. 19 (1967), 147-171.

[10] L. Hörmander: The analysis of linear partial differential operators II and III, Springer, Berlin-Heidelberg-New York-Tokyo, 1983.

[11] V. Iftimie: Opérateurs hypoelliptiques dans des éspaces de Gevrey, Bull. Math. Soc. Sci. Math. Rouanie 27 (1983), 317-333.

[12] C. Iwasaki: Gevrey-hypoellipticity and pseudo-differential operators on Gevrey class, Springer Lecture Notes 1256, 281-293.

[13] K. Kajitani: Wellposedness in Gevrey class of the Cauchy problem for hyperbolic operators, Bull. Sci. Math. 2^e serie, 111 (1987), 415-438.

[14] K. Kajitani and S. Wakabayashi: Microhyperbolic operators in Gevrey classes, to appear in Publ. RIMS Kyoto Univ.

[15] M. Koike: On the microlocal hypoellipticity of pseudodifferential operators, Tsukuba J. Mth. 8 (1984), 55-68.

[16] O. Liess and L. Rodino: Inhomogeneous Gevrey classes and related pseudo-differential operators, Boll. UN. Math. It., Sez. 6, 3-C (1984), 233-323.

[17] Y. Matsumoto: Theories of pseudo-differential operators of ultra-differential class, to appear.

[18] T. Matsuzawa: Gevrey hypoellipticity of a class of pseudo-differential operators, Tohoku Math. J. 39 (1987), 447-464.

[19] A. Melin and J. Sjöstrand: Fourier integral operators with complex valued phase functions, Springer Lecture Notes 459, 120-223.

[20] G. Métivier: Propriété des itérés et ellipticité, Comm. in P.D.E. 3 (1978), 827-876.

[21] Y. Morimoto: Criteria for hypoellipticity of differential operators, Pbul. RIMS, Kyoto Univ. 22 (1986), 1129-1154.

[22] T. Ōkaji: On the lowest index for semi-elliptic operators to be Gevrey hypoelliptic, J. Math. Kyoto Univ., 25 (1985), 693-701.

[23] C. Parenti and L. Rodino: Parametrices for a class of pseudo-differential operators I and II, Annali Mat. Pura ed Appl., 125 (1980), 221-278.

[24] L. Rodino: Gevrey hypoellipticity for a class of operators with multiple characteristics, Asterisque, 89-90 (1984), 249-262.

[25] L. Rodino and L. Zanghirati: Pseudo differential operators with multiple characteristics and Gevrey singularities, Comm. in P.D.E. II (1986), 673-711.

[26] T. Sakurai: Gevrey hypoellipticity for operators with non symplectic characteristics, to appear

[27] K. Taniguti: On the hypoellipticity and the global analytic-hypo-ellipticity of pseudo differential operators, Osaka J. Math. 11 (1974), 221-238.

[28] K. Taniguti: On multi-products of pseudo-differential operators in Gevrey classes and its application to Gevrey hypoellipticity, Proc. Japan Acad. 61 (1985), 291-293.

[29] F. Trèves: An invariant criterion of hypoellipticity, Amer. J. Math. 83 (1961), 645-668.

[30] S. Wakabayashi: Singularities of solutions of the Cauchy problem for hyperbolic systems in Gevrey classes, Japan J. Math. 11 (1985), 157-201.

[31] L. Zanghirati: Iterati di operatori e regolarita Gevrey microlocale anisotropa, Rend. Sem. Mat. Padova, 17 (1982), 85-104.

[32] C. Parenti and L. Rodino: Examples of hypoelliptic operators which are not microhypoelliptic, Bollettino U.M.I. (5) 17-B (1980), 390-409.

Kunihiko Kajitani
Institute of Mathematics
University of Tsukuba
Ibaraki 305
Japan

Seiichiro Wakabayashi
Institute of Mathematics
University of Tsukuba
Ibaraki 305
Japan.

H. KOMATSU

Laplace transforms of hyperfunctions

1. INTRODUCTION

We will present a very elementary theory of Laplace transforms of hyperfunctions of one variable. It has little to do with the main subject of this workshop but we hope it will help one to understand some of the works by Bronshtein [4], Sjöstand [19], Kataoka, Aoki [1], Ōuchi [15, 16], Komatsu [11] and others.

Actually our theory is not quite independent of the hyperbolic equations because the origin is in the Cauchy problem in the zero-dimensional case. We start with the constant coefficient case. Let

$$P = P(d/dx) = a_m(d/dx)^m + \dots + a_0 \tag{1}$$

be a linear ordinary differential operator with $a_i \in \underline{C}$. We want to solve

$$\begin{cases} P(d/dx)u(x) = f(x), & x > 0, \\ u^{(j)}(0) = g_j, & j = 0,\dots,m-1, \end{cases} \tag{2}$$

where f(x) is a (hyper-)function and $g_j \in \underline{C}$.

The classical method of Laplace transformation gives us the following solution (cf. Doetsch [5]).

Assume that the solution u(x) and its derivatives $u^{(j)}(x)$ up to order m are measurable functions of exponential type, i.e. there are constants H and C such that

$$|u^{(j)}(x)| \le Ce^{Hx}, \quad x > 0, \ j = 0,\dots,m. \tag{3}$$

Then the Laplace transform

$$\hat{u}(\lambda) = \int_0^\infty e^{-\lambda x}\, u(x)dx \tag{4}$$

is a holomorphic function on the half plane Re $\lambda >$ H and we have by integration by parts

$$\begin{aligned}\hat{f}(\lambda) &= (P(d/dx)u)^{\wedge}(\lambda) \\ &= P(\lambda)\hat{u}(\lambda) \\ &- (a_m u^{(m-1)}(0) + \dots + a_1 u(0)) \\ &- (a_m u^{(m-2)}(0) + \dots + a_2 u'(0))\lambda \\ &- \dots - a_m u(0)\lambda^{m-1}.\end{aligned}$$

Hence

$$\hat{u}(\lambda) = P(\lambda)^{-1} \{\hat{f}(\lambda) + (a_m g_{m-1} + \dots + a_1 g_0) + \dots + a_m g_0 \lambda^{m-1}\}, \tag{5}$$

so that the solution is obtained by the inversion formula

$$u(x) = \frac{1}{2\pi i} \int_{c-i\infty}^{c+i\infty} e^{\lambda x} \hat{u}(\lambda) d\lambda. \tag{6}$$

Although this method is very useful in practice, it has been believed to have the following three theoretical defects:

1) The datum f(x) must be of exponential type;

2) No good characterization is known of the Laplace image of the functions u of exponential type, so that we are not sure whether or not the right-hand side of (5) is in the image;

3) The integral (6) does not necessarily converge absolutely.

For example, consider the simplest problem

$$\begin{cases} (d/dx - \alpha)u(x) = 0, \\ u(0) = g. \end{cases}$$

The solution

$$u(x) = \frac{1}{2\pi i} \int_{c+i\infty}^{c+i\infty} \frac{g}{\lambda-\alpha} e^{\lambda x} d\lambda, \quad x > 0,$$

converges only as Cauchy's principal value.

The above method was introduced as a justification of the Heaviside operational calculus. To avoid these defects Mikusinski [14] invented an algebraic foundation based on the Titchimarsh theorem on convolutions of continuous functions (cf. also Yosida [20]). His method is very good to justify known results but the operational calculus seems to lose its computational character.

We will show that we can get rid of the defects only by extending the definition of Laplace transforms to hyperfunctions.

We recall that the space $\mathcal{B}_{[a,\infty)}$ of hyperfunctions with support in $[a,\infty)$ is defined by

$$\mathcal{B}_{[a,\infty)} = \mathcal{O}(\underline{C} \setminus [a,\infty)) \,/\, \mathcal{O}(\underline{C}), \tag{7}$$

where $\mathcal{O}(V)$ denotes the space of holomorphic functions on V. If F(z) is in $\mathcal{O}(\underline{C} \setminus [a,\infty))$, the hyperfunction f(x) represented by F(z) is denoted as

$$f(x) = F(x + io) - F(x - io). \tag{8}$$

The holomorphic function F is called a *defining function* of f.

<u>DEFINITION 1</u>: We define the space $\mathcal{B}^{\exp}_{[a,\infty]}$ of *Laplace hyperfunctions* with support in $[a,\infty]$ by

$$\mathcal{B}^{\exp}_{[a,\infty]} = \mathcal{O}^{\exp}(\underline{C} \setminus [a,\infty)) \,/\, \mathcal{O}^{\exp}(\underline{C}), \tag{9}$$

where $\mathcal{O}^{\exp}(V)$ denotes the space of all holomorphic functions on V of exponential type, i.e., V is a (Riemann) domain represented as a union of closed sectors Σ of the form

$$\Sigma = \{z \in \underline{C};\ \ \alpha \leq \arg(z-c) \leq \beta\}\ , \tag{10}$$

and $F \in \mathcal{O}^{\exp}(V)$ is a holomorphic function on V satisfying the estimates

$$|F(z)| \leq Ce^{H|z|}, \quad z \in \Sigma, \tag{11}$$

on each closed sector Σ in V with constants H and C.

The interval $[\alpha,\beta]$ is called the opening of Σ, and the union of openings $[\alpha,\beta]$ of all Σ in V is called the *opening* of V.

DEFINITION 2: We define the *Laplace transform* $\hat{f}(\lambda)$ of a Laplace hyperfunction $f \in \mathcal{B}^{\exp}_{[a,\infty]}$ represented by $F \in \mathcal{O}^{\exp}(\underline{C} \setminus [a,\infty))$ by the integral

$$\hat{f}(\lambda) = \int_{\Gamma} e^{-\lambda z} F(z)dz, \tag{12}$$

where Γ is a contour composed of the ray from $e^{i\beta}\infty$ to $c < a$ and the ray from c to $e^{i\alpha}\infty$ with $0 < \alpha < \pi/2$ and $3\pi/2 < \beta < 2\pi$.

The Laplace transform $\hat{f}(\lambda)$ does not depend on the choice of defining function F and the Laplace transformation gives a one-to-one correspondnece between the Laplace hyperfunctions $f(x) \in \mathcal{B}^{\exp}_{[a,\infty]}$ and the holomorphic functions $\hat{f}(\lambda)$ of exponential type on a domain R of opening $(-\pi/2, \pi/2)$ such that

$$\overline{\lim_{r\to\infty}} \frac{\log|\hat{f}(re^{i\theta})|}{r} \leq -a \cos\theta, \quad |\theta| < \pi/2, \tag{13}$$

(Theorems 2 and 5).

Moreover, the inclusion mapping $\mathcal{O}^{\exp}(\underline{C} \setminus [a,\infty)) \to \mathcal{O}(\underline{C} \setminus [a,\infty))$ induces the surjection

$$\rho : \mathcal{B}^{\exp}_{[a,\infty]} \to \mathcal{B}_{[a,\infty)}$$

(Theorem 6). Therefore we can extend any hyperfunction f(x) in $\mathcal{B}_{[a,\infty)}$ and in particular any continuous function f(x) on $[a,\infty)$ to a Laplace hyperfunction $\tilde{f}(x)$ in $\mathcal{B}^{\exp}_{[a,\infty]}$ and consider its Laplace transform $\hat{f}(\lambda)$.

2. LAPLACE TRANSFORMS OF HOLOMORPHIC FUNCTIONS OF EXPONENTIAL TYPE

Except for the surjectivity of ρ, the whole theory depends only on the classical works of Borel [3], Pólya [17] and Macintyre [13] on Laplace transforms of holomorphic functions of exponential type (see Boas [2]).

We consider a function $F(z) \in \mathcal{O}^{\exp}(V)$ on a domain V with connected opening

(A,B). The function

$$h(\theta) = \overline{\lim_{r\to\infty}} \frac{\log|F(re^{i\theta} + c)}{r} \tag{14}$$

on (A,B) is called the *indicator function* of F. It is shown that $h(\theta)$ is a continuous function independent of the origin c.

Suppose Σ of (10) is in V and $\theta \in [\alpha,\beta]$. We define the Laplace transform $\hat{F}_c(\lambda)$ of F with origin at c by the integral

$$\hat{F}_c(\lambda) = \int_c^{c+e^{i\theta}\infty} e^{-\lambda z} F(z)dz \tag{15}$$

and its analytic continuation.

Originally this is a holomorphic function of exponential type on the half plane $\mathrm{Re}(\lambda e^{i\theta}) > h(\theta)$. Since F(z) is of exponential type, we can deform the path of integral to a curve which eventually coincides with the ray from c to $e^{i\phi}\infty$ for any $\phi \in (A,B)$. Hence we obtain the following.

<u>THEOREM 1</u>: Suppose F(z) is a function in $\mathcal{O}^{\exp}(V)$ on a domain V of opening (A,B). Then its Laplace transform $\hat{F}_c(\lambda)$ with origin at c is a holomorphic function of exponential type on the Riemann domain

$$\rho(F) = \bigcup_{A<\theta<B} \{\lambda \in \underline{C};\ \ \mathrm{Re}(\lambda e^{i\theta}) > h(\theta)\} \tag{16}$$

of opening $(-\pi/2 - B,\ \pi/2 - A)$.

We call the complement

$$\sigma(F) = \bigcap_{A<\theta<B} \{\lambda \in \underline{C};\ \ \mathrm{Re}(\lambda e^{i\theta}) \leqq h(\theta)\} \tag{17}$$

the *convex spectrum* of F. (It is customarily called the conjugate indicator diagram.) Both $\rho(F)$ and $\sigma(F)$ should be regarded as sets in a suitable Riemann surface spread over $\underline{C}$ because the analytic continuation may have different branches on the same point in $\underline{C}$.

<u>THEOREM 2</u> (Borel, Pólya): If F(z) is an entire function of exponential type,

then $\sigma(F)$ is a compact convex set, i.e. the analytic continuation of $\hat{F}_c(\lambda)$ coincides of the overlapping domains.

We have only to expand F into the Taylor series and integrate (15) termwise.

THEOREM 3 (Macintyre): If $z \in V$ is in the subdomain of V that is the union of all convex curves in V homotopic to the original ray $c + R^+e^{i\theta}$, then

$$F(z) = \frac{1}{2\pi i} \int_\Gamma e^{\lambda z} \hat{F}_c(\lambda)d\lambda \qquad (18)$$

for a suitable curve Γ.

This is proved by the change of order of integrals and Cauchy's integral formula.

THEOREM 4: Let V be a domain of opening (A,B) with $\pi < B - A < 2\pi$. An $F(z) \in \mathcal{O}^{\exp}(V)$ can be continued to an entire function in $\mathcal{O}^{\exp}(\underline{C})$ if and only if the convex spectrum $\sigma(F)$ is compact.

The only if part is Theorem 2. The if part is proved by deforming the path Γ of (18) into a closed curve encircling the spectrum $\sigma(F)$.

3. LAPLACE HYPERFUNCTIONS AND THEIR LAPLACE TRANSFORMS

Let

$$f(x) = F(x + io) - F(x - io) \qquad (19)$$

be a Laplace hyperfunction with support in $[a,\infty]$ represented by an $F \in \mathcal{O}^{\exp}(\underline{C} \setminus [a,\infty))$. Since $\underline{C} \setminus [a,\infty)$ is a domain of opening $(0, 2\pi)$, the Laplace transform $\hat{F}_c(\lambda)$ is a holomorphic function of exponential type on the Riemann domain $\rho(F)$ of opening $(-5\pi/2, \pi/2)$. The overlapping domain R is clearly a domain of opening $(-\pi/2, \pi/2)$. We denote by $F_c^+(\lambda)$ (resp. $F_c^-(\lambda)$) the branch of $F_c(\lambda)$ on R with opening $(-\pi/2, \pi/2)$ (resp. $(-5\pi/2, -3\pi/2)$). Then the Laplace transform $\hat{f}(\lambda)$ of f(x) defined by Definition 2 is identified with

$$\hat{f}(\lambda) = \hat{F}_c^+(\lambda) - \hat{F}_c^-(\lambda). \qquad (20)$$

Clearly the right-hand side does not depend on the origin c. It follows from Theorem 4 that $\hat{f}(\lambda)$ does not depend on the defining function F either though the domain R may, and that $\hat{f}(\lambda) = 0$ if and only if $f(x) = 0$.

THEOREM 5: A holomorphic function $\hat{f}(\lambda)$ on a domain R of opening $(-\pi/2, \pi/2)$ is the Laplace transform of a Laplace hyperfunction $f(x) \in \mathcal{B}^{\exp}_{[a,\infty]}$ if and only if $\hat{f}(\lambda)$ is of exponential type on R and its indicator function $\hat{h}(\theta)$ satisfies (13).

Then for any Λ in R

$$F(z) = \frac{1}{2\pi i} \int_\Lambda^\infty e^{\lambda z} \hat{f}(\lambda)d\lambda \qquad (21)$$

is a defining function in $\mathcal{O}^{\exp}(\underline{C} \setminus [a,\infty))$ of the unique Laplace hyperfunction $f(x) \in \mathcal{B}^{\exp}_{[a,\infty]}$.

PROOF: Let $f(x) = G(x + io) - G(x - io)$ be a Laplace hyperfunction in $\mathcal{B}^{\exp}_{[a,\infty]}$ represented by $G(z) \in \mathcal{O}^{\exp}(\underline{C} \setminus [a,\infty))$. We take a point $c < a$ and consider the Laplace transform $\hat{G}_c(\lambda)$. By a simple estimate we see that its indicator function $\hat{H}(\theta)$ satisfies

$$\hat{H}(\theta) \leqq -c \cos \theta, \ -5\,\pi/2 < \theta < \pi/2.$$

Hence, it follows from (20) that the indicator function $\hat{h}(\theta)$ of $\hat{f}(\lambda)$ satisfies

$$\hat{h}(\theta) \leqq -c \cos \theta, \qquad |\theta| < \pi/2.$$

Since $c < a$ is arbitrary we have (13).

In order to prove that F(z) defined by (21) is a defining function of f(x), we make use of the Macintyre inversion formula

$$G(z) = \frac{1}{2\pi i} \int_\Gamma e^{\lambda z} \hat{G}_c(\lambda)d\lambda.$$

We deform the contour Γ in $\rho(G)$ into a double curve from Λ to ∞ with opposite directions plus a closed curve Γ_0 starting from Λ. Then we have

$$G(z) = \frac{1}{2\pi i} \int_{\Gamma_0} e^{\lambda z} G_c(\lambda)d\lambda + \frac{1}{2\pi i} \int_{\Lambda}^{\infty} e^{\lambda z} \hat{f}(\lambda)d\lambda.$$

The first term is clearly in $\mathcal{O}^{exp}(\underline{C})$.

Conversely suppose $\hat{f}(\lambda) \in \mathcal{O}^{exp}(\underline{R})$ satisfies (13). Then (21) defines a function $F(z) \in \mathcal{O}^{exp}(\underline{C} \setminus [a,\infty))$ by Theorem 1. Let $g(x) = F(x + io) - F(x - io)$ be the Laplace hyperfunction with defining function $F(z)$. We have by Theorem 3

$$\hat{f}(\hat{\lambda}) = \int_{\Gamma} e^{-\lambda z} F(z)dz.$$

On the other hand, the right-hand side is the Laplace transform $\hat{g}(\lambda)$ of $g(x)$ by the definition.

<u>THEOREM 6</u>: The canonical mapping

$$\rho: \mathcal{B}^{exp}_{[a,\infty]} \to \mathcal{B}_{[a,\infty)} \tag{22}$$

is surjective, i.e. every hyperfunction $f(x) \in \mathcal{B}_{[a,\infty)}$ has a defining function $F(z)$ in $\mathcal{O}^{exp}(\underline{C} \setminus [a,\infty))$.

This is the only theorem whose proof is not elementary.

On the analogy with the usual theory of hyperfunction (see e.g. [7]) a natural proof would be obtained if we could prove the cohomology vanishing theorem

$$H^1(V, \mathcal{O}^{exp}) = 0 \tag{23}$$

for any open set V in the radial compactification

$$\underline{D} = \underline{C} \cup S^1\infty$$

of $\underline{C}$, where $\mathcal{O}^{exp}$ is regarded as a sheaf on $\underline{D}$ (actually it is enough to prove (23) for $V = \underline{D} \setminus \{+\infty\}$). However, since the topological structure of $\mathcal{O}^{exp}$ was as complicated as the real analytic functions, we were unable to prove it. Instead we employed in [12] Saburi's theorem [18]

$$H^1(V, \mathcal{O}^{\text{infexp}}) = 0, \qquad (24)$$

which holds for a restricted class of open sets V in $\underline{D}$, where $\mathcal{O}^{\text{infexp}}$ denotes the sheaf of holomorphic functions on $\underline{D}$ of minimal exponential type.

We remark that the space $\mathcal{O}^{\exp}(\underline{C})$ of entire functions of exponential type is strictly smaller than the space $\mathcal{O}(\underline{C}) \cap \mathcal{O}^{\exp}(\underline{C} \setminus [a,\infty))$ of entire functions which is of exponential type except for the positive direction. Hence the mapping ρ is not injective. The elements in the kernel are called Laplace hyperfunctions with support only at ∞.

More generally we say that ∞ is not included in the support supp f of an $f(x) \in \mathcal{B}^{\exp}_{[a,\infty]}$ if its defining function F(z) can be continued to a function in $\mathcal{O}^{\exp}(\{z;\ |z| > M\})$ for an M. We define the support supp f on an $f \in \mathcal{B}^{\exp}_{[a,\infty]}$ by

$$\operatorname{supp} f = \begin{cases} \operatorname{supp} \rho(f), & \text{if } \infty \notin \operatorname{supp} f, \\ \operatorname{supp} \rho(f) \cup \{\infty\} & \text{, otherwise.} \end{cases} \qquad (25)$$

Then $\mathcal{B}^{\exp}$ forms a flabby sheaf on $[a,\infty]$ and we have

$$\operatorname{supp} \rho(f) = \operatorname{supp} f \cap \underline{R}. \qquad (26)$$

In particular, if $-\infty < a < b < \infty$, then we have the canonical isomorphism

$$\mathcal{B}_{[a,b]} = \mathcal{B}^{\exp}_{[a,b]}, \qquad (27)$$

where the left-hand side (resp. right-hand side) denotes the space of all hyperfunctions (resp. Laplace hyperfunctions) with support in [a,b].

4. ACTION OF PSEUDO-DIFFERENTIAL OPERATORS

For a Laplace hyperfunction

$$f(x) = F(x + io) - F(x - io)$$

in $\mathcal{B}^{\exp}_{[a,\infty]}$ we define its derivative f'(x) by

$$f'(x) = F'(x + io) - F'(x - io). \tag{28}$$

Clearly $f'(x)$ is a Laplace hyperfunction in $\mathcal{B}^{exp}_{[a,\infty]}$ independent of the defining function, and we have

$$(f')^{\wedge}(\lambda) = \lambda\hat{f}(\lambda) \tag{29}$$

for its Laplace transform.

Hence, if $P(d/dx)$ is a differential operator of the form (1), we have

$$(P(d/dx)f)^{\wedge}(\lambda) = P(\lambda)\hat{f}(\lambda).$$

More generally, suppose $Q(\lambda)$ is a function in the Laplace image $L\mathcal{B}^{exp}_{[c,\infty]}$ of $\mathcal{B}^{exp}_{[c,\infty]}$. Then the multiplication by $Q(\lambda)$ is a linear mapping

$$Q(\lambda) : L\mathcal{B}^{exp}_{[a,\infty]} \to L\mathcal{B}^{exp}_{[a+c,\infty]}.$$

Therefore we can define the action

$$Q(d/dx) : \mathcal{B}^{exp}_{[a,\infty]} \to \mathcal{B}^{exp}_{[a+c,\infty]} \tag{30}$$

by

$$Q(d/dx)f(x) = G(x + io) - G(x - io), \tag{31}$$

where

$$G(z) = \frac{1}{2\pi i}\int_{\Lambda}^{\infty} e^{\lambda z}\, Q(\lambda)\hat{f}(\lambda)d\lambda. \tag{32}$$

This is sufficient for the operational calculus in the constant coefficient case.

If $P(\lambda)$ is a non zero polynomial, then $Q(\lambda) = P(\lambda)^{-1}$ is clearly in $L\mathcal{B}^{exp}_{[0,\infty]}$, so that we have the inverse mapping

$$P(d/dx)^{-1} : \mathcal{B}^{exp}_{[a,\infty]} \to \mathcal{B}^{exp}_{[a,\infty]} \tag{33}$$

of $P(d/dx)$ for any $-\infty < a \leqq \infty$.

Let $0 < b \leqq \infty$. Suppose we are given a hyperfunction or, $(-\infty,b)$ having support in $[0,b)$. By the flabbiness of the hyperfunctions it can be continued to a hyperfunction in $\mathcal{B}_{[0,\infty)}$ and then by Theorem 6 to a Laplace hyperfunction $\tilde{f}(x)$ in $\mathcal{B}^{\exp}_{[0,\infty]}$. Since the difference of two extensions is in $\mathcal{B}^{\exp}_{[b,\infty]}$ and the action of $P(d/dx)^{-1}$ in (33) is independent of a, it follows that there is a unique solution $u(x) \in \mathcal{B}_{[0,b)}$ of

$$P(d/dx)u(x) = f(x), \tag{34}$$

which is obtained as the restriction to $(-\infty,b)$ of $P(d/dx)^{-1}\tilde{f}(x)$.

To take care of initial values, we have only to note the Green formula

$$\begin{aligned} &P(d/dx)(\theta(x)u(x)) \\ &= \theta(x)(P(d/dx)u(x)) \\ &+ (a_m u^{(m-1)}(a) + \dots + a_1 u(0))\delta(x) \\ &+ \dots + a_m u(0)\delta^{(m-1)}(x), \end{aligned} \tag{35}$$

where $\theta(x)$ is the Heaviside function.

We can also define for symbols $Q(x,\lambda)$ with variable coefficients its action on $\mathcal{B}_{[a,b)}$ by

$$Q(x, d/dx)f(x) = G(x + io) - G(x - io), \tag{36}$$

$$G(z) = \frac{1}{2\pi i} \int_{\Lambda}^{\infty} e^{\lambda z} Q(z,\lambda)\hat{f}(\lambda)d\lambda, \tag{37}$$

where $\hat{f}(\lambda)$ is the Laplace transform of an extension of $f(x)$ to $\mathcal{B}^{\exp}_{[a,\infty]}$.

To solve the Cauchy problem for the differential operator

$$P(x, d/dx) = a_m(x)(d/dx)^m + \dots + a_0(x) \tag{38}$$

with analytic coefficients, we construct a formal symbol

$$Q(x,\lambda) = \sum_{j=0}^{\infty} Q_j(x,\lambda) \tag{39}$$

so that

$$P(x,\lambda) \circ Q(x,\lambda) = 1 \tag{40}$$

in the sense of pseudo-differential operators in the usual way. Namely we expand $Q(x,\lambda)$ into a formal power series in λ^{-1} and solve

$$P(x, \lambda + d/dx)Q(x,\lambda) = 1$$

termwise comparing the coefficients.

Since $Q_j(x,\lambda)$ is of order $O(C^j j!)$, the sum (39) diverges. However, if we take a sufficiently large Λ and define

$$G_j(z) = \frac{1}{2\pi i} \int_{\Lambda_j}^{\infty} e^{\lambda z} Q_j(z,\lambda)\hat{f}(\lambda)d\lambda, \tag{41}$$

then

$$G(z) = \sum_{j=0}^{\infty} G_j(z) \tag{42}$$

is shown to converge in a neighbourhood of a outside $[a,\infty)$. Therefore we can define the action of $Q(x, d/dx)$ by

$$Q(x, d/dx)f(x) = G(x + io) - G(x - io) \tag{43}$$

in a neighbourhood of a.

Ōuchi [15, 16] introduced this type of method to analyze Hamada's solution [6] to the Cauchy problem of partial differential equations with meromorphic data in the complex domain.

We have considered in [11] the case where the Cauchy data are more complicated functions and have proved that the irregularity condition introduced in [8,9] is also necessary in order that the Cauchy problem for a formally hyperbolic operator be correctly posed in a Gevrey class of functions or ultradistributions.

References

[1] T. Aoki, Symbols and formal symbols of pseudodifferential operators, Advanced Studies in Pure Mathematics 4, Group Representations and Systems of Differential Equations, Kinokuniya, Tokyo, 1984. pp. 181-208.

[2] R.P. Boas, Jr., Entire Functions, Academic Press, New York, 1954.

[3] E. Borel, Leçons sur les Séries Divergents, 2éd., Gauthier-Villars, Paris, 1928.

[4] M.D. Bronshtein, The parametrix of the Cauchy problem for hyperbolic operators with characteristics of variable multiplicity, Trudy Moscov. Math. Obšč., 41 (1980), 83-99.

[5] G. Doetsch, Theorie und Anwendung der Laplace-transformation, Springer, Berlin, 1937.

[6] Y. Hamada, Problème analytique de Cauchy à caractéristiques multiples dont les données de Cauchy ont des singularitiés polaires, C.R. Acad. Sci. Paris, Sér. A, 276 (1973), 1681-1684.

[7] H. Komatsu, An introduction to the theory of hyperfunctions, Hyperfunctions and Pseudo-Differential Equations, Lecture Notes in Math., 287 (1973), 3-40.

[8] H. Komatsu, Irregularity of characteristic elements and construction of null-solutions, J. Fac. Sci., Univ. Tokyo, Sec. IA, 23 (1976), 297-342.

[9] H. Komatsu, Irregularity of characteristic elements and hyperbolicity, Publ. RIMS, Kyoto Univ., 12 Suppl. (1977), 233-245.

[10] H. Komatsu, Linear hyperbolic equations with Gevrey coefficients, J. Math. Pures Appl., 59 (1980), 145-185.

[11] H. Komatsu, Irregularity of hyperbolic operators, Hyperbolic Equations and Related Topics, Kinokuniya, Tokyo, 1986, pp. 155-179.

[12] H. Komatsu, Laplace transforms of hyperfunctions – A new foundation of the Heaviside calculus, J. Fac. Sci., Univ. Tokyo, Sec. IA, 34 (1987), to appear.

[13] A.J. Macintyre, Laplace's transformation and integral functions, Proc. London Math. Soc., (2) 45 (1938), 1-20.

[14] J. Mikusiński, Rachunek Operatorow, Warszawa, 1953.

[15] S. Ōuchi, Asymptotic behaviour of singular solutions of linear partial differential equations in the complex domain, J. Fac. Sci., Univ. Tokyo, Sec. IA, 27 (1980), 1-36.

[16] S. Ōuchi, An integral representation of singular solutions of linear partial differential equations in the complex domain, J. Fac. Sci., Univ. Tokyo, Sec. IA, 27 (1980), 37-85.

[17] G. Pólya, Untersuchungen über Lücken und Singularitäten von Potenzreihen, Math. Z., 29 (1929), 549-640.

[18] Y. Saburi, Fundamental properties of modified Fourier hyperfunctions, Tokyo J. Math., 8 (1985), 231-273.

[19] J. Sjöstrand, Singularités analytiques microlocales, Astérisque, 95 (1982), 1-166.

[20] K. Yosida, Operational Calculus A Theory of Hyperfunctions, Springer, New York - Berlin - Heidelberg - Tokyo, 1984.

Hikosaburo Komatsu
University of Tokyo,
Faculty of Sciences,
Department of Mathematics,
Hongo, Tokyo 113,
Japan.

G. LEBEAU

Équation des ondes semi-linéaires en dimension 3 d'espace. Un resultat de finitude

ABSTRACT: The semilinear wave equation in 3-space dimension: a finiteness result

Let $u(x_0,x')$ be a function in the Sobolev space $H^s_{loc}(\Omega)$, $(s > 2, \Omega \subset \mathbb{R}^4)$, solution of the semi-linear wave equation

$$(\partial^2_{x_0} - \Delta_{x'})u = \sum_{j=0}^{d} p_j(x)u^j \qquad p_j \in C^\infty(\Omega).$$

Suppose that the Cauchy data of u, $u_0 = u(0,x') \in H^s_{loc}(\omega)$, $u_1 = \partial u/\partial x_0(0,x') \in H^{s-1}_{loc}(\omega)$ $(\omega = \Omega \cap x_0 = 0$, Ω influenced by $\omega)$ are Lagrangian distributions of Hörmander on some analytic Lagrangian submanifold Λ of $T^*\omega$. We prove that for every real σ, there exists a closed, subanalytic, homogeneous and isotropic subset L_σ of $T^*\Omega$ depending only on Λ such that $WF^\sigma(u) \subset L_\sigma$, where WF^σ denotes the Sobolev H^σ microlocal singularities. As a consequence, we obtain that, for every integer k, there exists an open, dense subset Ω_k of Ω such that $u|_{\Omega_k} \in C^k$; examples, (such as in Beals [1]) where $u \notin H^\sigma_x$ at any points of an open subset U of Ω are therefore excluded if the Cauchy data of u are conormal distributions on analytic submanifolds of Ω.

1. INTRODUCTION

Soit $x = (x_0,x')$ le point courant de l'espace-temps $\mathbb{R}^4$, $x' = (x_1,x_2,x_3)$, $\square$ l'opérateur des ondes, $\square = \partial^2_{x_0} - \Delta_{x'}$, Ω un ouvert de $\mathbb{R}^4$ et $\omega = \{\Omega \cap x_0 = 0\}$ la trace de Ω à l'instant $t = 0$. On suppose que Ω est un domaine d'influence pour ω, c'est-à-dire que pour tout $x \in \Omega$ l'intersection du cône d'onde de sommet x avec l'hyperplan $x_0 = 0$ est contenu dans ω.

Soit $u(x_0,x')$ un élément de $H^s_{loc}(\Omega)$, $s > 2$, qui vérifie l'équation des ondes semi-linéaires:

$$\begin{aligned} &\square u = p(u) \text{ dans } \Omega \\ &u|_{x_0 = 0} = u_0 \in H^s_{loc}(\omega) \qquad\qquad (*) \\ &\frac{\partial u}{\partial x_0}\Big|_{x_0 = 0} = u_1 \in H^{s-1}_{loc}(\omega) \end{aligned}$$

où $p(u) = \sum_{j=0}^{d} p_j(x)u^j$ est un polynôme de u, les fonctions $p_j(x)$ étant C^∞ sur Ω.

Le problème qu'on se pose est d'estimer les singularitiés de u, connaissant les singularités des données de Cauchy u_0 et u_1; plus précisément, on cherche des estimations sur les différents fronts d'ondes Sobolev de u, $WF^\sigma(u)$, connaissant u_0 et u_1 (on sait que si $u \in H^s_{loc}(\Omega)$, $s > 2$ et $u_0,u_1 \in C^\infty(\omega)$, alors $u \in C^\infty(\Omega)$).

Un problème voisin du précédent consiste à déterminer les singularités de u dans l'avenir $\Omega_+ = \Omega \cap \{x_0 > 0\}$ connaissant les singularités de u dans le passé $\Omega_- = \Omega \cap (x_0 < 0)$, en supposant que Ω_+ est influencé par Ω_-.

Ce type de problème a été résolu par J-M. Bony [2] pour des régularités Sobolev $\sigma \leq 2s - s_0$, (comportement linéaire des singularités) puis étudié par M. Beals [1] et J-Y. Chemin [4] jusqu'a $\sigma \leq 3s-s_1$ (interaction-contrôlée), le cas de la dimension un d'espace ayant été élucidé par Rauch et Reed [7]. Au delà de la régularité 3s, on ne peut espérer contrôler la géométrie du front d'onde de u à partir de la géométrie du front d'onde des données, comme le montre un exemple de M. Beals pour lequel on a $u_0,u_1 \in C^\infty(\omega)\setminus\{x' = 0\})$ et le support singulier (u) remplit le cône d'onde $|x'| \leq |x_0|$ [1].

Si on veut contrôler la géométrie du front d'onde de u au-delà de la régularité 3s, on doit donc faire des hypothèses sur la nature des singularités des données de Cauchy, qui permettent de contrôler le front d'onde des produits au delà de la régularité 2s-n/2.

L'hypothèse supplémentaire est celle de conormalité: si V est une sous-variété de $\mathbb{R}^n$, une distribution $f \in H^s_{loc}(\mathbb{R}^n)$ est conormale sur V ssi pour tout $N \in \mathbb{N}$ et tous champs de vecteurs $X_1,\dots,X_N$ tangents à V on a $X_1,\dots,X_N f \in H^s_{loc}(\mathbb{R}^n)$.

L'interaction de trois ondes conormales non linéaires a été étudiée par J-M. Bony [3] et R. Melrose-N. Ritter [6] en dimension 2 d'espace pour l'équation $\square u = F(u)$: les singularités de la solution restent confinées sur les 3 ondes et sur le cône d'onde émis par leur point d'intersection. Ce résultat a été étendu par J-Y. Chemin [4] a l'équation $\square u = F(u,\nabla u)$.

Le résultat qui suit affirme que si les données de Cauchy sont de type conormal, pour tout réel σ, le front d'onde $WF^\sigma(u)$ est "petit".

<u>THÉORÈME</u>: Soit u solution de (*). On suppose que les traces u_0 et u_1 sont des distributions intégrales de Fourier de Hörmander, à symbole C^∞, sur une lagrangienne Λ de $T^*\omega$, analytique réelle.

Alors pour tout $\sigma \in \mathbb{R}$, il existe un ensemble sous-analytique homogène isotrope L_σ dans $T^*\Omega$, ne dépendant que de Λ, tel que

$$WF^\sigma(u) \subset L_\sigma.$$

<u>COROLLAIRE</u>: Pour tout entier k, il existe un ouvert dense Ω_k, dans Ω, tel que $u|_{\Omega_k}$ soit de classe C^k.

<u>Rappels</u>: 1) Un sous-analytique est localement une projection propre d'un ensemble analytique.

2) Si L est un sous-analytique homogène dans $T^*\Omega$, il est isotrope ssi la 1-forme canonique $\xi\cdot dx$ est nulle sur toute courbe analytique tracée dans L. On a alors nécessairement $\dim L \leq \dim \Omega$, d'où le corollaire.

Dans le II, on donne un exemple géométrique où Λ est simple, mais $\overline{UL_\sigma}$ ne peut pas être un sous-analytique isotrope, même en temps petit.

Dans le III, on donne une idée de la preuve du théorème (voir [5]).

II. UN EXEMPLE GEOMETRIQUE EN DIMENSION 3D'ESPACE: LE DIVISEUR A CROISEMENT NORMAL

On choisit pour lagrangienne initiale $\Lambda = \bigcup_{j=1}^{3} T^*_{x_j=0}$ et on va itérer (partiellement) les deux procédes naturels associés à l'équation semi-linéaire: propager les singularités puis prendre l'enveloppe convexe en ξ.

On note D_α la droite de $\mathbb{R}^4$, pour $\alpha \in \mathbb{R}^3$

$$D_\alpha = \{x' = \alpha x_0\}$$

et $T^*_{D_\alpha}$ son fibré conormal:

$$T^*_{D_\alpha} = \{(x',x_0;\xi',\xi_0);\quad x' = \alpha x_0 \quad \xi'\alpha + \xi_0 = 0\}.$$

Pour $|\alpha| \geq 1$, $T^*_{D_\alpha} \setminus \xi = 0$ intersecte le cône caractéristique $|\xi'|^2 = \xi_0^2$ et on note Γ_α le cône, projection en x des bicaractéristiques issues de $(T^*_{D_\alpha} \setminus \xi = 0) \cap \{|\xi'|^2 = \xi_0^2\}$. On a:

$$\Gamma_\alpha = \{(x',x_0);\ (x'-\alpha x_0)^2\ (\alpha^2-1) = (x'\cdot\alpha - x_0\alpha^2)^2\}.$$

L'intersection du cône Γ_α et de l'hyperplan $x_0 = 1$ est le cône γ_α de $\mathbb{R}^3$, de sommet α et tangent à la sphère de rayon 1, centrée à l'origine.

<u>1EME ETAPE</u>: En propageant linéairement Λ on obtient les 6 hypersurfaces caractéristiques $x_j \pm x_0 = 0$, $j = 1,\dots,3$.

<u>2EME ETAPE</u>: En prenant la fermeture convexe en ξ on obtient en particulier les fibrés conormaux aux 8 droites d'intersection 3 à 3: $T^*_{D_\varepsilon}$ avec $\varepsilon = (\pm1, \pm1, \pm1)$.

<u>3EME ETAPE</u>: Choisissons $\alpha_1 = (-1, +1, +1)$, $\alpha_2 = (+1, -1, +1)$, $\alpha_3(+1,+1, -1)$. En propageant lineairement $T^*_{D_{\alpha j}}$, on obtient les fibré conormaux aux cônes Γ_{α_j}.

<u>4EME ETAPE</u>: Les cônes γ_{α_j} sont tangents à la sphère unité le long de cercles C_{α_j} 2 à 2 tangents. Soit C_{α_4}, cercle tangent au 3 cercles C_{α_j}, $j=1,\dots,3$. Alors $\Gamma_{\alpha_1} \cap \Gamma_{\alpha_2} \cap \Gamma_{\alpha_3}$ contient D_{α_4} donc par enveloppe convexe en ξ on

obtient $T^*_{D_{\alpha_4}}$.

Etc ... A partir des points $(\alpha_2,\alpha_3,\alpha_4)$ on définit de même un point α_5 puis α_6 ...

REMARQUE: Une description complète des singularités émises par le diviseur à croisement normal devrait prendre en compte *toutes* les intéractions 3 à 3, y compris celles avec les cylindres issus des droites $x_j = x_i = 0$.

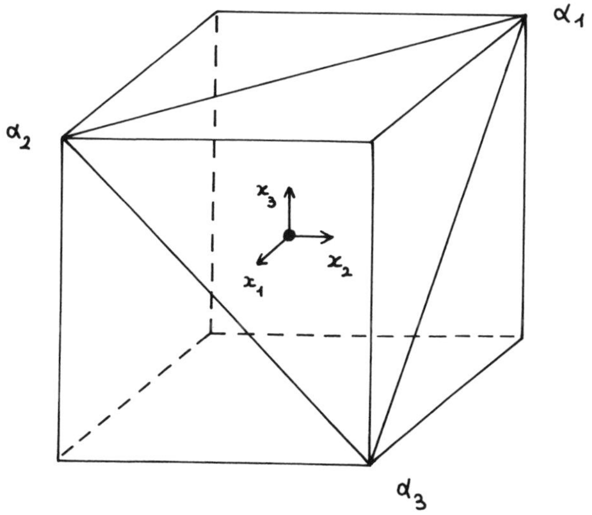

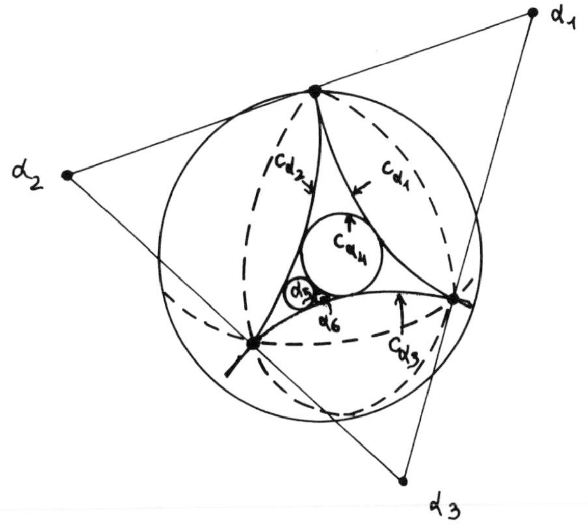

Schéma de la construction des cercles C_{α_j}, intersection avec la sphère unité des traces des cônes Γ_{α_j} à l'instant $x_o = 1$.

III. IDEE DE LA PREUVE DU THEOREME

DEFINITION: Une distribution $T \in \mathcal{D}(\mathbb{R}^n)$ vérifie l'hypothèse (I) ssi le front d'onde C^∞ de T est contenu dans un ensemble homogène, sous-analytique, isotrope.

L'hypothèse (I) est stable par image directe propre et produit tensoriel:

PROPOSITION 1: a) Si T(x) et S(y) verifient l'hypothèse (I), il en est de même de $T \otimes S$.

b) Si T(x,y) vérifie l'hypothese (I) et est à support compact en x, $\int T(x,y)dx$ vérifie l'hypothèse (I).

Le point b) résulte du fait que si $Z \subset T^*(\mathbb{R}^n_x \times \mathbb{R}^p_y)$ est sous-analytique, homogène et isotrope, contenu dans $|x| \leq M$, (y,η), $\exists x$, $(x,0,y,\eta) \in Z\}$ est sous-analytique, homogène isotrope.

La proposition suivante contient la partie non-linéaire de la preuve du théorème et est conséquence du théorème de désingularisation analytique réel de Hironaka.

PROPOSITION 2: Soient $f_1,\ldots,f_N$, $g_1,\ldots,g_N$ des fonctions analytiques réelles $\sigma_1(x,\lambda_1),\ldots,\sigma_N(x,\lambda_N)$ des symboles C^∞ de degré (strict) < -1. Alors

$$u(x) = 1_{\{g_1 \geq 0,\ldots,g_N \geq 0\}} \prod_{j=1}^{N} \int e^{i\lambda_j f_j(x)} \sigma_j(x,\lambda_j)d\lambda_j$$

verifie l'hypothese (I).

Soit v la solution du problème linéaire:

$$\square v = 0 \qquad \text{dans } \Omega$$

$$v|_{x_o=0} = u_o \qquad \frac{\partial v}{\partial x_o}\Big|_{x_o=0} = u_1.$$

Alors $v \in H^s_{loc}(\Omega)$ et est une distribution de Hörmander sur la lagrangienne L, réunion des bicaractéristiques de $\square$ passant au dessus de Λ. On pose

$$u = v + f;\ f = f_+ + f_-,\ f_\pm = f \cdot 1_{\pm x_o \geq 0}.$$

On est donc ramené à prouver le théorème pour f_+. On désigne par E_+ (resp. E_-) la paramétrix de $\square$ propageant vers le futur, $x_0 \geq 0$, (resp. le passé). On suppose dans la suite qu'on a $s < 5/2$, de sorte que si $g \in H^s_{loc}(\Omega)$, on a $E_+(g.1_{x_0 \geq 0}) \in H^s_{loc}(\Omega)$. Pour tout entier ℓ, on décompose f_+ sous la forme

$$f_+ = a_\ell + s_\ell$$

où a_ℓ, s_ℓ sont des éléments de $H^s_{loc}(\Omega)$, à support dans $x_0 \geq 0$, définis par récurrence par:

$$a_0 = 0 \qquad s_0 = f_+$$

$$a_{\ell+1} = E_+ \Big(\sum_{j,k} p_j C^k_j v^{j-k}\, 1_{x_0 \geq 0}\, a^k_\ell \Big)$$

$$s_{\ell+1} = E_+ \Big(\sum_{j,k,m \geq 1} p_j C^k_j v^{j-k} C^m_k\, a^{k-m}_\ell\, s^m_\ell \Big)$$

$a_{\ell+1}$ et $s_{\ell+1}$ sont obtenus en remplaçant f_+ par $a_\ell + s_\ell$ dans l'identité $f_+ = E_+(p(v\, 1_{x_0 \geq 0} + f_+))$.

Les a_ℓ sont calculables à partir de la solution v du problème linéaire. Pour étudier les s_ℓ, on utilise des intégrations par parties avec des espaces de Sobolev sur le produit $\mathbb{R}^4 \times \mathbb{R}^4 \times \ldots \times \mathbb{R}^4$ ($k+1$ facteurs).

<u>DEFINITION</u>: Soit $\underline{X}^k = (x^1, \ldots, x^k) \in \mathbb{R}^{4K}$, $\bar{y}^* = (\bar{y}, \bar{\eta}) \in T^*\mathbb{R}^4 \setminus 0$ et $\sigma \geq 0$. Une distribution $b(x^1, \ldots, x^k; y)$ appartient à B^σ_k au point $(\underline{X}^k, \bar{y}^*)$ s'il existe $\Phi \in C^\infty_0(\mathbb{R}^{4k})$ égale à 1 près de $\underline{X}^k$, $\phi \in C^\infty_0(\mathbb{R}^4)$ égale à 1 près de $\bar{y}$ et Γ voisinage conique de $\bar{\eta}$ tels que

$$\widehat{\Phi \cdot \phi \cdot b}(\xi_1, \ldots, \xi_k; \eta) \prod_{j=1}^{k} (1+|\xi_j|)^{-s} (1+|\eta|)^\sigma \in L^2(\mathbb{R}^{4k} \times \Gamma)$$

On note $\overset{+}{B}{}^\sigma_{k,\bar{y}^*} = \{u \in \mathcal{D}'(\Omega^{k+1})$, support $(u) \subset \{(X^k, y)\ \forall j\ x^j \in$ cône rétrograde issu de $y\}$ et $u \in B^\sigma_k$ en tout point $X^k, \bar{y}^*)\}$.

Ces espaces vérifient les propriétés suivantes:

1) Si $a(x^j) \in H^s(\mathbb{R}^4)$ près de $\underline{x}^j$ et $b \in B^\sigma_k$ près de $(\underline{X}^k, \bar{y}^*)$, alors $a.b \in B^\sigma_k$ au point $(\underline{X}^k, \bar{y}^*)$.

2) Si $b \in \overset{+}{B}{}^\sigma_{k,\bar{y}^*}$ est à support compact en x

$$\int b(x^1,\dots,x^k;y)dx^1 \dots dx^k \in H^{\sigma}_{y*}(\mathbb{R}^4)$$

où l'espace de droite est le Sobolev microlocal usuel.

3) Si $b \in \dot{B}^{\sigma}_{k,\bar{y}*}$, on note $E_-^{\otimes k}(b)$ l'unique distribution qui vérifie

$$\{\Box_{x_1} \dots \Box_{x_k} E_-^{\otimes k}(b) = b$$

Support $E_-^{\otimes k}(b) \subset \{(X^k,y);\ \forall j\ x^j \in$ cône rétrograde issu de $y\}$.

Alors $E_-^{\otimes k}(b) \in \dot{B}^{\sigma}_{k,\bar{y}*}$. Si de plus

$$\forall X^k;\ (X^k,\xi^1 = \dots = \xi^k = 0,\ \bar{y}*) \notin WF(b)$$

alors $E_-^{\otimes k}(b) \in \dot{B}^{\sigma+\frac{1}{}}_{k,\ \bar{y}*}$.

4) Soit $\prod_{j=1}^{k} \delta_{x^j,\ell_j}$ l'opérateur de $\mathcal{D}'(\Omega^{k+1})$ dans $\mathcal{D}'(\Omega^{\ell_1+\dots+\ell_k+1})$

défini par:

$$\forall \phi \in C_o^{\infty}(\Omega^{\ell_1+\dots+\ell_k+1}):$$

$$\prod_{j=1}^{k} \delta_{x^j,\ell_j}(b)(\phi) = \int b(x^1,\dots,x^k,y)\phi(\underbrace{x^1,\dots,x^1}_{\ell_1 \text{ fois}};\dots;\underbrace{x^k,\dots,x^k}_{\ell_k \text{ fois}};y)$$

Alors si $b \in \dot{B}^{\sigma}_{k,\bar{y}*}$, on à $\prod_{j=1}^{k} \delta_{x^j,\ell_j}(b) \in \dot{B}^{\sigma}_{\ell_1+\dots+\ell_k;\bar{y}*}$.

Posons a present $B^{\sigma}_{y*} = \bigoplus_{k=1}^{\infty} \dot{B}^{\sigma}_{k,\bar{y}*}$. On définit par récurrence des sous $\mathbb{C}$-espaces vectoriels V^{ν}_{ℓ}, $0 \leqq \nu \leqq \ell$ de B^{o}_{y*}, indépendants de $\bar{y}*$ en posant

$$V^{o}_{\ell} = \mathbb{C}\ \delta_{x=y}$$

$V^{\nu}_{\ell} = \mathbb{C}$. ev engendré par les distributions de la forme

$$\Pi\ \delta_{x^j,\ell_j} \cdot \prod_{j=1}^{k} a_{\ell-\nu}^{k_j-\ell_j}(x^j)\ v^{n_j-k_j}(x^j)\ p_{n_j}(x^j)\ E_-^{\otimes k}(b)$$

où $b(x^1,\dots,x^k;y) \in V^{\nu-1}_{\ell}$ et $\ell_j \leqq k_j \leqq n_j \leqq d$.

En utilisant des intégrations par parties successives à partir de l'identité $f_+(y) = \int f_+(x)\cdot\delta_{x=y}\, dx$, on montre:

LEMMA 1: $\forall i \in \{0,\dots,\ell\}$, $s_\ell(y)$ est combinaison linéaire finie de fonctions de la forme:

$$\int s_{\ell-i}(x^1) \dots s_{\ell-i}(x^k) b(x^1,\dots,x^k;y)\, dx^1 \dots dx^k$$

avec $b \in V_\ell^i$.

Soit alors Z_ℓ^i les sous-ensembles de $T^*\Omega$:

$$Z_\ell^i = \{\bar{y}^* \in T^*\Omega,\ \exists b(x^1,\dots,x^k;y) \in V_\ell^i,\ \exists(\underline{x}^1,\dots,\underline{x}^k) \text{ tel que } \forall j,\ \underline{x}_0^j \geqq 0 \text{ et } (\underline{x}^1,\dots,\underline{x}^k;\ \xi^1 = \dots = \xi^k = 0, \bar{y}^*) \in WF(b)\}.$$

En utilisant le lemme précédent et les propriétés des espaces B_k^σ on obtient alors

LEMME 2: Si $\bar{y}^* \notin Z_\ell^0 \cup \dots \cup Z_\ell^{\ell-1}$ on a $s_\ell \in H_{y^*}^\ell(\Omega)$.

Pour démontrer le théorème, il suffit alors de prouver que les a_ℓ et les éléments des V_ℓ^i vérifient l'hypothèse (I), ce qui résulte essentiellement des propositions 1 et 2.

Bibliographie

[1] M. Beals: "Self spreading and strength of singularities for solutions to semi-linear wave equations". Ann. of Math. 118 (1983), 187-214.

[2] J-M. Bony: "Calcul symbolique et propagation des singularités pour les équations aux dérivées partielles non linéaires". Ann. Sci. Ec. Norm. Sup., 4ème série, 14 (1981), 209-246.

[3] J-M. Bony: "Interaction des singularités...". Sém. Goulaouic-Meyer-Schwartz, 1981/82, n°2 et 1983/84, n°10.

[4] J-Y. Chemin: "Thèse de Doctorat. Université d'Orsay, 1986.

[5] G. Lebeau: "Problème de Cauchy semi-linéaire en 3 dimensions d'espace: un résultat de finitude". A paraître au Journal Fonct. Analysis.

[6] R. Melrose et N. Ritter: "Interaction of non linear progressing waves".

[7] J. Rauch et M. Reed: "Non-linear microlocal analysis of semi-linear hyperbolic-systems in one space dimension". Duke. Math. Journal, 49-2, 1982, 397-475.

Gilles Lebeau
Dept. de Mathématiques
Bat 425.
Université Paris Sud
91405 Orsay Cedex
France.

G. MÉTIVIER

Ondes discontinués pour les systèmes hyperboliques semi-lineaires

ABSTRACT: Discontinuous waves for first order semi-linear hyperbolic systems

This lecture is devoted to the study of discontinuous waves for first order strictly hyperbolic systems $Lu(t,x) = F(t,x,u(t,x))$ with $L = \partial_t + \Sigma A_j(t,x)\partial_{x_j}$. We consider solutions u which are smooth away from a union $\Sigma = \Sigma_1 \cup \ldots \cup \Sigma_N$ of characteristic manifolds (the fronts of the waves under consideration) and more precisely the solutions u are looked for in spaces of functions which are bounded and L^2-conormal with respect to Σ ($u \in L^\infty \cap H^{0,k}_\Sigma$, k large), so that jumps over Σ are allowed. Different patterns are considered: 1) Σ is reduced to one smooth manifold (propagation of one wave); 2) Σ is the bunch of the characteristic manifolds issued from a manifold Δ of codimension two contained in a space-like manifold (interaction of two waves Σ_1 and Σ_2 with $\Sigma_1 \cap \Sigma_2 = \emptyset$ in the past, or creation of waves by solving the Cauchy problem with data which are discontinuous over Δ); 3) when a boundary value problem is considered, let Σ be as in 2) but now with Δ contained in the boundary $\partial\Omega$ (reflection of one wave Σ_1 on an obstacle $\partial\Omega$, when $\Sigma_1 \cap \partial\Omega = \emptyset$ in the past).

The results are stated in terms of existence and continuation of solutions in the spaces $L^\infty \cap H^{0,k}_\Sigma$: if a solution is given in this space in the past $t < 0$, then it can be extended in the same space to positive times, and this extension can be continued as long as its L^∞-norm does not blow up. Similar results are given for the Cauchy problem and the mixed Cauchy problem. At last, if the data are small, the time of existence is large, so that interactions and reflections can be actually created.

1. SINGULARITES

Considérons un système $N \times N$ semi-linéaire, de la forme:

$$Lu(t,x) = F(t,x,u(t,x)) \tag{1.1}$$

où $L = \partial_t + \sum_{j=1}^{n} A_j(t,x)\partial_{x_j}$ est un système à coefficients réels C^∞, strictement hyperbolique dans la direction du temps, et où F est une fonction C^∞ de ses arguments. Dans toute la suite nous ne nous intéresserons qu'à des solutions localement bornées (1.1), pour lesquelles les deux membres de l'équation ont un sens.

De manière classique, une "onde" est une solution de (1.1) singulière sur une hypersurface $\Sigma \subset \mathbb{R}_t \times \mathbb{R}_x^n$, et on veut ici s'intéresser à des singularités fortes, incluant des discontinuités le long de Σ. Très schématiquement on peut formuler le problème de la manière suivante: trouver des espaces $E(\Omega)$ de fonctions bornées sur Ω, pouvant être singulières sur une configuration Σ constituée d'une ou plusieurs surfaces caractéristiques de L, possèdant les propriétés suivantes:

A). Propagation linéaire

Si $f \in E(\Omega)$, si u est solution de $Lu = f$ et si $u \in E(\Omega \cap \{t < 0\})$, alors $u \in E(\Omega)$. (En supposant que $\Omega \cap \{t \geq 0\}$ est contenu dans le domaine de détermination du passé $\Omega \cap \{t < 0\}$).

B). Propriété d'algèbre

Si F est une fonction C^∞ des ses arguments, et si $u \in E(\Omega)$ alors $F(u) \in E(\Omega)$.

Nous reviendrons sur les configurations Σ que nous voulons traiter; disons d'abord un mot sur le type de singularités que l'on peut considérer. D'une part, lorsque Σ est une hypersurface caractéristique C^∞, on veut inclure le cas où u est régulière (jusqu'au bord) de part et d'autre de Σ. On parle alors d'onde C^∞ (ou H^s) par morceaux, et on note $u \in p - C_\Sigma^\infty$(ou $u \in p - H_\Sigma^s$). C'est ce type de singularités qu'ont étudié J. Rauch et M. Reed [15] dans le cas de la dimension $n = 1$ d'espace, en utilisant les méthodes d'intégration le long des caractéristiques de L.

Dans le cas multidimensionnel ($n > 1$) les espaces $p - H^s$ ne sont plus adaptés (il n'y a pas propagation linéaire, cf. [9], [14]), et l'étude des singularités p - C^∞ est beaucoup plus délicate (cf. J. Rauch - M. Reed [16], [17] pour les systèmes 2 × 2, et Chen Shu Xing [7], Gu Ben [8] pour des systèmes 3 × 3). Par contre, l'idée fondamentale introduite par J.M. Bony [3], [4], est que les espaces de fonctions à singularité conormale par rapport à Σ, sont bien adaptés à la fois pour les propriétés A) et B). Rappelons que cette information "conormale" est celle, utilisée pour avoir B), qui précise le fait que le front d'onde des fonctions de E reste confiné sur le fibré conormal à Σ, propriété nécessaire pour A).

En effet, J.M. Bony [3], [4] a montré que pour $s > \frac{n+1}{2}$, les espaces $H^{s,k}_\Sigma$ (voir définition ci-dessous) vérifient effectivement les propriétés A) et B) lorsque Σ est constitué d'une seule surface caractéristique ou d'une union de surfaces caractéristiques qui se coupent deux à deux transversalement sur la même variété Δ de codimension deux. Ces résultats ont ensuite été étendus à des géométries plus compliquées (J.M. Bony [5], [6], R. Melrose-N. Ritter [10], [11]) et aussi à des problèmes de réflexion (M. Beals - G. Métivier [1], [2]).

Neanmoins la limitation $s > \frac{n+1}{2}$ est très sévère: lorsque Σ est une simple hypersurface, les fonctions $H^{s,k}_\Sigma$ sont, pour k assez grand, de régularité $C^{\frac{n}{2}-\frac{1}{2}}$. En fait, la propriété de propagation linéaire A) n'impose aucune restriction sur s et k, alors que la propriété B) est vraie dès que $s > \frac{1}{2}$ et $s + k > (n+1)/2$ dans le cas d'une surface, et dès que $s > 1$ et $s+k > (n+1)/2$ dans le cas de plusieurs surfaces se coupant sur Δ comme indiqué ci-dessus. Signalons que N. Ritter [18] a travaillé dans de tels espaces pour les équations du second ordre. Malheureusement, ces espaces n'autorisent toujours pas les discontinuités.

Le but de cet exposé est donc d'indiquer que les propriétés A) et B) sont vraies dans des espaces $L^\infty \cap H^{0,k}_\Sigma$, qui autorisent eux, des discontinuités (sauts) sur Σ. On en déduit des théorèmes de propagation et d'existence dans ces espaces.

2. LES CONFIGURATIONS

Nous allons maintenant décrire différentes configurations Σ, (désormais classiques) qui correspondent à des problèmes simples de propagation, d'interaction et de réflexion d'ondes. On se placera toujours sur un ouvert

Ω, assez petit, de la forme

$$\Omega = \{(t,x) \in \mathbb{R}^{n-1} / -T_0 < t < T_0 - \varepsilon|x|\} \tag{2.1}$$

avec ε et T_0 assez petits.

(Pb I) - Stabilité d'une onde isolée

On suppose que

$$\Sigma \text{ est une surface caractéristique } C^\infty \text{ de } L. \tag{2.2}$$

Typiquement, le résultat obtenu est que pour une solution bornée de (1.1), une singularité $H^{0,k}_\Sigma$ dans le passé $t < 0$, reste confinée sur Σ et de même nature dans $t \geqq 0$.

(Pb II) - Interaction de deux ondes

On se donne deux surfaces caractéristiques Σ_1 et Σ_2 telles que:

$$\Sigma_1 \cap \Sigma_2 \cap \Omega \cap \{t < 0\} = \emptyset \tag{2.3}$$

On suppose que Σ_1 et Σ_2 se coupent transversalement le long de Δ. Si $t^0 \geqq 0$ désigne le premier temps ou l'intersection est non vide, alors au voisinage de t^0, Δ est contenue dans une surface spatiale, et, restreignant Ω si besoin est on suppose donc

$$\Delta \text{ est contenu dans une surface spatiale.} \tag{2.4}$$

On désigne alors par $\Sigma_1,\Sigma_2,\ldots,\Sigma_N$ les N surfaces caractéristiques, 2 à 2 transverses, issues de Δ.

Typiquement, le résultat exprime que si une solution u n'a que des singularités conormales portées par Σ_1 et Σ_2 dans $t < 0$, alors les singularités de u dans $t \geqq 0$, sont conormales et portées par $\Sigma_1 \cup \Sigma_2 \cup \Sigma_3 \cup \Sigma_N$.

(Pb III) - Réflexion d'une onde sur un obstacle

On considere dans ce cas un problème aux limites

$$\begin{cases} L\ u = F(t,x,u) & \text{dans} \quad x_n > 0 \\ B\ u = g & \text{sur} \quad x_n = 0 \end{cases} \tag{2.5}$$

où B est un système C^∞ de conditions aux limites linéaires. On suppose que (L,B) satisfait une hypothèse de Lopatinski uniforme. On se donne alors une surface caractéristique Σ_1 telle que:

$$\Sigma_1 \cap \Omega \cap \{t < 0\} \cap \{x_n = 0\} = \emptyset \tag{2.6}$$

et que Σ coupe le bord $\{x_n = 0\}$ transversalement le long de Δ. Alors, en restreignant Ω si nécessaire, (2.4) est encore vrai et on définit $\Sigma = \Sigma_1 \cup \ldots \cup \Sigma_N$ comme précédemment.

Parce qu'on ne sait pas construire de solutions bornées de façon très générale, pour le système (1.1), on ne peut se contenter de poser les problèmes en terme de propagation, et aux trois problèmes énoncés ci-dessus, on peut ajouter:

(Pb IV) - <u>Problème de Cauchy</u>

On se donne

$$\Delta \text{ une variété de codimension 1 dans } \{t = 0\} \tag{2.7}$$

et on note $\Sigma = \Sigma_1 \cup \ldots \cup \Sigma_N$ la réunion des surfaces caractéristiques issues de Δ. Etant donnée h(x) conormale et singulière sur Δ, on cherche une solution u, ayant ses singularités sur Σ, du problème de Cauchy:

$$\begin{cases} L\ u = F(t,x,u) & \text{dans } t > 0 \\ u_{|t=0} = h \end{cases} \tag{2.8}$$

(Pb V) - <u>Problème mixte</u>

On cherche ici à résoudre le problème:

$$\begin{cases} L\ u = F(t,x,u) & \text{dans } t > 0, \quad x_n > 0 \\ B\ u = g & \text{dans } t > 0, \quad x_n = 0 \\ u_{|t=0} = h & \text{dans } t = 0, \quad x_n > 0 \end{cases} \tag{2.9}$$

h et g étant données conormales par rapport à Δ (par exemple régulières mais sans vérifier de compatibilités). Alors la solution u de (2.9) développe des singularités le long des surfaces Σ_j issues de Δ.

3. ENONCE DES RESULTATS

Pour simplifier l'exposé nous allons restreindre notre attention aux problèmes II et IV ci-dessus. Le problème I est en fait beaucoup plus facile et on renvoie à [13] pour des énoncés concernant les problèmes III et V.

On se donne donc au voisinage de l'origine dans R^{n+1} une variété Δ de codimension 2, contenue dans une surface spatiale et on note $\Sigma_1,\ldots,\Sigma_N$ les surfaces caractéristiques issues de Δ. On pose $\Sigma = \Sigma_1 \cup \ldots \cup \Sigma_N$ et Ω désigne un petit voisinage de l'origine, de la forme (2.1). Pour $T \in]-T_0,T_0[$, on pose $\Omega_T = \Omega \cap \{t < T\}$.

Soit M_j l'algèbre de Lie des champs de vecteurs C^∞ tangents à la fois à Σ_j et à Δ. Pour $k \in N$ on désigne par $H^{o,k}_{M_j}(U)$ l'espace des fonctions $u \in L^2(U)$ telles que pour toute famille $(M_1,\ldots,M_\ell)$ de $\ell \leqq k$ champs pris dans M_j on ait:

$$M_1 \circ M_2 \ldots \circ M_\ell u \in L^2(U) \tag{3.1}$$

Enfin on désigne par $H^{o,k}_\Sigma(U)$ l'espace somme des $H^{o,k}_{M_j}(U)$.

Dans toute la suite k est un entier fixé, $k > \frac{n+5}{2}$.

La première chose à faire est de construire des solutions discontinues. Pour cela on résout le problème de Cauchy, en supposant que:

$$\Delta \subset \{t = 0\} \tag{3.2}$$

Désignant par $\omega = \Omega \cap \{t = 0\}$, l'espace $H^{o,k}_\Delta(\omega)$ est défini de façon semblable à (3.1). On peut alors énoncer:

<u>THEOREME 3.1</u>: Si Δ vérifie (3.2) et si $h \in L^\infty \cap H^{o,k}_\Delta(\omega)$ alors il existe $T \in]0,T_0[$ et $u \in L^\infty \cap H^{o,k}_\Sigma(\Omega \cap \{0 < t < T\})$ solution du problème de Cauchy (2.8).

Pour ce qui concerne le problème de l'interaction on peut énoncer un résultat de propagation de régularité conormale, où l'on ne suppose plus (3.2):

THEOREM 3.2: Si $u \in L^\infty(\Omega)$ est solution de (1.1) et si pour un $T_1 \in]-T_0,T_0[$, $u \in H^{0,k}_\Sigma(\Omega_{T_1})$, alors, pour tout $U \subset\subset \Omega$, $u \in H^{0,k}_\Sigma(U)$.

Ce résultat est intéressant mais ne répond pas totalement à la question car il ne donne pas d'information quant à l'existence de la solution u. On veut donc aussi disposer de théorèmes de prolongement de solutions données sur Ω_{T_1}. Malheureusement, il apparait alors une difficulté technique importante: $L^\infty \cap H^{0,k}_\Sigma(\Omega_T)$ n'est pas toujours une algèbre (cf. propriété B) du §1); néanmoins, c'est le cas si Δ est transverse aux surfaces t = constantes, ou si Δ est contenue dans $\{t = 0\}$, (d'autres cas intermédiaires sont aussi envisageables, cf. [13]). Pour fixer les idées, nous ne considérons maintenant que le cas où (3.2) a lieu, en renvoyant à [13] pour les autres cas. On a alors:

THEOREME 3.3: On suppose à nouveau (3.2). Soit $u \in L^\infty \cap H^{0,k}_\Sigma(\Omega_{T_1})$ une solution de (1.1). Alors il existe $T \in]T_1,T_0[$ tel que u se prolonge en solution de (1.1) dans $L^\infty \cap H^{0,k}_\Sigma(\Omega_T)$.

L'unicité locale des solutions bornées de (1.1) étant triviale, on peut alors définir le temps maximum d'existence T_*, borne supérieure des $T > 0$ tel que u se prolonge à Ω_T. Pour "créer" l'interaction, on voudrait bien, au moins dans certains cas, que, T_1 étant donné < 0, on ait $T_* > 0$. Cela est possible grâce au:

THEOREME 3.4: On se place sous les hypothèses du théorème 3.3. On suppose que $F(t,x,0) = 0$ et on se donne $T_2 \in]T_1,T_0[$. Alors il existe $\delta > 0$ tel que si $\|u\|_{L^\infty \cap H^{0,k}_\Sigma(\Omega_{T_1})} \leq \delta$, on a $T_* \geq T_2$.

(La condition $F(t,x,o) = 0$ exprime que 0 est solution (1.1); on peut énoncer un résultat analogue pour tout solution voisine d'une solution donnée).

Pour finir, on peut énoncer un résultat qualitatif pour étudier les phénomènes qui apparaissent au temps T_*:

THEOREME 3.5: Sous les hypothèses du théorème 3.3 (ou du théorème 3.1) on a ou bien $T_* = T_0$, ou bien $T_* < T_0$ et alors:

$$\limsup_{t \to T_*} \|u(t)\|_{L^\infty} = +\infty$$

4. ETAPES DES PREUVES

Comme indiqué au §1, le point crucial consiste à verifier les propriétés A) et B) pour les espaces $E = L^\infty \cap H_\Sigma^{0,k}$. Pour montrer B), on prolonge les fonctions de $L^\infty \cap H_\Sigma^{0,k}(\Omega_T)$ en fonctions de $L^\infty \cap H_\Sigma^{0,k}(\mathbb{R}^{n+1})$, pour lesquelles la propriété d'algèbre est connue. Pour obtenir A) on considère le problème linéaire:

$$L\,u = f \tag{4.1}$$

La propagation de la régularité $H_\Sigma^{0,k}$ est connue (J.M. Bony [4]) et le point nouveau consiste à propager des estimations L^∞ pour les solutions conormales de (4.1).

PROPOSITION 4.1: Soit $f \in L^\infty \cap H_\Sigma^{0,k}(\Omega_T)$, et soit $u \in H_\Sigma^{0,k}(\Omega_T)$ solution de (4.1). Alors si $u \in L^\infty$ sur Ω_{T_1} $(-T_0 < T_1 < T < T_0)$, u est encore bornée sur Ω_T.

Notons Λ^0, l'espace des fonctions a, C^∞ sur $\mathbb{R}^{n+1}\setminus\Delta$, telles que pour toute suite $M_1,\ldots,M_\ell$ de champs tangents à Δ on ait $M_1 \circ \ldots \circ M_\ell a \in L^\infty_{loc}$. Notons Λ^1 l'espace des fonctions C^∞ sur $\mathbb{R}^{n+1}\setminus\Delta$ dont les dérivées premières sont dans Λ^0.

Introduisons alors une base $M_1,\ldots,M_{n+1}$ de l'algèbre de Lie des champs de vecteurs à coefficients Λ^1, tangents à Σ. Après diagonalisation, et changement de variables dépendantes, on peut mettre le système (4.1) sous la forme:

$$(X_j + c_j)u_j = f_j + \sum_{k \neq j} \sum_{\ell=0}^{n+1} b_{j,k}^\ell M_\ell u_k \tag{4.2}$$

où X_j est un champ tangent à Σ_j, à coefficients Λ^1, $c_j \in \Lambda^0$, les $b_{j,k}^\ell \in \Lambda^0$, $M_0 = Id$ et $M_1,\ldots,M_{n+1}$ est la base des champs tangents à Σ choisie précédemment.

LEMME 4.2: Si $f \in H_\Sigma^{o,k}(\Omega_T)$ avec $m > \frac{n+1}{2}$ alors

$$|f(t,x)| \leqq C\ \delta_0^{-1/2} \sum_{j=1}^{N} \delta_j^{-1/2} \tag{4.3}$$

où $\delta_0 = \delta_0(t,x)$ désigne la distance du point (t,x) à Δ et $\delta_j(t,x)$ la distance de (t,x) à Σ_j.

Puisque $u \in H_\Sigma^{o,k}$, on voit que le membre de droite de (4.2) appartient à $H_\Sigma^{o,k-1}$ et qu'il vérifie donc des estimations (4.3). Intégrant le long des caractéristiques de X_j, qui sont transverses aux Σ_ℓ pour $\ell \neq j$, on en déduit que

$$|u_j(t,x)| \leqq C_1\ \delta_j^{-1/2}(t,x) \tag{4.4}$$

En commutant (4.2) aux champs M_ℓ, on obtient de même:

$$|M_\ell u_j(t,x)| \leqq C_2\ \delta_j^{-1/2}(t,x)$$

d'où il résulte que le membre de droite de (4.2) est majoré par:

$$\|f_j\|_{L^\infty} + C_3 \sum_{k\neq j} \delta_k^{-1/2}(t,x) \tag{4.5}$$

Intégrant à nouveau (4.5) le long des caractéristiques de X_j, et puisque dans la sommation de (4.5), $k \neq j$, on en déduit que $u_j \in L^\infty$, ce qui prouve la proposition 4.1.

Une fois établies les propriétés A) et B) pour les espaces $L^\infty \cap H_\Sigma^{o,k}(\Omega_T)$ on obtient les théorèmes d'existence 3.1 et 3.3 à l'aide de schémas itératifs:

$$\begin{cases} L\ u_{\nu+1} = F(t,x,u_\nu) \\ u_{\nu+1}|_{t=0} = h \quad \text{ou} \quad u_{\nu+1}|_{t<T_1} = u \end{cases}$$

Les propriétés A) et B) garantissent en effet l'existence de la suite

$u_\nu \in L^\infty \cap H^{o,k}_\Sigma(\Omega_T)$. En même temps qu'on montre A) et B) on obtient des estimations, qui permettent d'en déduire que la suite u_ν reste bornée dans $L^\infty \cap H^{o,k}_\Sigma(\Omega_T)$ si T est assez petit (ou si $T - T_1$ est assez petit). On voit alors facilement que cette suite converge, et la limite est solution de (1.1). Les théorèmes 3.4 et 3.5 résultent d'une étude précisée des estimations.

Enfin le théorème 3.2 résulte du théorème 3.5 par déformation de surfaces spatiales.

Les démonstration complètes se trouvent dans [12], [13].

References

[1] M. Beals, G. Métivier, Progressing wave solution to certain non linear mixed problems Duke Math. J. 53 (1986), p. 125-137.

[2] M. Beals, G. Métivier, Reflexion of transversal progressing waves in non linear strictly hyperbolic mixed problems. Amer. J. Math., 109 (1987), p. 335-360.

[3] J.M. Bony, Calcul symbolique et propagation des singularités pour les équations aux dérivées partielles non linéaires. Ann. Sc. de l(Ecole Norm. Sup. 14 (1981), p. 209-246.

[4] J.M. Bony, Interaction des singularités pour les équations aux dérivées partielles non linéaires, Sém. Goulaouic-Meyer-Schwartz, Ecole Polytechnique, exposé n° 22 (1979-80) et exposé n° 2 (1981-82).

[5] J.M. Bony, Interaction des singularités pour les équations de Klein-Gordon non linéaires. Sém. Goulaouic-Meyer-Schwartz, Ecole Polytechnique, exp. n° 10 (1983-84).

[6] J.M. Bony, Second microlocalisation and propagation of singularities for semilinear hyperbolic equations. Preprint-Orsay.

[7] Chen Shu Xing, Piecewise smooth solutions of semilinear hyperbolic systems in higher space dimension, Preprint-Fudan's university (Shanghai).

[8] Gu Ben, Reflection of singularities at the boundary for semilinear hyperbolic system, Preprint - Fudan's university (Shanghai).

[9] A. Majda, S. Osher, Initial boundary value problems for hyperbolic equations with uniformly characteristic boundary, Comm. Pure Appl. Math., 28 (1975), p. 607-676.

[10] R. Melrose, N. Ritter, Interaction of non linear progressing waves for semilinear wave equations;Ann. Math., 121 (1985), p. 187-213.

[11] R. Melrose, N. Ritter, Interaction of non linear progressing waves for semilinear wave equations II; (Preprint).

[12] G. Métivier, The Cauchy problem for semilinear hyperbolic systems with discontinuous data; Duke Math. J. 53 (1986), p. 983-1011.

[13] G. Métivier, Propagation, interaction and reflection of discontinuous progressing waves for semilinear systems, Preprint.

[14] G. Métivier, Problèmes de Cauchy et ondes non linéaires, Journées Equations aux Dérivées Partielles, St. Jean de Monts (1986).

[15] J. Rauch, M. Reed, Jump discontinuities of semilinear strictly hyperbolic systems in two variables: creation and propagation. Comm. Math. Phys. 81 (1981), p. 203-227.

[16] J. Rauch, M. Reed, Discontinuous progressing waves for semilinear systems, Comm. Partial Diff. Equa. 10 (1985).

[17] J. Rauch, M. Reed, Bounded, stratified and striated solutions of hyperbolic systems, Preprint.

[18] N. Ritter, Progressing wave solutions to non linear hyperbolic Cauchy problems, Ph. D. Thesis, M.I.T. (1984).

Guy Métivier
Universite de Rennes I,
Mathematique et Informatique,
Campus de Beaulieu,
F 35042 Rennes Cedex
France.

A.J. MILANI

Global existence for nonlinear dissipative hyperbolic equations with a small parameter

1. We are concerned with the question of global existence of smooth solutions of the initial value problem for the quasi-linear dissipative wave equation

$$(H\varepsilon)\begin{cases} \varepsilon u_{tt} + \sigma u_t - \sum_{i,j=1}^{n} a_{ij}(\nabla u)\partial_i\partial_j u = 0 \\ u(0) = u_0, \quad u_t(0) = u_1, \end{cases}$$

under the uniformly strong ellipticity condition

$$\exists \nu > 0, \forall p, \forall q \in \mathbf{R}^n, \quad \Sigma a_{ij}(p)q^i q^j \geq \nu \; |q|^2. \qquad (1)$$

In (H_ε), $u = u(x,t) \in \mathbf{R}^1$, $x \in \mathbf{R}^n$, $t \geq 0$, $\sigma \in \mathbf{R}^+$; $\partial_j = \partial/\partial x_j$ and ∇ is the gradient with respect to the space variables only; ε is a small positive parameter, and the coefficients a_{ij} are smooth and symmetric.

2. We consider solutions of (H_ε) that are smooth in the sense of Kato, [1]: more precisely, we consider $s \in \mathbf{N}$ such that $s > 1 + n/2$, assume that $u_0 \in H^{s+1}$ $(= H^{s+1}(\mathbf{R}^n))$ and $u_1 \in H^s$, and look for solutions in the space

$$X(T) = \bigcap_{j=0}^{s+1} C^j(0,T;H^{s+1-j}),$$

where $T > 0$ is a given number. A direct application of Kato's results yields

THEOREM 1 (local existence): $\forall u_0 \in H^{s+1}$, $\forall u_1 \in H^s$, $\forall \varepsilon > 0$, $\exists \tau > 0$, $\exists!$ $u \in X(\tau)$, solution of (H_ε).

We remark that this value of τ is not the life span of the solution, but rather a minimum guaranteed time of existence of the solution, which can very well be defined in later times. In [3] we proved

THEOREM 2 (uniformly local existence): $\exists \varepsilon_0 > 0$, $\forall u_0 \in H^{s+1}$, $\forall u_1 \in H^s$,

$\exists \tau_0 > 0, \quad \forall \varepsilon \leq \varepsilon_0, \exists!\ u \in (\tau_0)$, solution of (H_ε).

This result is uniformly local, in the sense that τ_0 does not depend on ε; we look now for conditions to ensure that $\tau_0 = +\infty$ or, at least, $t_0 \geq T$. Adapting a result of Matsumura, [2], which exploits the presence of the dissipation term u_t, we can prove that $\tau_0 = +\infty$ if the initial datum u_0 is sufficiently small; more precisely, in [3] we established

<u>THEOREM 3</u> (global existence): $\exists \varepsilon_1 \leq \varepsilon_0$, $\exists \lambda = \lambda(\nu,\sigma) > 0$ such that, if

$$\|u_0\|_{s+1} + \sqrt{\varepsilon}\,\|u_1\|_s \leq \lambda\ , \text{ then } \tau_0 = +\infty \quad \forall \varepsilon \leq \varepsilon_1.$$

<u>3</u>. The question naturally arises, whether the smallness requirements on u_0 can be relaxed; in [4] we obtained a first answer for more regular solutions, relating the global solvability of (H_ε) to that of the reduced parabolic problem

$$(P) \quad \begin{cases} \sigma v_t - \Sigma a_{ij}(\nabla v)\partial_i \partial_j v = 0 \\ (0) = u_0, \end{cases}$$

formally obtained from (H_ε) by putting $\varepsilon = 0$ and losing the initial condition on u_t. Roughly speaking, we obtain that solutions of (H_ε) exist globally for small ε if and only if the solution of (P) is defined globally; more precisely, assume that $u_0 \in H^{s+2}$ and $u_1 \in H^{s+1}$, and that (P) has a unique solution v in the space

$$V = C(0,T;H^{s+2}) \cap H^1(0,T;H^{s+1}).$$

Then

<u>THEOREM 4</u> (almost global existence): $\exists \varepsilon_2 > 0$ such that $\forall \varepsilon \leq \varepsilon_2$, (H_ε) has a unique solution u in the space

$$X_+(T) = \bigcap_{j=0}^{s+2} C^j(0,T;H^{s+2-j});$$

moreover $\exists \Lambda = \Lambda(T; \|v\|_V) > 0$ such that $\forall \varepsilon \leq \varepsilon_2$,

$$\sup_{0\leq t\leq T}\{\|u(t)\|^2_{s+2}+\varepsilon\|u_t(t)\|^2_{s+1}+\int_0^t[\|\nabla u\|^2_{s+1}+\|u_t\|^2_{s+1}]\}\leq\Lambda\ . \tag{2}$$

This result is almost global, in that T can be arbitrary, but must be finite. We remark that Λ does not depend on ε, so that, because of (2), we can recover the solution of (P) as a weak limit of solutions of (H_ε) as $\varepsilon \downarrow 0$; this convergence is singular, because of the loss of one initial condition (see [3]). Indeed, the "only if" part of our result is

THEOREM 5 (singular convergence): assume that $\exists\varepsilon_3 > 0$, $\exists\Lambda > 0$ such that $\forall\varepsilon \leq \varepsilon_3$, $(H\varepsilon)$ has a solution $u = u^\varepsilon \in X_+(T)$, satisfying (2). Then there exists a unique $v \in V$, solution of (P), and $v = \text{w-}\lim_{\varepsilon\downarrow 0} u^\varepsilon$ in $X_+(T)$.

Finally, solutions of (H_ε) and of (P) are related by

THEOREM 6 (difference estimates):

$$u = v + O(\varepsilon^{1/4}) \qquad \text{in } C(0,T;H^{s+1})$$

$$u = v + O(\varepsilon^{1/2}) \qquad \text{in } C(0,T;H^s)\cap H^1(0,T;H^{s-1}),$$

$$u_t = v_t + (u_1 - v_t(0))e^{-t/\varepsilon} + O(\varepsilon^{1/8}) \quad \text{in } C(0,T;H^{s-1}).$$

The last of these estimates describes the boundary layer in time, due to the loss of the initial condition on u_t; in particular, convergence of u_t to v_t is uniform in $C(\tau,T;H^{s-1})$, $\forall\tau \in]0,T]$. These results are the quasi-linear analogous of the linear estimates established by Zlamal, [5].

4. We briefly sketch the idea of the proof of Theorem 4, referring to [4] for all other details.

Step 1: we directly look for solutions of (H_ε) of the form $u = v + y$, so that the new unknown y solves

$$(D_\varepsilon)\quad \begin{cases} \varepsilon y_{tt} + \sigma y_t - \Sigma a_{ij}(\nabla v + \nabla y)\partial_i\partial_j y = f(v,y) \equiv \\ \equiv \Sigma[a_{ij}(\nabla v + \nabla y) - a_{ij}(\nabla\ )]\partial_i\partial_j v - \varepsilon v_{tt}, \\ y(0) = 0,\ y_t(0) = y_1 \equiv u_1 - v_t(0) = u_j - \Sigma a_{ij}(\nabla u_o)\partial_i\partial_j u_o. \end{cases}$$

In (D_ε), the inhomogeneous term f is semilinear, and can be estimated in terms of ∇y and of ε; the initial datum $y_1 \in H^s$ and y_0 is small, so that by Theorem 3 we can establish the global solvability for (D_ε) in X(T), with y satisfying (in particular) the uniform estimate

$$\sup_{0\leq t\leq T} \{\|\nabla y(t)\|_s^2 + \int_0^t \|y_t\|_s^T\} \leq K, \tag{3}$$

where $K = K(T; \|v\|_V) > 0$ is independent of ε. Consequently, (H_ε) is globally solvable in X(T) as well, and its solutions u satisfy an estimate similar to (3).

Step 2: calling E(u,t) the "energy" norm appearing at the left side of (2), we establish the following a priori estimate on the solutions of (H_ε) (which are uniformly locally defined, because of Theorem 2):

$$E(u,t) \leq E(u,0)\exp \int_0^t \{h(\|\nabla u(\tau)\|_s) + \|u_t(\tau)\|_s\}d\tau, \tag{4}$$

where $h:\mathbf{R}^+ \to \mathbf{R}^+$ is a continuous increasing function.

Step 3: we insert estimate (3), written for u, into (4) and obtain that

$$E(u,t) \leq E(u,0)K_1, \quad K_1 = K_1(T; \|v\|_V):$$

this a priori bound on the solutions of (H_ε) allows us to extend them to the whole of [0,T], thus establishing global solvability of (H_ε) in $X_+(T)$.

ACKNOWLEDGEMENTS

We would like to thank Professors Cattabriga, Colombini, Murthy and Spagnolo for their kind invitation to the Workshop on "Hyperbolic Equations", and for the friendly and exciting atmosphere they were able, as usual, to create in Pisa.

References

[1] T. Kato, Quasi-linear Equations of Evolutions, with Application to Partial Differential Equations. Lect. Notes Math. 448, 25-70; Springer-Verlag, Berlin, 1975.

[2] A. Matsumura, Global Existence and Asymptotics of the Solutions of Second Order Quasi-linear Hyperbolic Equations with First Order Dissipation. Publ. R.I.M.S. Kyoto Univ., 13 (1977), 349-379.

[3] A. Milani, Long Time Existence and Singular Perturbations Results for Quasi-linear Hyperbolic Equations with Small Parameter and Dissipation Term. Non Linear Analysis, 10/11 (1986), 1237/1248.

[4] A. Milani, Long Time Existence and Singular Perturbation Results for Quasi-linear Hyperbolic Equations with Small Parameter and Dissipation Term, II. To appear in Non Linear Analysis.

[5] M. Zlamal, The Parabolic Equations as a Limiting Case of Hyperbolic and Elliptic Equations. Proc. Equadiff I, Prague 1962 (1963), 243-247.

Albert J. Milani
Department of Mathematics
University of Wisconsin
Milwaukee
WI 53201
U.S.A.

S. MIZOHATA

On propagation of singularities

1. INTRODUCTION

We are concerned with the following problem. Let

$$au = (\partial_t + i\lambda(t,x;D_x) + c(t,x;D_x))u(t,x) = f(t,x) \qquad (1.1)$$

$$(t,x) \in [-T,T] \times R^{\ell},$$

where $\lambda(t,x;\xi)$ is real-valued C^{∞}-function in (t,x) and homogeneous of degree 1 in ξ, and $c(t,x;\xi)$ is C^{∞}-function in (t,x), and as symbol of pseudo-differential operator acting on R^{ℓ}, belongs to $S^{\rho}_{1,0}$, where $0 \leq \rho < 1$. Let Γ be the null bicharacteristic strip passing through $(t,x;\tau,\xi) = (0,x_0;\tau^0,\xi^0)$, where for simplicity we assume $|\xi^0| = 1$. Let $\Gamma(t)$ be the point of Γ corresponding to t. Namely denote

$$\Gamma(t) = (t,x(t);\tau(t),\xi(t)),$$

where

$$\dot{x}(t) = \lambda_{\xi}(t,x(t);\xi(t)),$$

$$\dot{\tau}(t) = -\lambda_t(t,x(t);\xi(t)), \qquad \text{for} |t| \leq T. \qquad (1.2)$$

$$\dot{\xi}(t) = -\lambda_x(t,x(t);\xi(t)).$$

Observe that the second relation is satisfied if we put $\tau(t) = -\lambda(t,x(t);\xi(t))$.

The propagation of singularities in C^{∞} class means that

THEOREM 1: Let $c(t,x;\xi) \in S^0_{1,0}$. Assume that

$$(0,x_0;\tau^0,\xi^0) \notin WF(u), \text{ and } \Gamma(t) \notin WF(f) \text{ for } |t| \leq T. \qquad (1.3)$$

Then it holds $\Gamma(t) \notin WF(u)$ for $|t| < T$.

In the case $c(t,x;\xi) \in S^{\rho}_{1,0}$ with $\rho > 0$, we must assume that $\lambda(t,x;\xi)$ and $c(t,x;\xi)$ belong to some Gevrey class $\gamma^{(s)}$ $(s > 1)$ with $1/s > \rho$ with respect to (t,x). Precise conditions are given in Section 4. Then Theorem 1 takes the following form, where $WF_s(u)$ denotes the wave front set in the sense of Gevrey class s.

THEOREM 2: Let $\rho < 1/s$. Then

$$\Gamma(0) \notin WF_s(u), \text{ and } \Gamma(t) \notin WF_s(f) \text{ for } |t| \leq T \tag{1.4}$$

implies $\Gamma(t) \notin WF_s(u)$ for $|t| < T$.

Finally, if $\lambda(t,x;\xi)$ and $c(t,x;\xi)$ are analytic in $(t,x;\xi)$ (precise definition is given in Section 5), we obtain

THEOREM 3: Let $\rho < 1$. Then

$$\Gamma(0) \notin WF_A(u) \text{ and } \Gamma(t) \notin WF_A(f) \quad \text{for } |t| < T \tag{1.5}$$

implies $\Gamma(t) \notin WF_A(u)$ for $|t| < T$.

These three theorems are already known.(*) However, the proof of Theorems 2 and 3 seems to be fairly complicated. Namely, a standard method is to transform (1.1) by Fourier integral operator into $\partial_t + c'(t,x;D_x)$, however this method needs fairly delicate analysis especially in the case of Gevrey class (see Taniguchi [6]). The purpose of this paper is to show that one can treat these three cases by the same direct method (microlocal energy method). Our method is to use a fairly simple, microlocalizer around Γ of the form $k_n(t,x;D)$ where

$$k_n(t,x;\tau,\xi) = \beta_0(t-x(t))\chi(\tfrac{\tau}{n} - \tau(t), \tfrac{\xi}{n} - \xi(t)) = \beta(t,x)\alpha_n(t;\tau,\xi) \tag{1.6}$$

for $|t| \leq T$,

(*) We cite only representative works in the References.

where $\beta_0(x)$, $\chi(\tau,\xi)$ are cut off functions of small size around the origin.

From (1.2) we see easily that

$$(\partial_t + \lambda_\xi(t)\partial_x - n\lambda_t(t)\partial_\tau - n\lambda_x(t)\partial_\xi)k_n(t,x;\tau,\xi) = 0, \tag{1.7}$$

where $\lambda_\xi(t) = \lambda_\xi(t,x;\xi)|_{\Gamma(t)}$, and the same convention is made for $\lambda_t(t)$, $\lambda_x(t)$. This means that $k_n(t,x;\tau,\xi)$ is constant along the parallels to Γ.

Let us explain a little concretely our method in C^∞-case. We argue for $t \geqq 0$. By introducing a suitable cut off function $\chi_0(t)$, and introducing large parameter M, we consider the following microlocal energy (L^2 norm) around $\Gamma(t)$ or rather around $\Gamma_n(t) = (t,x(t);\ n\tau(t),\ n\xi(t))$

$$\|n^{-Mt}\ \chi_0(t)k_nu\|.$$

This introduction of n^{-Mt} is convenient or even necessary to our argument because we have no information on u in a neighbourhood of $t = T$. Similarly, in Gevrey case we consider

$$\|\exp(-r_0n^{1/2}t)\chi_0(t)k_nu\| .$$

Our method is a microlocal version of the method used in the problem of hypoellipticity. More strictly, from view point of methodology, our method could be compared with the treatment of hypoellipticity with *auxiliary conditions*, because we assume that $\Gamma(0) \notin WF(u)$. One of such examples is the following:

<u>AGRANOVIČ'S RESULT</u>. If $u \in \mathcal{D}'(\Omega)$ (Ω being a bounded domain) satisfies $P(D)u = f \in C^\infty(\Omega)$, and that $u(x) \in C^\infty$ in a neighbourhood of $\partial\Omega$, then $u(x) \in C^\infty(\Omega)$.

Finally we explain how to apply this method to general situation. We follow the standard way. Let $P(x,D)$ be a differential operator of order m. We are concerned with u satisfying

$$Pu = f. \tag{1.8}$$

We assume that, by suitable change of notations and coordinates, P can be expressed as $P(t,x;D_t,D_x)$ and whose principal symbol P_m essentially has the form

$$P_m(t,x;\tau,\xi) = (\tau + \lambda(t,x;\xi))^k\, Q_{m-k}(t,x;\tau,\xi), \tag{1.9}$$

where

$$Q_{m-k}\big|_{(0,x_0;\tau^0;\xi^0)} \neq 0. \tag{1.10}$$

Then we use the perfect factorization of P

$$P = P_k \circ Q + R, \tag{1.11}$$

where

$$P_k(t,x;D_t,D_x) = (D_t + \lambda(t,x;D_x))^k + \sum_{j=1}^{k} a_j(t,x;D_x)(D_t+\lambda(t,x;D_x))^{k-j}, \tag{1.12}$$

and order $a_j < j$ $(1 \leqq j \leqq k)$, and R is a regularing operator.

In view of (1.11), we are merely concerned with the problem concerning

$$P_k u = f. \tag{1.13}$$

Now this equation is reduced to an equivalent system. Let $(D_t + \lambda)^j u = u_j$ $(0 \leqq j \leqq k-1)$. Denote

$$\max_j \text{ order } a_j/j = \rho. \tag{1.14}$$

Then putting

$$\tilde{u}_j = \langle\Lambda\rangle^{(k-1-j)\rho}\, u_j \ (0 \leqq j \leqq k-1),\ (\widehat{\langle\Lambda\rangle^s v} = \langle\xi\rangle^s \hat{v} = (1 + |\xi|^2)^{s/2}\hat{v}),$$

(1.13) can be expressed in the form

$$(D_t + \lambda(t,x;D_x))I\,\tilde{U} + A(t,x;D_x)\tilde{U} = F, \tag{1.15}$$

where $A = (a_{ij})$, order $a_{ij}(t,x;\xi) \leq \rho$; $F = {}^t(0,\dots,0,f)$, $\tilde{U} = {}^t(\tilde{u}_0,\dots,\tilde{u}_{k-1})$.
Observe that $\Gamma(0) \notin WF(u)$ is equivalent to

$$\Gamma(0) \notin \bigcup_{j=0}^{k-1} WF(\tilde{u}_j). \tag{1.16}$$

In view of the form (1.15), and taking account of (1.16), and in view of the argument that we shall explain hereafter, we see that Theorems 1, 2 and 3 yield the corresponding results for (1.8).

We give here only the proof of Theorems 1 and 2. The main result is Theorem 2.

2. PRELIMINARIES

Let

$$a(t,x;\, D_t,D_x) = \partial_t + i\lambda(t,x;D_x) + c(t,x;D_x), \tag{2.1}$$

and put its principal part by $\mathring{a}$

$$\mathring{a}(t,x;D_t,D_x) = \mathring{a}(t,x;D) = \partial_t + i\lambda(t,x,D_x). \tag{2.2}$$

Let Γ be the null bicharacteristic of operator a, issued from $(0,x_0;\tau^0,\xi^0)$:

$$\begin{aligned} \dot{x}_j(t) &= \lambda_{\xi_j}(t) \quad (1 \leq j \leq \ell). \\ \dot{\tau}(t) &= -\lambda_t(t), \\ \dot{\xi}_j(t) &= -\lambda_{x_j}(t) \quad (1 \leq j \leq \ell), \end{aligned} \tag{2.3}$$

where $\lambda_{\xi_j}(t) = \lambda_{\xi_j}(t,x,\xi)|_\Gamma = \lambda_{\xi_j}(t,x(t);\xi(t))$, and the same convention is made on $\lambda_t(t)$, $\lambda_{x_j}(t)$.

Observe that

$$\tau(t) + \lambda(t,x,\xi(t)) = 0. \tag{2.4}$$

First, in order to consider a micro-localizer around Γ, we consider two cut off functions.

$\chi_N(\Xi) = \chi_N(\tau,\xi)$ is a localizer of size r_0 around the origin satisfying the following conditions.

$$\chi_N(\Xi) \in C_0^\infty, \quad 0 \leqq \chi_N(\Xi) \leqq 1, \; = 1 \text{ for } |\Xi| \leqq r_0/2, \text{ and } = 0$$

for $|\Xi| \geqq r_0$.

$$|\chi_N^{(\mu+\nu)}(\Xi)| \leqq (Ncr_0^{-1})^{|\mu|} c_\nu r_0^{-|\nu|}, \text{ for } |\mu| \leqq N. \tag{2.5}$$

$$\beta_0(x) \in C_0^\infty, \; 0 \leqq \beta_0(x) \leqq 1, \; = 1 \text{ for } |x| \leqq r_0/2, \text{ and } = 0 \text{ for } |x| \geqq r_0.$$

$$|\beta_0{}^{(\mu+\nu)}(x)| \leqq (Ncr_0^{-1})^{|\mu|} c_\nu r_0^{-|\nu|} \text{ for } |\mu| \leqq N. \tag{2.6}$$

Next we define the following localizer around Γ, or rather $\Gamma_n = (t,x(t); n\tau(t), n\xi(t))$.

$$\alpha_n(t;\tau,\xi) = \chi_N((\tau-n\tau(t))/n, \quad (\xi-n\xi(t))/n)$$

$$= \chi_N \left(\frac{\tau}{n} - \tau(t), \frac{\xi}{n} - \xi(t)\right). \tag{2.7}$$

$$\beta(t,x) = \beta_0(x-x(t)).$$

We use the micro-localizer of the form

$$k_n(t,x;D) = \beta(t,x)\alpha_n(t;D), \tag{2.8}$$

and call it a micro-localizer of size r_0 around Γ or rather Γ_n. In fact, notice that (2.8) can be written

$$\beta_0(x-x(t))\chi_N(D_t/n - \tau(t), D_x/n - \xi(t)). \tag{2.9}$$

We consider also

$$k_n{}^{(p)}{}_{(q)}(t,x;\tau,\xi) = \beta_{(q)}(t,x)\alpha_n^{(p)}(t;\tau,\xi). \tag{2.10}$$

Observe that

$$k_{n(q)}^{(p)} \circ \mathring{a}(v) = k_{n(q)}^{(p)} \circ (\partial_t + i\lambda)$$

$$(\partial_t + i\lambda) \circ (k_{n(q)}^{(p)} v) - (\partial_t k_{n(q)}^{(p)} + \sum_{j=1}^{\ell} i\lambda^{(e_j)} \circ k_{n(q+e_j)}^{(p)}$$

$$- \sum_{j=0}^{\ell} i\lambda_{(e_j)} \circ k_{n(q)}^{(p+e_j)})(v) \tag{2.11}$$

$$+ \sum_{|\mu+\nu|\geq 2} \mu!^{-1}\, \nu!^{-1} (-1)^{|\nu|}\, \lambda_{(\mu)}^{(\nu)} \circ k_{n(q+\nu)}^{(p+\mu)}\, (v).$$

In the sequel, our concern is concentrated to the second term of the right-hand side. This term is small in some sense. In fact, we have

LEMMA 2.1: The following holds

$$(\partial_t + \lambda_\xi(t)\partial_x - n\lambda_t(t)\partial_\tau - n\lambda_x(t)\partial_\xi)\, k_n(t,x;\tau,\xi) = 0. \tag{2.12}$$

PROOF:

$$\partial_t k_n(t,x;\tau,\xi) = \partial_t \beta(t,x)\alpha_n(t;\tau,\xi) + \beta(t,x)\partial_t \alpha_n(t;\tau,\xi).$$

In view of (2.3) and (2.4),

$$\partial_t \beta(t,x) = - \partial_x \beta_0(x-x(t))\dot{x}(t) = - \partial_x \beta\cdot\lambda_\xi(t),$$

$$\partial_t \alpha_n(t;\tau,\xi) = \partial_t \{\chi_N(\tau/n - \tau(t),\ \xi/n - \xi(t))\}$$

$$= -\partial_{\tau'} \chi_N(t;\tau',\xi')\dot{\tau}(t) - \partial_{\xi'}\chi_N(t,\tau',\xi')\dot{\xi}(t)$$

$$= \partial_{\tau'}\chi_N(t;\tau',\xi')\lambda_t(t) + \partial_{\xi'}\chi_N(t;\tau',\xi')\lambda_x(t),$$

where $\partial_{\tau'}\chi_N(t;\tau',\xi')$ means $\partial_{\tau'} \chi_N(t;\tau',\xi')|_{\tau' = \tau/n-\tau(t),\ \xi' = \xi/n - \xi(t)}$.

Hence this is equal to $n\partial_\tau\, \alpha_n(t;\tau,\xi)$. In the same way $\partial_{\xi'}\, \chi_N(t,\tau'\xi') = n\, \partial_\tau\, \alpha_n(t;\tau,\xi)$. Summing up:

$$\partial_t \alpha_n(t;\tau,\xi) = n\,\partial_\tau \alpha_n(t;\tau,\xi)\lambda_t(t) + n\partial_\xi\, \alpha_n(t;\tau,\xi)\lambda_x(t).$$

Thus we see that

$$\partial_t k_n = -\,\partial_x\beta\, \lambda_\xi(t)\alpha_n + \beta(t,x)\,(n\lambda_t(t)\partial_\tau\, \alpha_n + n\lambda_x(t)\partial_\xi\alpha_n)$$

$$= -\,\lambda_\xi(t)\partial_x k_n + n\lambda_t(t)\partial_\tau\, k_n + n\lambda_x(t)\partial_\xi k_n. \qquad \square$$

NOTE 1. (2.12) is fairly simple relation. In fact, for any $\mu = (\mu_0,\mu_1,\ldots,\mu_2)$, and $\nu = (0,\nu_1,\ldots,\nu_\ell)$, we obtain

$$(\partial_t + \lambda_\xi(t)\partial_x - n\lambda_t(t)\partial_\tau - n\lambda_x(t)\partial_\xi)\, k^{(\mu)}_{n(\nu)}(t,x;\tau,\xi) = 0. \tag{2.13}$$

This relation says that

$$\partial_t k_n{}^{(p)}{}_{(q)}(t,x;D) + \sum_{j=1}^{\ell} i\lambda^{(e_j)}(t)\, k_n{}^{(p)}{}_{(q+e_j)} - n\sum_{j=0}^{\ell} i\lambda_{(e_j)}(t)(k_{n(q)}^{(p+e_j)}) = 0. \qquad \square \tag{2.14}$$

In order to estimate the remainder term of the asymptotic expansion in (2.11), it would be necessary to modify symbols $\lambda(t,x;\xi)$ and $c(t,x;\xi)$ from the beginning.

Let $\chi_n(\xi)$ be a cut off function, $\chi_n(\xi) \in C_0^\infty$, $0 \leqq \chi_n(\xi) \leqq 1$, $\chi_n(\xi) = 1$ for $n/R \leqq |\xi| \leqq Rn$, and $= 0$ for $|\xi| \geqq 2Rn$ or $|\xi| \leqq n/2R$, satisfying

$$|\chi_n{}^{(\mu+\nu)}(\xi)| \leqq (\tfrac{N}{n}\, cR^{-1})^{|\mu|}\, c'_\nu\, n^{-|\nu|} \quad \text{for } |\mu| \leqq N. \tag{2.15}$$

Observe that, if we choose R large, then

$$\mathrm{supp}_\xi\, [\alpha_n(t;\tau,\xi)] \Subset \chi_n(\xi) \text{ for all } t \in [0,T]. \tag{2.16}$$

Let

$$\begin{aligned} &\lambda_{n,loc}(t,x;\xi) = \lambda(t,x;\xi)\chi_n(\xi),\quad c_{n,loc}(t,x;\xi) = c(t,x;\xi)\chi_n(\xi),\\ &b_{n,loc}(t,x;\xi) = i\lambda_{n,loc}(t,x;\xi) + c_{n,loc}(t,x;\xi). \end{aligned} \tag{2.17}$$

Hence

$$b(t,x;D)\alpha_n(t;D) = b_{n,loc}(t,x;D)\alpha_n(t;D). \tag{2.18}$$

This relation is used later for the estimate

$$\chi_0(t)\beta_{(q)}(t,x)\alpha_n^{(p)}(t;D)(b-b_{n,loc})v.$$

We also use the following cut off function:

$$\tilde{\chi}_n(\tau,\xi) \in C_0^\infty,\ 0 \leqq \tilde{\chi}_n(\tau,\xi) \leqq 1,\ = 1 \text{ for } (n/R)^2 \leqq \tau^2 + |\xi|^2 \leqq (Rn)^2,$$

and $= 0$ for $\tau^2 + |\xi|^2 \leqq (n/2R)^2$ or $\geqq (2Rn)^2$. Hence

$$k_n{}^{(p)}{}_{(q)}(t,x;D)\ (1-\tilde{\chi}_n(D)) = 0 \text{ for all } t \in [0,T]. \tag{2.19}$$

NOTE 2. Let us remark that

$$i^{-1}\ H_{\mathring{a}}(k_n) = i^{-1}\{\mathring{a}(t,x;\tau,\xi),\ k_n(t,x;\tau,\xi)\}$$

$$= (\partial_t + \lambda_\xi(t,x;\xi)\partial_x - \lambda_t(t,x;\xi)\partial_\tau - \lambda_x(t,x;\xi)\partial_\xi)k_n(t,x;\tau,\xi),$$

where $\{\ ,\ \}$ stands for Poisson bracket. From the definition, it is clear that $k_n(t,x;\tau,\xi) = 1$ along Γ. (2.12) says that k_n are constant along the parallels of Γ. Concerning (2.11), let us notice that if we denote

$$\{a(x,D),\ b(x,D)\}' = i^{-1}(a_\xi \circ b_x - a_x \circ b_\xi)$$

$$= \sum_j (a^{(e_j)} \circ b_{(e_j)} - a_{(e_j)} \circ b^{(e_j)}),$$

the second term of the right-hand side can be expressed by

$$- \{\tau + \lambda(t,x;\xi),\ k_n(t,x;\tau,\xi)\}'. \quad \square$$

In the next section, we are led to estimate (for notations, see (3.10)),

$$\|(\lambda_{n,loc}^{(e_j)} - \lambda^{(e_j)}(t))\chi_0(t)\beta_{(q)}\,\alpha_n^{(p)}v_n\|_M \tag{2.20}$$

$$+\|(\lambda_{n,loc\,(e_j)}-n\,\lambda_{(e_j)}(t))\chi_0(t)\beta_{(q)}\alpha_n^{(p)}\,v_n\|_M.$$

In order to estimate this, we use the sharp Gårding inequality. However, the localization $\lambda_{n,loc}$ is not enough for it. For this purpose, we use the localization of size $2r_0$ around Γ.

Let $p(t,x;\xi)$ be one of $\lambda_{n,loc}^{(e_j)}$, $\lambda_{n,loc(e_j)}$. We introduce C^∞-maps:

$(t,x) \to \tilde{x}(t,x)$ for $(t,x) \in [0,T] \times R^\ell_x$, $(t,\xi) \to \tilde{\xi}_n(t,\xi)$ for $(t,\xi) \in [0,T] \times R^\ell$, satisfying

i) $\tilde{x}(t,x) = x$ for $|x-x(t)| \leqq r_0$, and $= x(t)$ for $|x-x(t)| \geqq 2r_0$,

$$|\tilde{x}(t,x) - x(t)| \leqq 2r_0, \tag{2.21}$$

ii) $\tilde{\xi}_n(t,\xi) = \xi$ for $|\xi-n\xi(t)| \leqq nr_0$, and $= n\xi(t)$ for

$|\xi-n\xi(t)| \geqq 2nr_0$, $|\tilde{\xi}_n(t,\xi)-n\xi(t)| \leqq 2nr_0$, $|\partial_\xi^\mu\, \tilde{\xi}_n(t,\xi)| \leqq c''_\mu\, n^{-|\mu|}$.

Then

$$\tilde{p}_{n,loc}(t,x;\xi) = p(t,\, \tilde{x}(t,x);\, \tilde{\xi}_n(t,\xi)). \tag{2.22}$$

Observe that

$$|\tilde{p}_{n,loc}(t,x;\xi) - p(t,x(t);n\xi(t))| \tag{2.23}$$

$$\leqq (\sup|\partial p(t,x;\xi)/\partial x|)|\tilde{x}(t,x)-x(t)|+(\sup|\partial p(t,x;\xi)/\partial\xi|)|\tilde{\xi}_n(t,\xi)-n\xi(t)|,$$

where sup is taken over $(t,x;\xi) \in [0,T] \times \{x;|x-x(t)| \leqq 2r_0\} \times \{\xi;|\xi-n\xi(t)|\leqq 2nr_0\}$.

Now we denote the micro-localization by $\tilde{\lambda}_{n,loc}^{(e_j)}$, etc, (2.20) is estimated by

$$\|(\tilde{\lambda}_{n,loc}^{(e_j)} -\lambda^{(e_j)}(t))\chi_0(t)k_{n(q)}^{(p)}v_n\|_M + \|(\tilde{\lambda}_{n,loc(e_j)}-n\lambda_{(e_j)}(t))\chi_0(t)k_{n(q)}^{(p)}v_n\|$$

$$+\|(\lambda_{n,loc}^{(e_j)} -\lambda_{n,loc}^{(e_j)})\chi_0(t)\beta_{(q)}\alpha_n^{(p)}v_n\|+ \tag{2.24}$$

$$+\|(\tilde{\lambda}_{n,loc(e_j)}-\lambda_{n,loc(e_j)})\chi_0(t)\beta_{(q)}\alpha_n^{(p)}v_n\|.$$

Let us notice that to the first two terms, we can apply sharp Gårding inequality. Hence these are estimated in the form

$$A' r_0 \| k_n{}^{(p)}{}_{(q)} v_n \|_M + nA' r_0 \| k_n{}^{(p)}{}_{(q)} v_n \|_M, \tag{2.25}$$

where A' depends only on $\lambda(t,x;\xi)$.

Concerning the last two terms, let us notice that

$$\begin{aligned} &(\lambda_{n,loc}^{(e_j)} - \tilde{\lambda}_{n,loc}^{(e_j)})(t,x;\xi) = 0, \\ &(\lambda_{n,loc(e_j)} - \tilde{\lambda}_{n,loc(e_j)})(t,x;\xi) = 0, \end{aligned} \tag{2.26}$$

for $(x,\xi) \in \mathrm{supp}_x[\beta(t,x)] \times \mathrm{supp}_\xi[\alpha_n(t;\tau,\xi)]$.

3. PROOF OF THEOREM 1

Let $\chi_0(t) \in C_0^\infty$, $\chi_0(t) = 1$ for $t \in [\delta, T-2\delta/3]$, and $= 0$ for $t \leqq 0$ or $t \geqq T-\delta/3$, where we assume δ to be small.

Then operating $\chi_0(t)$ to

$$au = (\partial_t + b(t,x;D_x))u = f, \tag{3.1}$$

we have

$$a(v) = \chi_0'(t)u + \chi_0(t)f, \tag{3.2}$$

where

$$v = \chi_0(t)u, \tag{3.3}$$

and we assume that

$$v \in H^{-k}(R^{\ell+1}), \ (^\exists\, k \geqq 0). \tag{3.4}$$

In (3.2) we replace a by $a_{n,loc} = \partial_t + b_{n,loc}$ (see (2.17)) and v by $v_n = \tilde{\chi}_n(D)v$ (see (2.19)), and operate $\tilde{\chi}_0(t)k_n(t,x;D)$ ($= \chi_0(t)\beta(t,x)\alpha_n(t;D)$) from left. Then

$$\chi_0(t)k_n\, a_{n,loc}(v_n) = \chi_0(t)k_n(\chi_0'(t)u + \chi_0(t)f + (a_{n,loc}-a)v_n - a(1-\tilde{\chi}_n(D))v). \tag{3.5}$$

The left-hand side is written in the form

$$a_{n,loc}(\chi_0(t)k_n\, v_n) + [\chi_0(t)k_n,\, a_{n,loc}]v_n,$$

and the second term is written in the form

$$[\chi_0(t)k_n,\, \partial_t]v_n + [\chi_0(t)k_n,\, b_{n,loc}]v_n$$

$$= -\chi_0'(t)k_n(v_n) - \chi_0(t)\,\partial_t k_n(v_n) + [\chi_0(t)k_n,\, b_{n,loc}]v_n.$$

Hence

$$a_{n,loc}(\chi_0(t)k_n v_n) = \chi_0(t)\partial_t k_n\, v_n + [b_{n,loc}, \chi_0 k_n]v_n + g, \tag{3.6}$$

where

$$g = (\chi_0'(t)k_n v_n + \chi_0(t)k_n\chi_0'(t)u) + \chi_0(t)k_n\, \chi_0 f \tag{3.7}$$

$$+ \chi_0(t)k_n(b_{n,loc}-b)v_n - \chi_0(t)k_n\, b(1-\tilde{\chi}_n(D))u.$$

Now taking account of (2.11), asymptotic expansion up to k_0 terms gives

$$\chi_0(t)\partial_t k_n\, v_n + [b_{n,loc}, \chi_0 k_n]v_n \tag{3.8}$$

$$= (\chi_0(t)\partial_t k_n v_n + \sum_{j=1}^{\ell} i\lambda_{n,loc}^{(e_j)} \circ \chi_0(t)k_{n(e_j)} - \sum_{j=0}^{\ell} i\lambda_{n,loc(e_j)}\chi_0(t)k_n^{(e_j)} v_n)$$

$$+ i \sum_{2\leq|\mu+\nu|\leq k_0} \mu!^{-1}\nu!^{-1}(-1)^{|\nu|}\lambda_{n,loc}{}^{(\nu)}{}_{(\mu)} \circ \chi_0(t)k_n^{(\mu)}{}_{(\nu)}v_n$$

$$+ \sum_{1\leq|\mu+\nu|\leq k_0} \mu!^{-1}\nu!^{-1}(-1)^{|\nu|}c_{n,loc}{}^{(\nu)}{}_{(\mu)} \circ \chi_0(t)k_{n(\nu)}^{(\mu)}v_n + r_{n,k_0}(v_n).$$

Taking account of (2.12), the first term of the right-hand side becomes

$$h_n(v_n) = \sum_{j=1}^{\ell} i\,(\lambda_{n,loc}^{(e_j)} - \lambda^{(e_j)}(t))(\chi_0(t)k_{n(e_j)}v_n)$$
$$- \sum_{j=0}^{\ell} i\,(\lambda_{n,loc(e_j)} - n\lambda_{(e_j)}(t))(\chi_0(t)k_n^{(e_j)}v_n). \tag{3.9}$$

The estimate of this term is crucial to our argument. On the other hand, in order to efface the effect of g, more precisely of the first term of the right-hand side of (3.7), we introduce the following norm.

$$\|v\|_M = \|n^{-Mt}\, v\|, \tag{3.10}$$

where M is a large parameter. Let us notice that the support of v is contained in $t \in [0,T]$.

<u>LEMMA 3.1</u> (Energy inequality): Let

$$(\partial_t + b_{n,loc})(\chi_0(t)v) = f \quad (\text{or } a_{n,loc}\,(\chi_0(t)v) = f).$$

Then

$$\|a_{n,loc}(\chi_0(t)v\|_M \geqq (M \log n - c)\, \|\chi_0(t)v\|_M, \tag{3.11}$$

where c is a constant independent of n.

<u>PROOF</u>: This is well-known in view of $n^{-Mt} = \exp(-Mt \log n)$.

Put $n^{-Mt}(\chi_0(t)v) = w$. w has its support in [0,T]. First

$n^{-Mt}a_{n,loc}(\chi_0(t)v) = (a_{n,loc} + M\log n)(w)$. Next since

$$\mathrm{Re}(a_{n,loc}w,\, w) \geqq -c\,\|w\|^2,$$

$$\|n^{-Mt}f\|\,\|w\| \geqq \mathrm{Re}(a_{n,loc}w,\, w) + M\log n\,\|w\|^2 \geqq (M\log n - c)\,\|w\|^2. \quad \square$$

LEMMA 3.2:

$$\|h_n(v_n)\|_M \leqq Ar_0 \sum_{j=1}^{\ell} \|\chi_0(t)k_{n(e_j)}v_n\|_M + nAr_0 \sum_{j=0}^{\ell} \|\chi_0(t)k_n^{(e_j)} v_n\|_M + (\text{negligible})$$

where A depends only on $\lambda(t,x;\xi)$.

PROOF: Here negligible means that it is estimated by any negative power of n. The proof was explained in previous section (see (2.25)). A detailed account will be given later. □

From the above two lemmas, we obtain

$$\begin{aligned}(\log n-c)\,\|\chi_0(t)k_n v_n\|_M &\leqq Ar_0 \sum_j \|\chi_0(t)k_{n(e_j)}v_n\|_M + \\ &+ nAr_0 \sum_j \|\chi_0(t)k_n^{(e_j)} v_n\|_M \\ + c(k_0)\Big(\sum_{2\leqq|\mu+\nu|\leqq k_0} n^{1-|\nu|} &\|\chi_0(t)k_{n(\nu)}^{(\mu)} v_n\|_M + \\ + \sum_{1\leqq|\mu+\nu|\leqq k_0} n^{-|\nu|} \|\chi_0(t)k_{n(\nu)}^{(\mu)} v_n\|_M\Big) &+ \|r_{n,k_0}(v_n)\|_M + \|g\|_M + (\text{negligible})\end{aligned} \tag{3.12}$$

We are led to consider $\chi_0(t)k_{n(q)}^{(p)}$ $(= \chi_0(t)\beta_{(q)}(t,x)\alpha_n^{(p)}(t;D))$ instead of $\chi_0(t)k_n$. We have the same type inequality as above. Notice that $q = (0,q_1,\ldots,q_\ell)$.
Put

$$\chi_0(t)k_{n(q)}^{(p)} v_n = u_{p,q}. \tag{3.13}$$

Then

$$\begin{aligned}(M\log n-c)\,\|u_{pq}\|_M &\leqq Ar_0 \sum_j \|u_{p,q+e_j}\|_M + Anr_0 \sum_j \|u_{p+e_j,q}\|_M \\ + c(k_0)\Big(\sum_{2\leqq|\mu+\nu|\leqq k_0} n^{1-|\nu|}\|u_{p+\mu,q+\nu}\|_M\Big) &+ c' \sum_j n^{-1} \|u_{p,q+e_j}\|_M + c'\|u_{p+e_j,q}\|_M \\ + \|r_{n,pq}(v_n)\|_M &+ \|g_{n,pq}\|_M + \|h_{n,pq}\|_M,\end{aligned} \tag{3.14}$$

where

$$r_{n,pq}(v_n) = [\chi_0 k_{n(q)}^{(p)}, b_{n,loc}]v_n - \sum_{1\leq|\mu+\nu|\leq k_0} \mu!^{-1}\nu!^{-1}(-1)^{|\nu|} \times b_{n,loc(\mu)}^{(\nu)}(u_{p+\mu,q+\nu}),$$

$$g_{n,pq} = (\chi_0'(t)k_{n(q)}^{(p)}v_n + \chi_0(t)k_{n(q)}^{(p)}\chi_0'(t)u) + \chi_0(t)k_{n(q)}^{(p)}\chi_0(t)f + \chi_0(t)k_{n(p)}^{(q)}((b_{n,loc}-b)v_n - b(1-\tilde{\chi}_n(D))v), \tag{3.15}$$

$$h_{n,pq} = \sum_j i(\lambda_{n,loc}^{(e_j)} - \tilde{\lambda}_{n,loc}^{(e_j)})(u_{p,q+e_j}) - \sum_j i(\lambda_{n,loc(e_j)} - \tilde{\lambda}_{n,loc(e_j)})(u_{p+e_j,q}).$$

Here we consider all (p,q) satisfying $|p+q| \leq N$. This number N is the same one appearing in (2.5), (2.6). The choice of N is connected with the estimate of $\|r_{n,pq}(v_n)\|_M$.

We add (for $|p+q| \leq N$) with weight c_{pq}^n the both sides of $(3.14)_{pq}$. This weight is determined in such a way that all terms of the form $\|u_{p+\mu,q+\nu}\|_M$ with $|p+q+\mu+\nu| \leq N$ are absorbed in the left hand side. As one might observe, only the first terms in the right-hand side of $(3.14)_{pq}$ play decisive role. We choose c_{pq}^n by

$$c_{pq}^n = ((2\ell+2)Ar_0/M\log n)^{|p+q|}\, n^{|p|} = c(n;A,r_0,M)^{|p+q|}\, n^{|p|}. \tag{3.16}$$

Next we choose N in such a way that

$$c(n;a,r_0,M)\, Ncr_0^{-1} = e^{-1}. \tag{3.17}$$

Namely

$$N = c(A)M\log n, \text{ where } \quad c(A) = ((2\ell + 2)Ace)^{-1}. \tag{3.18}$$

Hence

$$e^{-N} = n^{-c(A)M}. \tag{3.19}$$

Observe that $c^n_{pq} \,/\, c^n_{p+\mu,q+\nu} = c(n;A,r_0,M)^{-|\mu+\nu|} n^{-|\mu|}$.

Denoting

$$c^n_{pq}\, u_{pq} = v_{pq}, \tag{3.20}$$

we obtain

$$(M\log n - c)\,\|u_{00}\|_M \leq c(k_0) \sum_{|p+q+\mu+\nu|>N} c(n;A,r_0,M)^{-|\mu+\nu|} n^{1-|\mu+\nu|} \|v_{p+\mu,q+\nu}\|_M$$

$$+ \sum_{|p+q|\leq N} c^n_{pq} \|r_{n,pq}(v_n)\|_M + \sum_{|p+q|\leq N} c^n_{pq} \|g_{n,pq}\|_M + \sum_{|p+q|\leq N} c^n_{pq} \|h_{n,pq}\|_M. \tag{3.21}$$

Look at the first term of the right-hand side. Since (see (2.5), (2.6))

$$\begin{aligned} |\alpha_n^{(p+\mu)}|_0 &\leq (\tfrac{N}{n}\, c r_0^{-1})^{|p|} c_\mu r_0^{-|\mu|}\, n^{-|\mu|} && \text{for } |p| \leq N, \\ |\beta_{(q+\nu)}|_0 &\leq (Ncr_0^{-1})^{|q|}\, c_\nu r_0^{-|\nu|} && \text{for } |q| \leq N, \end{aligned} \tag{3.22}$$

we obtain, taking account of (3.17),

$$c(n;A,r_0,M)^{-|\mu+\nu|} n^{1-|\mu+\nu|} \|v_{p+\mu,q+\nu}\| \leq e^{-|p+q|} (c' r_0^{-1})^{|\mu+\nu|} n^{1-|\mu+\nu|} \times \text{const. } n^k \|v\|_{-k}.$$

Now since $|\mu+\nu| \leq k_0$, and $|p+q+\mu+\nu| > N$, it holds $|p+q| > N-k_0$. Then the last quantity is estimated by $e^{-N} c'(k_0) n^k \|v\|_{-k}$.

Next we consider the second term in (3.21).

$$\|r_{n,pq}(v_n)\| \leq c(k_0) \sum_{|\mu+\nu|=k_0+1} n^{1-|\nu|}\, |\alpha_n^{(p+\mu)}|_0 |\beta_{(q+\nu)}|_0 \text{ const. } n^k \|v_n\|_{-k}$$

$$\leq \text{const. } c(k_0)\, (\tfrac{N}{n}\, c r_0^{-1})^{|p|} (Ncr_0^{-1})^{|q|} n^k \|v\|_{-k} \sum_{|\mu+\nu|=k_0+1} n^{1-|\mu+\nu|}.$$

Hence

$$c^n_{pq} \|r_{n,pq}(v_n)\| \leqq e^{-|p+q|} \; c''(k_0) \; n^{-k_0+k} \; \|v\|_{-k} \; .$$

Summing up:

$$(\text{first + second terms}) \leqq (c'(k_0) \; n^{-c(A)M} + c''(k_0) \; n^{-k_0}) \; n^k \; \|v\|_{-k} \; . \quad (3.23)$$

After this preparation, let us show that

PROPOSITION 3.3:

$$\|\chi_0(t)k_n v\|_M \leqq (c'(k_0) \; n^{-c(A)M} + c''(k_0) \; n^{-k_0}) \; n^k \; \|v\|_{-k}$$

$$+ \sum_{|p+q|\leqq N} c^n_{pq} \|\chi_0'(t) \; k_n^{(p)}{}_{(q)} v_n\|_M + \sum_{|p+q|\leqq N} c^n_{pq} \|\chi_0(t)k_{n(q)}^{(p)}\chi_0'(t)u\|_M$$

$$+ \sum_{|p+q|\leqq N} c^n_{pq} \|\chi_0(t)k_{n(q)}^{(p)} \; (\chi_0(t)f)\| \; + \text{(rapidly decreasing)},$$

where rapidly decreasing means that a sequence which can be estimated in the form $c(L)n^{-L}$ for any large integer when n tends to infinity.

NOTE. c(L) is independent of M.

To prove Proposition 3.3, it is enough to show

LEMMA 3.4:

i) $c^n_{pq} \quad \|\chi_0(t)k_{n(q)}^{(p)} \; (b_{n,loc} - b)v_n\|$

ii) $c^n_{pq} \quad \|\chi_0(t)k_{n(q)}^{(p)} \; b \; (1-\tilde{\chi}_n(D))v\|$

iii) $c^n_{pq} \; \|(\lambda_{n,loc}^{(e_j)} - \tilde{\lambda}_{n,loc}^{(e_j)})u_{p,q+e_j}\| + c^n_{pq} \; \|(\lambda_{n,loc(e_j)} - \tilde{\lambda}_{n,loc(e_j)})u_{p+e_j,q}\|$

are all negligible. More precisely there exists a constant $C_0(L)$ such that they are all estimated by $C_0(L)n^{-L}$, where L can be taken as large as we wish, and $C_0(L)$ is independent of (p,q) provided that $|p+q| \leq N$.

<u>PROOF</u>: i) Recall that $k_{n\ (q)}^{(p)} = \beta_{(q)}(t,x)\ \alpha_n^{(p)}(t;D)$. We commute $\alpha_n^{(p)}$ with $(b_{n,loc}-b)$. Hence by expansion,

$$\beta_{(q)}\ \alpha_n^{(p)}(b_{n,loc}-b) = \sum_{|\mu|\leq L} \mu!^{-1}\ \beta_{(q)}(t,x)(b_{n,loc}-b)_{(\mu)}\alpha_n^{(p+\mu)} + r_{L,pq}.$$

By (2.18) the first terms under Σ vanish. Now

$$\| r_{L,pq}\|_{L(H^{-k},L^2)}\ \|v_n\|_{-k} \leq c_1(L)|\beta_{(q)}|_0\ |\alpha_n^{(p+\mu)}|_0 \text{ const. } n^{1+k}\ \|v_n\|_{-k},$$

where $|\mu| = L+1$.

Since $c_{pq}^n|\beta_{(q)}|_0|\alpha_n^{(p+\mu)}|_0 \leq e^{-|p+q|}\ c_\mu r_0^{-|\mu|}\ n^{-|\mu|}$, we see that

$$c_{pq}^n\ \|r_{L,pq}(v_n)\| \leq c_2(L)\ e^{-|p+q|}\ n^{-L+k} \text{ cont. } \|v\|_{-k}.$$

ii) We use the partition of unity in (τ,ξ)-space as explained in [4], p.52-.

$$0 \leq \phi_k(\rho) \leq 1, \quad \text{supp } [\phi_k] \subset [4^k/3,\ 3\cdot 4^k], \quad \sum_k \phi_k(\rho) = 1,$$

$$|\partial^m\phi_k(\rho)| \leq c_m/R_k^m \quad (m = 0,1,2,\ldots),\ R^k = 4^k.$$

Now denoting characteristic function of supp $[\phi_k]$ by $\chi_k(\tau,\xi)$, we write

$$v = \sum_k \phi_k(D)\chi_k(D)v = \sum_k \phi_k(D)v_k.$$

Then

$$k_{n\ (q)}^{(p)}\ b\ (1-\tilde{\chi}_n(D))v = \sum_k k_{n(q)}^{(p)}\ b(1-\tilde{\chi}_n(D))\phi_k(D)v_k.$$

We commute b with $(1-\tilde{\chi}_n(D))\phi_k$. Remarking that

$$\text{supp } [(1-\tilde{\chi}_n(\tau,\xi))\phi_k(\tau,\xi)] \cap \text{supp}_{\tau\xi}[\alpha_n(t;\tau,\xi)] = \phi,$$

we arrive at the conclusion. Of course we use

$$\|\chi_0(t)k_n^{(p)}{}_{(q)}b(1-\tilde{\chi}_n(D))v\| \leq \sum_k \|\chi_0(t)k_{n(q)}^{(p)}\, b(1-\tilde{\chi}_n(D))\phi_k(D)v_k\| .$$

iii) We consider only the first term. Let us recall $u_{p,q+e_j} = \chi_0(t)\beta_{(q+e_j)}\alpha_n^{(p)}v_n$. Since this first term can be written

$$\chi_0(t)\,(\lambda_{n,loc}^{(e_j)} - \tilde{\lambda}_{n,loc}^{(e_j)})\,\beta_{(q+e_j)}\,\alpha_n^{(p)}(t;D)v_n,$$

recalling (2.26), we can commute $(\lambda_{n,loc}^{(e_j)} - \tilde{\lambda}_{n,loc}^{(e_j)})$ with $\beta_{(q+e_j)}$ and estimate the remainder term in the same as in i). □

Now we go back to Proposition 3.3. We consider two terms in the second line. These terms are concerned with the considerations on the support of $\chi_0'(t)$. We separate this support into $[0,\delta]$ and $[T-2\delta/3,\ T-\delta/3]$. We write

$$\chi_0'(t) = \chi_\delta(t) + \chi_T(t), \quad \text{where supp } [\chi_\delta(t)] \subset [0,\delta], \quad \text{and}$$

$$\text{supp } [\chi_T(t)] \subset [T-2\delta/3,\ T-\delta/3].$$

Concerning the part corresponding to $\chi_\delta(t)$, this is estimated by

$$\sum_{|p+q|\leq N} c_{pq}^n\ \|\chi_\delta(t)k_n^{(p)}{}_{(q)}\ v_n\| + \sum_{|p+q|\leq N} c_{pq}^n \|\chi_0(t)k_{n(q)}^{(p)}\ (\chi_\delta(t)u)\| . \tag{3.24}$$

Incidentally, we consider the term

$$\sum_{|p+q|\leq N} c_{pq}^n\ \|\chi_0(t)k_n^{(p)}{}_{(q)}\ (\chi_0(t)f)\| . \tag{3.25}$$

As we shall explain in the Appendix, the assumptions

$$(0,x_0;\tau^0,\xi^0) \notin WF(u), \quad \Gamma(t) \notin WF(f) \text{ for } t \in [0,T] \tag{3.26}$$

imply that *both terms in* (3.24) *and* (3.25) *are rapidly decreasing*.

Now we state

<u>LEMMA 3.5:</u> The following holds

i) $\displaystyle\sum_{|p+q|\leq N} c_{pq}^n\ \|\chi_T(t)\ k_{n(q)}^{(p)}\ v_n\|_M \leq c_0\ n^{-M(T-\delta)}\ n^k\ \|v\|_{-k}$.

ii) $\sum_{|p+q|\leqq N} c^n_{pq} \|\chi_0(t)\, k^{(p)}_{n(q)}\, \chi_T(t)u\|_M \leqq c_1\, n^{-M(T-\delta)}\, n^k \|\tilde{\chi}_0(t)u\|_{-k}$

+ (rapidly decreasing),

where c_0 and c_1 are constants independent of n and M, and $\tilde{\chi}_0(t)$ is any cut off function with support in [0,T] and $\chi_T(t) \subset\!\subset \tilde{\chi}_0(t)$.

PROOF: i) There is no problem

$$c^n_{pq} \|n^{-Mt}\, \chi_T(t)\, k^{(p)}_{n(q)}\, v_n\| \leq n^{-M(T-\delta)}\, c^n_{pq} \|\chi_T(t)\, k^{(p)}_{n(q)}\, v_n\|$$

$$\leqq n^{-M(T-\delta)}\, e^{-|p+q|} \text{ const. } n^k \|v\|_{-k}.$$

ii) $\chi_0(t)\, k^{(p)}_{n(q)}\, \chi_T(t)u = \chi_0(t)\, k^{(p)}_{n(q)}\, \chi_T(t)(\tilde{\chi}_0(t)u)$.

Commuting $k^{(p)}_{n(q)}(t,x;D)$ with $\chi_T(t)$, this becomes

$$= \sum_{0\leqq\nu\leqq L} \nu!^{-1}\, \chi_{T(\nu)}(t) \times \partial^\nu_\tau\, k^{(p)}_{n(q)}\, (\tilde{\chi}_0(t)u) + s_{L,pq}(\tilde{\chi}_0 u).$$

Observe that

$$c^n_{pq} \|s_{L,pq}(\tilde{\chi}_0(t)u)\| \leq e^{-|p+q|}\, c(L)\, n^{-L} \text{ const. } n^k \|\tilde{\chi}_0(t)u\|_{-k}. \quad \square$$

Now let t_0 be a point in $[\delta, T-\delta]$. Let $\chi_{t_0}(t)$ $(\in C^\infty_0)$ be a cut off function around t_0 of size r_0, namely $\chi_{t_0}(t) = 1$ for $|t-t_0| \leq r_0/2$, and $= 0$ for $|t-t_0| \geqq r_0$. Then

$$\|\chi_{t_0}(t)k_n\, v\| \leq n^{M(t_0+r_0)} \|\chi_0(t)k_n\, v\|_M. \tag{3.27}$$

In fact, since $\chi_{t_0}(t)\chi_0(t) = \chi_{t_0}(t)$,

$$\|n^{-Mt}\, \chi_{t_0}(t)\chi_0(t)\, k_n v\| \leq \|\chi_0(t)k_n\, v\|_M.$$

On the other hand, $\|n^{-Mt}\, \chi_{t_0}(t)k_n\, v\| \leq n^{-M(t_0-r_0)}\|\chi_{t_0}(t)k_n\, v\|$.

Now we assume (3.26). Then taking account of Lemma 3.5 and (3.27), from Proposition 3.3, we obtain the inequality

$$\|\chi_{t_0}(t)k_n\, v\| \leq n^{M(t_0+r_0)}(c'(k_0)n^{-c(A)M}+c''(k_0)n^{-k_0}+cn^{-M(T-\delta)})n^k c_k(u) \\ + n^{M(t_0+r_0)}\text{ (rapidly decreasing)}, \tag{3.28}$$

where $c_k(u)$ is a constant depending only on $u(t,x)$.

Now we are in a position to prove

PROPOSITION 3.6: Let $0 < t_0 < T$. Assume (3.26). Then, provided that $t_0 < c(A)$, $\|\chi_{t_0}(t)k_n\, v\|$ is rapidly decreasing. Hence $\Gamma(t_0) \notin WF(u)$.

NOTE: $c(A)$ is defined by (3.18).

PROOF: We use the arbitrariness of two large parameters M and k_0 in (3.28). Let s be an arbitrary positive number. We may assume $t_0 + r_0 < T-\delta$, and $t_0 + r_0 < c(A)$, because we may assume r_0 and δ arbitrarily small. Then first we choose M in such a way that

$$M(T-\delta-t_0-r_0) \geq s+k,\ M(c(A)-(t_0+r_0)) \geq s+k.$$

Next, we choose k_0 in such a way that $k_0 \geq s+k+M(t_0+r_0)$.

Then we see that

$$\|\chi_{t_0}(t)k_n\, v\| \leq c_s\, n^{-s} \quad \text{if } n \text{ is large.}$$

4. PROOF OF THEOREM 2

In this section, we treat the case of Gevrey class $\gamma^{(s)}$ $(s > 1)$. We can carry out the proof of Theorem 2 almost along the same line as in C^∞-case.

We assume $\lambda(t,x;\xi)$ and $c(t,x;\xi)$ are of Gevrey class $\gamma^{(s)}$ in (t,x) and analytic in ξ. More precisely,

$$c(t,x;\xi) \in S^{\rho}_{1,0} \quad \text{with} \quad \rho < 1/s,$$

and

$$|\lambda_{(\mu)}^{(\nu)}(t,x;\xi)| \leq B\, C_0^{|\mu+\nu|}\mu!^{\,s}\,\nu!\,|\xi|^{1-|\nu|},$$
$$|c_{(\mu)}^{(\nu)}(t,x;\xi)| \leq B\, C_0^{|\mu+\nu|}\mu!^{\,s}\,\nu!\,|\xi|^{\rho-|\nu|}, \text{ for } |\nu| \leq R_0^{-1}|\xi|^{1/s}, \tag{4.1}$$

where R_0 is a suitable large number.

The condition on c is usually called Gevrey pseudo-analytic symbol. Let us write again

$$au = f, \text{ or } (\partial_t + b)u = (\partial_t + i\lambda + c)u = f. \tag{4.2}$$

We enumerate the places where we change previous arguments.

(1) Weight function. We replace n^{-Mt} by $\exp(-r_0\, n^{1/s}t)$, and we denote

$$\|v\|_E = \|\exp(-r_0\, n^{1/s}\, t)v\|.$$

(2) Weight c_{pq}^n. We replace $c_{pq}^n = ((2\ell+2)Ar_0/M\log n)^{|p+q|}\, n^{|p|}$ by

$$c_{pq}^n = \{(2\ell+2)A\}^{|p+q|}\, n^{|p|-|p+q|/s}. \tag{4.3}$$

We denote often

$$(2\ell+2)A = A'. \tag{4.4}$$

(3) Number N. We replace $N = c(A)M \log n$ by

$$N = c(A)r_0\, n^{1/s}, \tag{4.5}$$

where c(A) is the same as in previous section. Namely

$$c(A) = ((2\ell+2)Ace)^{-1} = (A'ce)^{-1}. \tag{4.6}$$

Let us notice the relation which we use often:

$$c_{pq}^n(Ncr_0^{-1})^{|p+q|}\, n^{-|p|} = e^{-|p+q|}, \text{ hence } c_{pq}^n|\alpha^{(p)}{}_n|_0|\beta_{(q)}|_0 \leq e^{-|p+q|}.$$

Next, we define $\lambda_{n,loc}(t,x;\xi)$ and $c_{n,loc}(t,x;\xi)$ by (2.17) by using $\chi_n(\xi)$. We make more precise the condition (2.15) in the following way

$$|\chi_n^{(\mu+\nu)}(\xi)| \leqq (\frac{N}{n} cR^{-1})^{|\mu|} \nu!^{1+\varepsilon'}(cR^{-1})^{|\nu|} \quad \text{for} \quad |\mu| \leqq N, \tag{4.7}$$

where ε' (> 0) is chosen small in such a way that $1 + 2\varepsilon' < s$. This implies in particular that $\lambda_{n,loc}$, $c_{n,loc}$ satisfy the following estimates, if necessary by changing C_0 to larger one,

$$\begin{aligned} &|\lambda_{n,loc}{}^{(\nu)}_{(\mu)}| \leqq B\, C_0^{|\mu+\nu|}\mu!^{\,s}\, \nu!^{\,1+\varepsilon'}\, n^{1-|\nu|}, \\ &|c_{n,loc}{}^{(\nu)}_{(\mu)}| \leqq B\, C_0^{|\mu+\nu|}\mu!^{\,s}\, \nu!^{\,1+\varepsilon'}\, n^{\rho-|\nu|} \quad \text{for} \quad |\nu| \leqq R_0^{-1} n^{1/s}. \end{aligned} \tag{4.8}$$

In accordance with this cut off function, we make more precise (2.5) and (2.6) as follows:

$$|\chi_N^{(\mu+\nu)} \leqq (Ncr_0^{-1})^{|\mu|} \nu!^{\,1+\varepsilon'} (cr_0^{-1})^{|\nu|} \text{ for } |\mu| \leqq N,$$

$$|\beta_{0(\mu+\nu)}| \leqq (Ncr_0^{-1})^{|\mu|} \nu!^{\,1+\varepsilon'} (cr_0^{-1})^{|\nu|} \text{ for } |\mu| \leqq N.$$

Thus

$$\begin{aligned} &|\alpha_n^{(\mu+\nu)}(t;\tau,\xi)| \leqq (Ncr_0^{-1}/n)^{|\mu|} \nu!^{\,1+\varepsilon'} (cr_0^{-1}/n)^{|\nu|} \quad \text{for} \quad |\mu| \leqq N, \\ &|\beta_{(\mu+\nu)}(t,x)| \leqq (Ncr_0^{-1})^{|\mu|}\nu!^{\,1+\varepsilon'}(cr_0^{-1})^{|\nu|} \quad \text{for} \quad |\mu| \leqq N. \end{aligned} \tag{4.9}$$

Let us notice that here *the derivatives of* $\beta(t,x)$ *are merely concerned with those in* x, whereas those of α_n are concerned with (τ,ξ).

Lemma 3.1 is replaced by

<u>LEMMA 4.1</u> (Energy inequality): Let $(\partial_t + b_{n,loc})(\chi_0(t)v) = f$. Then it holds

$$\|a_{n,loc}(\chi_0(t)v)\|_E \leqq (r_0\, n^{1/s}\text{-const. } n^{\rho})\| \chi_0(t)v\|_E.$$

Since the proof is almost the same as that of Lemma 3.1, we omit the proof. In an actual case, we use the following asymptotic expansion

$$[\chi_0(t)k^{(p)}_{n(q)}, b_{n,loc}]v_n = \sum_{1\leq|\mu+\nu|\leq N-|p+q|} \mu!^{-1}\ \nu!^{-1}(-1)^{|\nu|} b^{(\nu)}_{n,loc(\mu)}$$

$$(\chi_0(t)k^{(p+\mu)}_{(q+\nu)}v_n) + r_{n,pq}(v_n). \tag{4.10}$$

Then we obtain the following relation, corresponding to $(3.14)_{pq}$:

$$\begin{aligned}(r_0\, n^{1/s}\text{-const. } n^\rho)\, \|u_{pq}\|_E &\leq Ar_0 \sum_j \|u_{p,q+e_j}\|_E + Anr_0 \sum_j \|u_{p+e_j,q}\|_E \\ &+ \Big(\sum_{2\leq|\mu+\nu|\leq N-|p+q|} \mu!^{-1}\ \nu!^{-1} \|\lambda^{(\nu)}_{n,loc(\mu)}(u_{p+\mu,q+\nu})\|_E \\ &+ \sum_{1\leq|\mu+\nu|\leq N-|p+q|} \mu!^{-1}\ \nu!^{-1} \|c^{(\nu)}_{n,loc(\mu)}(u_{p+\mu,q+\nu})\|_E, \\ &+ \|r_{n,pq}(v_n)\|_E + \|g_{n,pq}\|_E + \|h_{n,pq}\|_E,\end{aligned} \tag{4.11$_{pq}$}$$

where $r_{n,pq}(v_n)$ is defined by (4.10) and $g_{n,pq}$ and $h_{n,pq}$ are the same as those in $(3.15)_{pq}$.

First of all, observe that, taking account of (4.3),

$$c^n_{pq} / c^n_{p+\mu,q+\nu} = A'^{\,-|\mu+\nu|}\ n^{-|\mu|+|\mu+\nu|/s}. \tag{4.12}$$

Next

$$c^n_{pq}\,(Ar_0 \sum_j \|u_{p,q+e_j}\|_E + Anr_0 \sum_j \|u_{p+e_j,q}\|_E)$$

$$= \frac{1}{2\ell+2}\, r_0\, n^{1/s} \Big(\sum_j \|v_{p,q+e_j}\|_E + \sum_j \|v_{p+e_j,q}\|_E\Big),$$

where $v_{pq} = c^n_{pq}\, u_{pq}$. Let us notice that, for $v_{p,q+e_j}$, j runs through from 1 to ℓ, and for $v_{p+e_j,q}$, j runs through from 0 to ℓ. Hence

$$\begin{aligned}(r_0\, n^{1/s}\text{-const. } n^\rho)\|v_{pq}\|_E &\leq \frac{1}{2\ell+2}\, r_0 n^{1/s}\Big(\sum_j \|v_{p,q+e_j}\|_E + \sum_j \|v_{p+e_j,q}\|_E\Big) \\ &+ J_{1,pq} + J_{2,pq} + c^n_{pq}\,\|r_{n,pq}(v_n)\|_E + c^n_{pq}\,\|g_{n,pq}\|_E + c^n_{pq}\,\|h_{n,pq}\|_E,\end{aligned} \tag{4.14$_{pq}$}$$

where

$$J_{1,pq} = \sum_{2\leq|\mu+\nu|\leq N-|p+q|} \mu!^{-1}\nu!^{-1} c^n_{pq} \| \lambda_{n,loc(\mu)}^{(\nu)}(u_{p+\mu,q+\nu}) \|_E,$$

$$J_{2,pq} = \sum_{1\leq|\mu+\nu|\leq N-|p+q|} \mu!^{-1}\nu!^{-1} c^n_{pq} \| c_{n,loc(\mu)}^{(\nu)}(u_{p+\mu,q+\nu}) \|_E$$

Let us show that

<u>LEMMA 4.2</u>: Provided that r_0 is sufficiently small, we have

$$\sum_{|p+q|\leq N} J_{1,pq} \leq n \frac{\text{const.}}{n^{2(1-1/s)}} \sum_{|p+q|\leq N} \|v_{pq}\|_E, \tag{4.15}$$

where const. is independent of n (but may depend on r_0).

<u>PROOF</u>: Taking account of (4.7) and assuming there $C_0 \geq 1$ and also of (4.12),

$$c^n_{pq}\ \mu!^{-1}\ \nu^{-1} \|\ \lambda_{n,loc(\mu)}^{(\nu)}(u_{p+\mu,q+\nu}) \|_E$$

$$\leq c(\ell)\ B\ C_0^{|\mu+\nu|} |\mu+\nu|!^{s-1}\, n^{1-|\mu+\nu|(1-1/s)}\ A'^{-|\mu+\nu|} \|v_{p+\mu,q+\nu}\|_E$$

$$\leq c(\ell)\ B\ n\ (C_0\ A'^{-1} |\mu+\nu|^{s-1}/ n^{1-1/s})^{|\mu+\nu|}\ \|v_{p+\mu,q+\nu}\|_E.$$

Hence

$$\sum_{|p+q|\leq N} J_{1,pq} \leq nc(\ell)B \sum_{|p+q|\leq N}\ \sum_{2\leq|\mu+\nu|\leq N-|p+q|} \left(\frac{|\mu+\nu|^{s-1}}{n^{1-1/s}} C_0 A'^{-1}\right)^{|\mu+\nu|}$$

$$\times\ \|v_{p+\mu,q+\nu}\|_E.$$

Now (p',q') and m being fixed, then the number of $(\mu,\nu) > 0$ such that $p+\mu = p'$, $q+\nu = q'$ is less than $(m+1)^{2\ell+1}$. Then, the right-hand side is estimated by

$$nc(\ell)B \sum_{|p+q|\leq N} \left(\sum_{2\leq m\leq N} \left(\frac{m^{s-1}}{n^{1-1/s}} C_0 A'^{-1}\right)^m (m+1)^{2\ell+1}\right) \|v_{pq}\|_E.$$

Since $C_0 A'^{-1}\ m^{s-1}/n^{1-1/s} \leq C_0 A'^{-1}\ N^{s-1}/n^{1-1/s} = C_0\ A'^{-1}(c(A)r_0)^{s-1}$.

Choosing r_0 small in such a way that

$$C_0 A'^{-1} (c(A)r_0)^{s-1} \leq 1/2, \tag{4.16}$$

the above sum is estimated by

$$\frac{2^{s-1}}{n^{2(1-1/s)}} C_0 A'^{-1} \sum_{m\geq 2} (1/2)^m (m+1)^{2\ell+1} \leq \frac{\text{const.}}{n^{2(1-1/s)}} . \qquad \square$$

The estimate of $\sum_{|p+q|\leq N} J_{2,pq}$ is done in the same way.

$$J_{2,pq} \leq \text{const. } n^{\rho-1+1/s} \sum_{|p+q|\leq N} (\sum_j \|v_{p,q+e_j}\|_E + \sum_j \|v_{p+q_j,q}\|_E)$$

$$+ \frac{\text{const.}}{n^{2(1-1/s)}} n^{\rho} \sum_{2\leq|p+q|\leq N} \|v_{pq}\|_E.$$

The last term is absorbed in $\Sigma\ J_{1,pq}$. Hence, observing $\rho < 1/s$, the addition of the both sides of $(4.14)_{pq}$ for $|p+q|\leq N$ yields

$$(r_0 n^{1/s} - \text{const.} n^{\rho}) \|u_{00}\|_E + \sum_{1\leq|p+q|\leq N} (\frac{1}{2\ell+2} - \text{const.} n^{-(1-1/s)}) r_0 n^{1/s} \|v_{pq}\|_E$$

$$\leq \sum_{|p+q|=N} (2\ell+2)^{-1} r_0 n^{1/s} (\sum_j \|v_{p,q+e_j}\|_E + \sum_j \|v_{p+e_j,q}\|_E) \tag{4.17}$$

$$+ \sum_{|p+q|\leq N} c^n_{pq} \|r_{n,pq}(v_n)\|_E + \sum_{|p+q|\leq N} c^n_{pq} \|g_{n,pq}\|_E + \sum_{|p+q|\leq N} c^n_{pq} \|h_{n,pq}\|_E.$$

Observe that in the right-hand side the first and second terms have the same character.

<u>LEMMA 4.3</u>: Provided that r_0 is sufficiently small, it holds

$$c^n_{pq} \|r_{n,pq}(v_n)\| \leq e^{-N} \ (= \exp(-c(A) r_0\, n^{1/s})).$$

<u>PROOF</u>: In view of the form (4.10) in $r_{n,pq}(v_n)$, $c^n_{pq} \|r_{n,pq}(v_n)\|$ is estimated by

$$c^n_{pq}\ \mu!^{-1}\ \nu!^{-1}\ \|b_{n,loc(\mu)}^{(\nu)}\|\ |\alpha_n^{(p+\mu)}|_0 |\beta_{(q+\nu)}|_0\ n^k\ \|v\|_{-k}\ \times(\text{some polynomial in } n), \quad (4.18)$$

where $|p+q+\mu+\nu| = N$. Notice that here polynomial in n can be taken independent of (p,q). A part of this estimate is already done in the proof of Lemma 4.2. Namely

$$\mu!^{-1}\ \nu!^{-1}\ \|b_{n,loc(\mu)}^{(\nu)}\| \leqq c(\ell)Bn\ (C_0|\mu+\nu|^{s-1}/n)^{|\mu+\nu|}\ n^{|\mu|}.$$

Next,

$$c^n_{pq}\ |\alpha_n^{(p+\mu)}|_0|\beta_{(q+\nu)}|_0 \leqq c^n_{pq}\ (Ncr_0^{-1})^{|p+q|}n^{-|p+\mu|}\ (Ncr_0^{-1})^{|\mu+\nu|}$$

$$= (N\ A'cr_0^{-1}/n^{1/s})^{|p+q|}\ n^{-|\mu|}(Ncr_0^{-1})^{|\mu+\nu|} = e^{-|p+q|}\ n^{-|\mu|}(Ncr_0^{-1})^{|\mu+\nu|}.$$

On the other hand, $C_0|\mu+\nu|^{s-1}\ Ncr_0^{-1}/n \leqq c_0\ N^s\ cr_0^{-1}/n$ (since $|\mu+\nu| \leqq N$)

$$\leqq C_0\ cr_0^{-1}\ (c(A)r_0)^s = C_0\ c\ c(A)^s\ r_0^{s-1}.$$

If r_0 is small, namely if

$$r_0^{s-1} \leqq C_0^{-1}\ (ce)^{-1}\ c(A)^{-s}, \quad (4.19)$$

the last quantity is less than e^{-1}. Hence

$$(4.18) \leqq e^{-N} \times (\text{some polynomial in } n). \quad \square$$

From (4.17), we obtain

$$\|u_{00}\|_E \leqq \exp(-c(A)r_0\ n^{1/s}) \times (\text{some polynomial in } n) \quad (4.20)$$

$$+ \sum_{|p+q|\leqq N} c^n_{pq}\ \|g_{n,pq}\|_E + \sum_{|p+q|\leqq N} c^n_{pq}\ \|h_{n,pq}\|_E.$$

<u>NOTE</u>:

$$\|u_{00}\|_E = \|\chi_0(t)k_nv_n\|_E = \|\chi_0(t)k_nv\|_E = \|\exp(-r_0\ n^{1/s}t)\chi_0(t)k_nv\|. \quad \square$$

It remains to consider the estimates of two terms in the right-hand side of (4.20). Namely we must show two lemmas corresponding to Lemmas 3.4 and 3.5.

LEMMA 4.4: Provided that r_0 is sufficiently small,

i) $\sum_{|p+q|\leq N} c^n_{pq} \|\chi_0(t)k^{(p)}_{n(q)}(b_{n,loc}-b)v_n\|$,

ii) $\sum_{|p+q|\leq N} c^n_{pq} \|\chi_0(t)k^{(p)}_{n(q)}b(1-\tilde{\chi}_n(D))v\|$,

iii) $\sum_{|p+q|\leq N} c^n_{pq} \|(\lambda^{(e_j)}_{n,loc}-\tilde{\lambda}^{(e_j)}_{n,loc})u_{p,q+e_j}\| + \sum_{|p+q|\leq N} c^n_{pq} \|(\lambda_{n,loc(e_j)} - \tilde{\lambda}_{n,loc(e_j)})u_{p+e_j,q}\|$

are all estimated by e^{-N} × (some polynomial in n). More strictly. they are *negligible*. (See Note after the proof).

PROOF: i) and ii) are estimated along the same line as the proof of Lemma 3.4. We show iii). To carry out this, we use the following micro-localization of $\lambda_{n,loc(e_j)}$ and $\lambda^{(e_j)}_{n,loc}$.

Let $\psi_0(\xi)$ be a localizer of size $2r_0$ around the origin. Namely $\psi_0(\xi) = 1$ for $|\xi| \leq r_0$, $= 1$ for $|\xi| \geq 2r_0$, $0 \leq \psi_0(\xi) \leq 1$;

$$|\psi_0^{(\mu)}(\xi)| \leq \mu!^{1+\varepsilon'} (cr_0^{-1})^{|\mu|} \quad \text{with } 1 + 2\varepsilon' < s.$$

Put

$$\psi_n(\xi) = \psi_0(\tfrac{\xi}{n} - \xi(t)).$$

We define also $\tilde{x}(x)$ (localizer of size $2r_0$ around the origin) as follows. $\tilde{x}(x) = x$ for $|x| \leq r_0$, and $= 0$ for $|x| \geq 2r_0$, $|\tilde{x}(x)| \leq 2r_0$. Denote

$$\lambda_{n,loc(e_j)}(t,x;\xi) = a(t,x;\xi).$$

First let $a_1(t,x;\xi) = a(t,x(t) + \tilde{x}(x-x(t));\xi)$.

Then we define

$$\tilde{a}_{n,loc}(t,x;\xi) = a_1(t,x;\xi)\psi_n(\xi) + a_1(t,x;n\xi(t))(1-\psi_n(\xi)).$$

We see that

$$\tilde{a}_{n,loc}(t,x;\xi) - a(t,x;\xi) = 0 \text{ for } (x,\xi) \in \operatorname{supp}_x[\beta(t,x)] \cap \operatorname{supp}_\xi[\alpha_n(t;\tau,\xi)].$$

More strictly

$$\tilde{a}_{n,loc}(t,x;\xi) - a(t,x;\xi) = (a_1(t,x;\xi) - a(t,x;\xi))\psi_n(\xi)$$

$$+ (a_1(t,x;n\xi(t)) - a(t,x;\xi))(1-\psi_n(\xi)).$$

So we are led to consider

$$c^n_{pq}\, \chi_0(t)(a_1-a)\psi_n(D_x)\beta_{(q)}\alpha_n^{(p+e_j)}v,\; c^n_{pq}\, \chi_0(t)(a_1-a(t,x;n\xi(t))(1-\psi_n(D_x))\times$$

$$\times\, \beta_{(q)}\, \alpha_n^{(p+e_j)}v.$$

Let us consider the first term. Commuting $\beta_{(q)}$ with $(a_1-a)\psi_n(D_x)$, we are led to consider

$$c^n_{pq}\, \nu!^{-1}\, \beta_{(q+\nu)}((a_1-a)\psi_n)^{(\nu)}\, \alpha_n^{(p+e_j)}v.$$

These are zero. The remainder term when we take its asymptotic expansion up to the L-th term is estimated, except for a trivial factor, by

$$e^{-|\bar{p}+q|}\, |\nu|!^{\varepsilon'}\, |(cr_0^{-1})^{|\nu|}((a_1-a)\psi_n)^{(\nu)}|_0\, \|v\|_{-k}, \quad \text{where } |\nu| = L. \tag{4.21}$$

Now since $|\cdot|_0 \leqq |\nu|!^{1+\varepsilon'}\, (cr_0^{-1}/n)^{|\nu|}$, we have the following type estimate (except for some polynomial in n):

$$e^{-|p+q|}(L^{1+2\varepsilon'}\, (cr_0^{-1})^2/n)^L. \tag{4.22}$$

This quantity is estimated by

$$e^{-|p+q|}\exp(-c'\, r_0^2\, n^{\,1/1+2\varepsilon'}). \tag{4.23}$$

Since $1/1+2\varepsilon' > 1/s$, this term is negligible compared with $\exp(-c(A)r_0 n^{1/s})$. It is the same with the second term. □

NOTE: We say that a sequence c_n is *negligible* when it is estimated in the form

$$c_n \leqq \exp(-L\, r_0\, n^{1/s}) \tag{4.24}$$

when n tends to infinity. Here L can be taken as large as we wish, when if necessary by taking r_0 small. The above proof shows that all the quantity in Lemma 4.4 are *negligible*.

Finally we show the following lemma, which corresponds to Lemma 3.5.

LEMMA 4.5: The following

i) $$\sum_{|p+q|\leqq N} c^n_{pq} \|\chi_T(t) k^{(p)}_{n(q)}\, v_n\|_E \leqq c_0 \exp(-(T-\delta)r_0\, n^{1/s}) n^k \|v\|_{-k} \times (\text{some polynomial in } n),$$

ii) $$\sum_{|p+q|\leqq N} c^n_{pq} \|\chi_0(t) k_n{}^{(p)}{}_{(q)}\, \chi_T(t) u\|_E \leqq c_1 \exp(-(T-\delta)r_0\, n^{1/s}) n^k \times (\text{some polynomial in } n) + (\text{negligible term}).$$

PROOF: Part i) is the same as that of Lemma 3.5. The second part is proved in the following way. We can assume $\chi_T(t)$ $(= \chi_0'(t))$ is of Gevrey class $1+\varepsilon'$.

$$|\chi_{T(\nu)}(t)| \leqq \nu!^{\,1+\varepsilon'} (c'\,\delta^{-1})^\nu \quad \text{with } 1 + 2\varepsilon' < s.$$

Without loss of generality, we can assume $\delta \geqq r_0$. Then we look at the proof of Lemma 3.5. Then for $|\nu| \leqq L$,

$$\|c^n_{pq}\, \nu!^{-1}\, \chi_{T(\nu)}(t)\, \partial^\nu_\tau k^{(p)}_{n(q)}\, (\tilde{\chi}_0(t)u)\|_E$$

is estimated by

$$\exp(-r_0(T-\delta)n^{1/s})\, \|c^n_{pq}\, \nu!^{-1}\, \chi_{T(\nu)}(t)\, \partial^\nu_\tau k^{(p)}_{n(q)}\, (\tilde{\chi}_0(t)u)\|,$$

and this is estimated by

$$e^{-|p+q|}\nu!^{\varepsilon'}\ (c'\delta^{-1})\ \nu!^{1+\varepsilon'}(cr_0^{-1}/n)^{\nu}\ n^k\ \|\tilde{\chi}_0(t)u\|_{-k}.$$

This is estimated again by

$$e^{-|p+q|}\ (c'\delta^{-1}\ cr_0^{-1}\ \nu^{1+2\varepsilon'}/n)^{\nu}\ n^k\ \|\tilde{\chi}_0(t)u\|_{-k}.$$

Finally, $c_{pq}^n \|s_{L,pq}(\tilde{\chi}_0(t)u)\|$ is estimated, in view of the above estimate, by

$$e^{-|p+q|}\ \exp(-\gamma n^{1/1+2\varepsilon'}),\quad (^{\exists}\gamma > 0). \qquad \square$$

Now from (4.20), and using (3.27), we arrived at our desired result:

$$\begin{aligned}
&\|\chi_{t_0}(t)k_n v\| \\
&\leqq \exp((t_0+r_0)r_0\ n^{1/s})[\exp(-c(A)r_0\ n^{1/s}) + \exp(-(T-\delta)r_0\ n^{1/s})] \times \\
&\qquad \times\ (\text{some polynomial in } n) + (\text{negligible term}) + \qquad\qquad (4.25)\\
&+ \exp((t_0+r_0)r_0\ n^{1/s})[\sum_{|p+q|\leqq N} c_{pq}^n \|\chi_\delta(t)k_{n(q)}^{(p)}\ v_n\| + \\
&\qquad + \sum_{|p+q|\leqq N} c_{pq}^n \|\chi_0(t)k_{n(q)}^{(p)}(\chi_0(t)f\|].
\end{aligned}$$

Let us recall the definition of $\chi_\delta(t)$ (see (3.24)). In the Appendix we shall show that the last term under Σ is negligible. Hence

<u>PROPOSITION 4.6</u>: Let $0 < t_0 < T$. Then provided that $t_0 < c(A)$, by taking r_0 small, it holds

$\|\chi_{t_0}(t)k_n v\| \leqq \exp(-\varepsilon n^{1/s})$ for n large $(^{\exists}\varepsilon > 0)$. Hence

$(t_0,x(t_0);\tau(t_0),\xi(t_0)) \notin WF_s(u)$.

APPENDIX

We explain briefly some basic facts about our microlocalizers.

1) C^∞ case. Let $(x_0,\xi^0) \in R^\ell \times S^{\ell-1}$, we say that $\{\gamma_n(\xi) = \gamma(\xi/n), \zeta(x)\}$ is a microlocalizer of size r around (x_0,ξ^0), if it satisfies

1) $\gamma(\xi) \in C_0^\infty$, $\gamma(\xi) = 1$ for $|\xi - \xi^0| \leq r/2$, and $= 0$ for $|\xi - \xi^0| \geqq r$.

2) $\zeta(x) = 1$ for $|x-x_0| \leq r/2$, and $= 0$ for $|x-x_0| \geqq r$.

2) Gevrey case. We say that $\{\gamma_n(\xi), \zeta_n(x)\}$ is a microlocalizer of size r around (x_0,ξ^0) in the sense of $\gamma^{(s)}$ $(1 < s < \infty)$, if it satisfies, besides (A.1),

1) $|\gamma_n^{(\nu+\nu')}(\xi)| \leq (\frac{N}{n} cr^{-1})^{|\nu|} \nu'!^{1+\varepsilon'} (cr^{-1}/n)^{|\nu'|}$ for $|\nu| \leq N$,

where $N = (ce)^{-1} r\, n^{1/s}$, and $\varepsilon'(> 0)$ satisfies $1 + 2\varepsilon' < s$. Let us remark that c can be considered as an absolute constant.

2) $|\zeta_{n(\mu+\mu')}| \leq (Ncr^{-1})^{|\mu|} \mu'!^{1+\varepsilon'} (cr^{-1})^{|\mu'|}$ for $|\mu| \leq N$.

3) Real analytic case. We say that $\{\gamma_n(\xi), \zeta_n(x)\}$ is a microlocalizer in analytic sense of size r around (x_0,ξ^0), if it satisfies, besides (A.1),

1) $|\gamma_n^{(\nu+\nu')}(\xi)| \leq (\frac{N}{n} cr^{-1})^{|\nu|} c_{\nu'} r^{-|\nu'|}$ for $|\nu| \leq N$.

2) $|\zeta_{n(\mu+\mu')}(x)| \leq (Ncr^{-1})^{|\mu|} c_{\mu'}$, for $|\mu| \leq N$,

where c is the same as in Gevrey case, and $N = \frac{1}{2} \ell^{-1} (ce)^{-2} r^2 n$.

Using these microlocalizers, we consider microlocal energy $\|\gamma_n(D)\zeta(x)u\|$ for $u \in \mathcal{D}'$, where $\| \cdot \|$ denotes L^2-norm. We can show the following relation of the microlocal energy with the wave front sets.

LEMMA A.1: Let $u \in \mathcal{D}'$. Then

1) $(x_0,\xi^0) \notin WF(u)$ is equivalent to the following assertion:
$\|\gamma_n(D)\zeta(x)u\|$ is rapidly decreasing when n tends to infinity, provided that

we choose the size r of the microlocalizer $\gamma_n(D)\zeta(x)$ sufficiently small.

2) $(x_0,\xi^0) \notin WF_s(u)$ is equivalent to the following: We have

$$\|\gamma_n(D)\zeta(x)u\| \leqq \exp(-\varepsilon n^{1/s}) \quad (^\exists\varepsilon > 0), \tag{A.2}$$

(A.2)

when n tends to infinity, provided that we choose the size of microlocalizer in Gevrey sense sufficiently small.

3) $(x_0,\xi^0) \notin WF_A(u)$ is equivalent to the following: We have

$$\|\gamma_n(D)\zeta(x)u\| \leqq \exp(-\varepsilon n) \quad (^\exists\varepsilon > 0), \tag{A.3}$$

when n tends to infinity, provided that we choose the size of microlocalizer in analytic sense sufficiently small.

<u>NOTE 1</u>: In [4], we showed these in indirect form. The above improvement is due essentially to Y. Takei.

<u>NOTE 2</u>: In applications, it is convenient to use generalized microlocalizers. This is defined replacing (A.1) by

$$\text{supp}[\gamma_n(\xi)] \subset \{\xi; |\xi-n\xi^0| \leqq nr\}, \quad \text{supp}[\zeta(x)] \subset \{x; |x-x_0| \leqq r\}. \tag{A.1$'$}$$

For generalized microlocalizers, we can claim only one direction of the assertion in Lemma A.1. Namely

$$(x_0,\xi^0) \notin WF(u) \text{ implies } \|\gamma_n \zeta u\| \text{ is rapidly decreasing,}$$

$$(x_0,\xi^0) \notin WF_s(u) \text{ implies } \|\gamma_n \zeta u\| \leqq \exp(-\varepsilon n^{1/s}),$$

$$(x_0,\xi^0) \notin WF_A(u) \text{ implies } \|\gamma_n \zeta u\| \leqq \exp(-\varepsilon n).$$

What about the smallness of the size r in the above lemma? This depends on each u. However we can claim the following semi-global version of Lemma A.1.

Let K be a compact set in $R^\ell \times S^{\ell-1}$ such that $K \cap WF(u) = \phi$.

LEMMA A.2: Under the above assumption, there exists $r_0(> 0)$ and $n(s)$ such that for any generalized microlocalizer $\{\gamma_n, \zeta\}$ of size r_0 around any $(x_0, \xi^0) \in K$ for any s, it holds

$$\|\gamma_n(D)\zeta(x)u\| \leq n^{-s} \text{ for } n \geq n(s),$$

provided that we restirct $\{\gamma_n(\xi), \zeta(x)\}$ to a family of microlocalizers satisfying the following type uniform estimates:

$$|\gamma_n^{(\mu)}(\xi)| \leq c_\mu r_0^{-|\mu|}/n^{|\mu|}, \quad |\zeta_{(\nu)}(x)| \leq c'_\nu r_0^{-|\nu|} \text{ for all } \mu, \nu \geq 0.$$

PROOF: We use the Heine-Borel theorem.

NOTE 3: Similar results hold in Gevrey and analytic cases.

We can generalize the above microlocalizers or generalized microlocalizers. In this article we used the local energy of the form $\|\Lambda_n(x,D)u\|$. In this case we assume $u \in H^{-k}$ ($\exists$ k). We call $\Lambda_n(x,D)$ is a microlocalizer of size r_0 around (x_0, ξ^0) if it satisfies

1) $\Lambda_n(x,\xi) \in C_0^\infty$ and $\Lambda_n(x,\xi) = 1$ for $(x,\xi) \in \{|x-x_0| \leq r_0/2\} \times \{|\xi - n\xi^0| \leq nr_0/2\}$.

2) $\text{supp}\, \Lambda_n(x,\xi) \subset \{|x-x_0| \leq r_0\} \times \{|\xi - n\xi^0| \leq nr_0\}$, (A.4)

$$|\Lambda_n{}^{(\alpha)}{}_{(\beta)}(x,\xi)| \leq c_{\alpha\beta}\, n^{-|\alpha|}.$$

In the case where we assume only 2), we call Λ_n generalized microlocalizer. We obtain

LEMMA A.3: Let $(x_0, \xi^0) \notin WF(u)$. Then there exists $\tilde{r}(> 0)$ such that, for any generalized microlocalizer $\Lambda_n(x,D)$ of size $r_0 \leq \tilde{r}$ around (x_0, ξ^0), it holds

$$\|\Lambda_n(x,D)u\| \text{ is rapidly decreasing.}$$

PROOF: By Lemma A.1, there exists a microlocalizer $\{\gamma_n(D), \zeta(x)\}$ around

(x_0,ξ^0) such that $\|\gamma_n(D)\zeta(x)u\|$ is rapidly decreasing. We choose $\Lambda_n(x,\xi)$ in such a way that $\Lambda_n(x,\xi) \subset\subset \zeta(x)\gamma_n(\xi)$. Then using the decomposition,

$$\Lambda_n u = \Lambda_n \gamma_n(D)u = \Lambda_n \gamma_n(D)\zeta(x)u + \Lambda_n\gamma_n(D)(1-\zeta(x))u, \tag{A.5}$$

we see the desired property.

Conversely

LEMMA A.4: Assume $\|\Lambda_n u\|$ is rapidly decreasing for some microlocalizer. Then $(x_0,\xi^0) \notin WF(u)$.

PROOF: We choose a microlocalizer $\{\zeta(x), \gamma_n(D)\}$ around (x_0,ξ^0) in such a way that $\zeta(x)\gamma_n(\xi) \subset\subset \Lambda_n(x,\xi)$. Then in view of the expression

$$\gamma_n(D)\zeta(x)u = \gamma_n(D)\zeta(x)\Lambda_n(x,D)u + \gamma_n(D)\zeta(x)\ (1-\Lambda_n(x,D))u, \tag{A.6}$$

we see easily the desired property.

Similar results hold in the case of Gevrey wave front sets. We assume for the derivatives, for instance,

$$|\Lambda_{n(\nu)}^{(\mu+\mu')}(x,\xi)| \leq (Ncr_0^{-1})^{|\mu|}\ \mu'!^{\ 1+\varepsilon'}(cr_0^{-1}/n)^{|\mu'|}\nu!^s\ C_0^{|\nu|}\ , \tag{A.7}$$

where N is the same number defined at the beginning in the case of Gevrey, it is the same with $1+\varepsilon'$.

LEMMA A.5: Let $u \in H^{-k}$ ($^\exists$ k), and $(x_0,\xi^0) \notin WF_s(u)$. Then there exists $\tilde{r}$ (> 0) such that for any generalized microlocalizer in Gevrey sense of size $r_0 \leq \tilde{r}$, it holds

$$\|\Lambda_n(x,D)u\| \leq \exp(-\varepsilon n^{1/s}) \quad (^\exists\ \varepsilon > 0). \tag{A.8}$$

PROOF: The proof is the same as that of Lemma A.3. Look at (A.5).

$$\Lambda_n\gamma_n(D)(1-\zeta(x)) = \sum_{|\mu|\leq N} \mu!^{-1}(1-\zeta(x))_{(\mu)}(\Lambda_n\gamma_n(D))^{(\mu)}+r_{N,n}(x,D). \tag{A.9}$$

The first term under Σ is zero operator, $\|r_{N,n}(x,D)u\|$ is estimated except for a trivial factor by

$$(Ncr_0^{-1})^N(ce\ell r_0^{-1}/n)^N \|u\|_{-k} = (\frac{N}{n} c^2 e\ell r_0^{-2})^N.$$

Since $N = (ce)^{-1} r_0 n^{1/s}$ and $1/s < 1$, we see that this term is negligible. In view of Lemma A.1,2), we see the desired property.

<u>LEMMA A.6</u>: Assume (A.8) for some microlocalizer Λ_n. Then $(x_0,\xi^0) \notin WF_s(u)$.

<u>PROOF</u>: The proof is the same as that of Lemma A.4. Look at (A.6).

$$\gamma_n(D)\zeta(x)(1-\Lambda_n(x,D))u = \sum_{|\mu|\leqq N} |\mu|^{-1}(\zeta(x)(1-\Lambda_n(x,D))_{(\mu)}\gamma_n^{(\mu)}u \qquad \text{(A.10)}$$

$$+ s_{N,n}(x,D)u.$$

The first term is zero. Now we make precise the sizes of two microlocalizers. Let r_0 be the size of Λ_n. We may choose that of $\{\zeta(x),\gamma_n(D)\}$ as $r_0/2$. Then $(\zeta(x)(1-\Lambda_n(x,\xi))_{(\mu)}$ is estimated by $C_0\ \mu!^s$. This can be seen in the following way. $|\zeta(x)_{(\mu)}| \leqq \mu!^{1+\varepsilon'}(2cr_0^{-1})^{|\mu|}$ $(1+\varepsilon' < s)$. $|\Lambda_{n(\mu)}| \leqq \mu!^s C_0^{|\mu|}$. Since $\mu!^{1+\varepsilon'}(2cr_0^{-1})^{|\mu|} << \mu!^s\ C_0^{|\mu|}$, by Leibniz we get the above estimate. Hence $\|s_{N,n}(x,D)u\|$ is estimated except for a trivial factor by

$(N^{s-1}C_0)^N (\frac{N}{n} 2cr_0^{-1})^N \|u\|_{-k} \leqq (N^s 2C_0 cr_0^{-1}/n)^N \|u\|_{-k}$. Since $N^s = (ce)^{-s} r_0^s\ n$,

the last term becomes $((ce)^{-s}2C_0 c\ r_0^{s-1})^N \|u\|_{-k}$.

Thus if we choose r_0 small, this term becomes negligible.

Finally we return to actual case. We use

$$\chi_0(t)k_n(t,x;D_t,D_x) = \chi_0(t)\beta(t,x)\alpha_n(t;D_t,D_x), \quad \text{where}$$

$$\beta\alpha_n(t,x;\tau,\xi) = \beta_0(x-x(t))\chi_N(\frac{\tau}{n} - \tau(t), \frac{\xi}{n} - \xi(t)).$$

Let us notice that t plays a specific role. Especially, in the case of Gevrey and analytic case, $(x(t),\tau(t),\xi(t))$ belongs to the same Gevrey class s and analytic class respectively, and its nature (for example, radius of convergence) depends only on $\lambda(t,x;\xi)$.

The above microlocalizer is not properly microlocalizer of the form $\Lambda_n(x,D)$ that we considered here. For this purpose, we use a partition of unity in t: $\sum_j \chi_{t_j}(t) = 1$ and consider

$$\chi_0(t)k_n = \sum_j (\chi_{t_j}(t)\chi_0(t))k_n.$$

Then for each term under Σ we can apply Lemmas stated above. We see that several facts itted in the proof of theorems can be deduced easily from the proofs of lemmas.

References

[1] J. Chazarain, Propagation des singularités pour une classe d'opérateurs à caractéristiques multiples et résulubilité locale, Ann. Inst. Fourier 24 (1974), 203-223.

[2] L. Hörmander, Uniqueness theorems and wave front sets for solutions of linear differential equations with analytic coefficients, Comm. Pure Appl. Math. 24 (1971), 671-704.

[3] S. Mizohata, Propagation des singularités au sense Gevrey pour les opérateurs différentiels à multiplicité constante, Séminaire sur les équations aux dérivées partielles de J. Vaillant (1982-83), 106-133, Hermann, 1984.

[4] S. Mizohata, On the Cauchy problem, Academic Press, 1986.

[5] L. Rodino and L. Zanghirati, Pseudodifferential operators with multiple characteristics and Gevrey singularities, Comm. in PDE, 11 (1986), 673-711.

[6] K. Taniguchi, Fourier integral operators in Gevrey class on R^n and the fundamental solution for a hyperbolic operator, Pub. RIMS, Kyoto Univ. 20 (1984), 491-542.

Sigeru Mizohata
Faculty of Engineering
Osaka Electro-Communication University
18-8 Hatsu-cho, Neyagawa,
Osaka 572
Japan.

Y. MORIMOTO

Propagation of wave front sets and hypoellipticity for degenerate elliptic operators

INTRODUCTION

In [K-S], Kusuoka-Strook gave a sufficient condition of hypoellipticity for degenerate elliptic operators of second order, as an application of the Malliavin calculus (see Theorem 8.13 of [K-S]). Their condition is applicable even to infinitely degenerate elliptic operators which do not satisfy the famous sufficient condition given by Hörmander [Hr]. One of the remarkable results by means of their method is as follows: A differential operator $L_0 \equiv D_t^2 + D_{x_1}^2 + \exp(-1/|x_1|^\sigma)D_{x_2}^2$. $\sigma > 0$, is hypoelliptic in R^3 if and only if $\sigma < 1$. Non-hypoellipticity of L_0 when $\sigma \geqq 1$ is quite interesting because it was already known by Fediǐ [Fd] that $A_0 \equiv D_{x_1}^2 + \exp(-1/|x_1|^\sigma)D_{x_2}^2$ $(= L_0 - D_t^2)$ is hypoelliptic in R^2 for any $\sigma > 0$.

In an aspect on the microlocal analysis, the difference of hypoellipticity between L_0 and A_0 seems to be interpreted from the fact that L_0 with $\sigma \geqq 1$ is hyperbolic with respect to D_t in a certain sense. Following "the uncertainty principle" discussed in Fefferman-Phong [F-P], [Ff], we consider a small box

$$B_0 = \{(x,\xi) \in T^*R_x^2;\ |x| \leqq \delta,\ \ |\xi-\xi_0| \leqq \delta^{-1}\},\ \delta > 0$$

where $\xi_0 = (0,\eta)$ and $|\eta|$ is sufficiently large. Set $\delta = (2\log|\eta|)^{-1/\sigma}$ and assume $\sigma \geqq 1$. Then the symbol $a_0(x,\xi)$ of A_C satisfies

$$0 \leqq a_0(x,\xi) \leqq 10(\log\langle\xi\rangle)^2 \text{ on } B_0,\ (\langle\xi\rangle^2 = 1 + |\xi|^2).$$

Note that the Cauchy problem for $\tilde{L}_0 = D_t^2 + 10(\log\langle D_x\rangle)^2$

$$\begin{cases} \tilde{L}_0 u = 0, & t \in [0,T] \\ u|_{t=0} = u_0(x), & \partial_t u|_{t=0} = u_1(x), \end{cases}$$

is H^∞ well-posed. Since L_0 is majorated by $\tilde{L}_0$ in a small tube neighbourhood of the null-bicharacteristic curve of D_t passing through $(0,(0,\xi_0))\in T^*(R^3_{t,x})\setminus 0$, one may expect the propagation of wave front sets of solutions of $L_0u = 0$ along this null-bicharacteristic curve.

The above argument is, of course, crude. In what follows, we shall show the existence of a singular solution of $L_0u = 0$ when $\sigma \geqq 1$, by using the similar method as in Métivier [Me], where nonanalytic hypoellipticity of a class of differential operators of the form $D_t^2 + A(x,D_x)$ was studied. (cf. Baouendi-Goulaouic [B-G]). Before constructing the singular solution, we also give a sufficient condition for hypoellipticity, in order to prove the hypoellipticity of L_0 when $0 < \sigma < 1$.

1. CRITERIA FOR HYPOELLIPTICITY

Let $P = p(x,D_x)$ be a differential operator of order $m \geqq 1$ with coefficients in $C^\infty(R^n_x)$, that is,

$$p(x,D_x) = \sum_{|\alpha|\leqq m} a_\alpha(x)D_x^\alpha, \quad a_\alpha(x) \in C^\infty(R^n_x),$$

where for multi-index $\alpha = (\alpha_1,\ldots,\alpha_n)$, $|\alpha| = \alpha_1 + \ldots + \alpha_n$, $D_x^\alpha = D_1^{\alpha_1} \ldots D_n^{\alpha_n}$ and $D_j = -i\partial_{x_j}$.

We say that P is *hypoellitpic* (C^∞-*hypoelliptic*) *in* R^n if for any $u\in\mathcal{D}'(R^n)$ and for any open set Ω of R^n, $Pu \in C^\infty(\Omega)$ implies $u \in C^\infty(\Omega)$. Let Λ and $\log \Lambda$ be pseudodifferential operators with symbols $\langle\xi\rangle$ and $\log\langle\xi\rangle$, respectively. We write $p^{(\alpha)}_{(\beta)}(x,\xi) = \partial_\xi^\alpha D_x^\beta p(x,\xi)$ for multi-indices α and β. We set $\|u\|_s = \|\Lambda^s u\|$ for real s and $u \in C_0^\infty(R^n)$, where $\|\cdot\|$ denotes the usual L^2 norm.

THEOREM 1: Assume that for any $x_0 \in R^n$ there exists a neighbourhood ω of x_0 satisfying the following property: For any $\varepsilon > 0$ estimates

$$\|(\log \Lambda)^m u\| \leqq \varepsilon\|Pu\| + C_\varepsilon\|u\|, \quad u \in C_0^\infty(\omega), \tag{1}$$

$$\sum_{0<|\alpha+\beta|<m} \|(\log \Lambda)^{|\alpha+\beta|} p^{(\alpha)}_{(\beta)} u\|_{-|\beta|} \leqq \varepsilon \|Pu\| + C'_\varepsilon \|u\|, \quad u \in C_0^\infty(\omega), \tag{2}$$

hold with constants C_ε and C'_ε. Here $P^{(\alpha)}_{(\beta)} = p^{(\alpha)}_{(\beta)}(x,D_x)$. Then P is hypoelliptic in R^n. Furthermore we have

$$WF\ Pv = WF\ v \text{ for any } v \in \mathcal{D}'(R^n). \tag{3}$$

Let L be a differential operator of second order with C^∞-coefficients, that is,

$$L = \sum_{j,k} a_{jk}(x)D_jD_k + \sum_j ib_j(x)D_j + c(x),\ D_j = -i\partial_{x_j}$$

We assume that

$$\begin{cases} a_{jk} \text{ and } b_j \text{ are real valued,} \\ \sum a_{jk}(x)\xi_j\xi_k \geqq 0 \text{ for all } (x,\xi) \in R^{2n} \end{cases}$$

COROLLARY 1: Let L be the above differential operator of second order. If for any $x_0 \in R^n$ there exists a neighbourhood ω of x_0 such that for any $\varepsilon > 0$ the estimate

$$\|(\log \Lambda)^2 u\| \leqq \varepsilon\|Lu\| + C_\varepsilon \|u\|, \quad u \in C_0^\infty(\omega) \tag{4}$$

holds with a constant C_ε then we have (3).

About the proof of Theorem 1 (and Corollary 1) we refer to [M_6].

COROLLARY 2: Let L be the above operator. If for any $x_0 \in R^n$ there exists a neighbourhood ω of x_0 such that for any $\varepsilon > 0$ the estimate

$$\|(\log \Lambda)u\|^2 \leqq \varepsilon\, \mathrm{Re}(Lu,u) + C_\varepsilon \|u\|^2, \quad u \in C_0^\infty(\omega) \tag{5}$$

holds with a constant C_ε then we have (3).

PROOF: The estimate (4) easily follows by substituting $(\log \Lambda)u$ into (5).
Q.E.D.

By this corollary, we can prove the hypoellipticity of L_0 with $\sigma < 1$ if we only check (5) (see the next section).

We remark that the estimate (4) is not always necessary for second order differential operator to be hypoelliptic. We can easily see that the hypoelliptic operator $A_0 \equiv D^2_{x_1} + \exp(-1/|x_1|^\sigma)D^2_{x_2}$, $\sigma > 0$, given by Fediĭ [Fd], does not satisfy (4) when $\sigma \geqq 1$. However, the estimate (4) is necessary to be hypoelliptic for a class of differential operators, (for example, $D_t^2 + A_0$). The result concerning the necessity of (4) can be discussed for some class of operators of higher order. Let m be even positive integer and let P_0 be a differential operator of the form

$$P_0 = D_t^m + A(x,D_x) \text{ in } R_t \times R_x^n, \tag{6}$$

where $A(x,D_x)$ is a differential operator of order m with C^∞-coefficients. We assume that $A(x,D_x)$ is formally self-adjoint in Ω and bounded from below, that is, there exists a real c_0 such that

$$(A(x,D_x)u,u) \geqq c_0 \,\|u\|^2 \tag{7}$$

for $u \in L^2(\Omega)$ satisfying $Au \in L^2(\Omega)$.

<u>THEOREM 2</u>: Let P_0 be the above operator. Assume that P_0 is hypoelliptic in $R_t \times \Omega$. Then for any $x_0 \in \Omega$ there exists a neighbourhood ω of x_0 such that for any $\varepsilon > 0$ the estimate

$$\|(\log \Lambda)^{m/2}u\|^2 \leqq \varepsilon \operatorname{Re}(P_0u,u) + C_\varepsilon \|u\|^2, \ u \in C_0^\infty(R_t \times \omega) \tag{8}$$

holds with a constant C_ε. Here Λ, of course, denotes $\langle D_t,D_x\rangle = (1+D_t^2+|D_x|^2)^{1/2}$.

When m = 2 the estimate (4) follows from (8). Using (8) (i.e. (5)) we can prove non-hypoellipticity of L_0 with $\sigma \geqq 1$ by means of the reduction to the absurdity.

From now on we shall mention the outline of the proof of Theorem 2, which is a slight modification of the method given in [Me]. Following [Me], for $s \geqq 1$ we introduce $G^s(\bar\Omega;A)$ the space of $L^2(\Omega)$ such that $A^ku \in L^2(\Omega)$ for $k = 1,2,\ldots$ and moreover there exists a constant M satisfying

$$\|A^k u\|_{L^2(\Omega)} \leqq M^{k+1}(k!)^{sm}, \quad k = 1,2,\dots . \tag{9}$$

We also introduce the space $G^s(\Omega;A)$ of $u \in L^2_{loc}(\Omega)$ whose restriction in any $\Omega_1 \subset\subset \Omega$ is in $G^s(\bar{\Omega}_1;A)$.

<u>PROPOSITION 1</u>: Assume that $G^1(\Omega;A) \not\subset C^\infty(\Omega)$. Then P_0 is not hypoelliptic in $R_t \times \Omega$.

<u>PROOF</u>: There exists a $u_0 \in G^1(\Omega;A)$ such that $u_0 \notin C^\infty(\Omega)$. The series

$$u(t,x) = \sum_{k=0}^{\infty} (it)^{mk}(-A)^k u_0(x)/(mk)! \tag{10}$$

is strongly convergent in $L^2(\tilde{\Omega})$ for some $\tilde{\Omega} = I_\delta \times \Omega_1$, where $I_\delta = (-\delta,\delta) \subset R_t$ and $\Omega_1 \subset\subset \Omega$. We have $P_0 u = 0$ and u is not C^∞ in $\tilde{\Omega}$ because $u_0 = u(0,x)$ is not C^∞ in Ω. Q.E.D.

Note that for any open set $\omega \subset\subset \Omega$

$$\mathrm{Re}(P_0 u,u) = \|D_t u\|^2 + (Au,u), \; u \in C_0^\infty(R_t \times \omega).$$

In view of Proposition 1, for the proof of Theorem 2 it suffices to show

<u>PROPOSITION 2</u>: Assume that $G^1(\Omega;A) \subset C^\infty(\Omega)$. Then for any $x_0 \in \Omega$ there exist a neighbourhood of ω of x_0 such that for any $\varepsilon > 0$ the estimate

$$\|(\log\langle D_x\rangle)^{m/2} u\|^2 \leqq \varepsilon(Au,u) + C_\varepsilon \|u\|^2, \quad u \in C_0^\infty(\omega) \tag{11}$$

holds with a constant C_ε (cf. Theorem 3.5 of [Me]).

We remark that almost the same result as this proposition was obtained independently by Hoshiro [Hs]. In the proof of Proposition 2 we may replace A by $A+\mu$ for any real μ because $G^1(\Omega;A) = G^1(\Omega;A+\mu)$. Taking a large $\mu > 0$, from (7) we may assume that $(Au,u) > 0$ for $u \in L^2(\Omega)$ satisfying $Au \in L^2(\Omega)$. Therefore, we have the Friedrichs extension $(A,D(A))$ in $L^2(\Omega)$ of $A(x,D_x)$, as a positive self-adjoint realization. For the proof of (11) it suffices to show that for any $\varepsilon > 0$ and any $r > 0$ there exists a $C_{\varepsilon,r}$ such that

$$\|(\log\langle D_x\rangle)^{mr}u\|^2 \leqq \varepsilon\|A^r u\|^2 + C_{\varepsilon,r}\|u\|^2, \quad u \in C_0^\infty(\omega). \tag{12}$$

The proof of (12) is performed by the same procedure as in [Me] p. 840-848, except for modifying Lemma 3.2 in [Me]. For the detail, we refer to Section 2 of [M_5].

2. APPLICATIONS

As stated in the preceding section, hypoellipticity and non-hypoellipticity of L_0 can be seen by Corollary 2 and Theorem 2, respectively, if we show that the condition $0 < \sigma < 1$ is sufficient and necessary for L_0 to satisfy (5). Instead of checking (5) for L_0 we shall consider a slightly more complicated example than L_0 and study its hypoellipticity.

Let P_1 be a differential operator of the form

$$P_1 = D_t^2 + D_x^2 + x^{2\ell}D_y^2 + D_z(g(y) + z^2)D_z \quad \text{in } R^4,$$

where $g \in C^\infty$ satisfies $g^{(j)}(0) = 0$ for any j and $g(y) > 0$ $(y \neq 0)$.

PROPOSITION 3: The above operator P_1 is hypoelliptic in R^4 if $g(y)$ satisfies

$$\lim_{y\to 0} |y|^{1/(\ell+1)}|\log g(y)| = 0. \tag{13}$$

Assume in addition that $yg'(y) \geqq 0$, that is, $g(y)$ and $g(-y)$ are non-decreasing in $[0,\infty)$. Then (13) is necessary for P_1 to be hypoellitpic in R^4.

For the first part of this proposition we use Corollary 2. Note

$$\begin{aligned}\mathrm{Re}(P_1u,u) &= \|D_tu\|^2 + \|D_xu\|^2 + \|x^\ell D_yu\|^2 \\ &\quad + (g(y)D_z^2u,u) + \|zD_zu\|^2, \quad u \in C_0^\infty.\end{aligned} \tag{14}$$

From Hörmander's classical theorem in [Hr] and its sharp version given in [R-S] and [F-P] we have

$$C(\mathrm{Re}(P_1u,u) + \|u\|^2) \tag{15}$$

$$\geq \|D_xu\|^2 + \|x^{\ell}D_yu\|^2 + (g(y)D_z^2u,u) + \|u\|^2$$

$$\geq \||D_y|^{1/(\ell+1)}\hat{u}\|^2 + (g(y)\zeta^2\hat{u},\hat{u}), \quad u \in C_0^\infty,$$

where $\hat{u}(\cdot,\zeta)$ denotes the Fourier transform of $u(\cdot,z)$. Set $M = |\zeta|$ and $V(y) = g(y)M^2$. Then, in view of (15), for the proof of (5) it suffices to show that for any integer $k > 0$ there exists a $M_k > 0$ such that for a $c > 0$

$$\||D_y|^{1/(\ell+1)}u\|^2 + (V(y)u,u) \tag{16}$$

$$\geq c(k \log M)^2 \|u\|^2, \quad u \in C_0^\infty(R^1),$$

$$\text{if } M \geq M_k.$$

To derive this we prepare the following theorem (cf. Theorem B in [M_7]). Let λ be $0 < \lambda \leq 1$. We consider a symbol of the form

$$a(x,\xi) = |\xi|^{2\lambda} + V(x), \ x \in R^n,$$

where $V(x)$ is a non-negative measurable function. Let C denote a set of boxes

$$B \equiv \{(x,\xi) \in R^{2n};\ |x_j - x_{0j}| \leq \delta/2,\ |\xi_j - \xi_{0j}| \leq \delta^{-1}/2\} \tag{17}$$

for all $(x_0,\xi_0) \in R^{2n}$ and all $\delta > 0$. Clearly, the volume of $B \in C$ is equal to 1.

THEOREM 3: Assume that there exists a $c > 0$ such that for any $B \in C$

$$m(\{(x,\xi) \in B;\ a(x,\xi) \geq R\}) \geq c, \tag{18}$$

where $m(.)$ denotes Lebesgue measure. Then we have

$$\||D_x|^{\lambda}u\|^2 + (V(x)u,u) \geq c'R\|u\|^2, \quad u \in C_0^\infty(R^n), \tag{19}$$

where $c' > 0$ depends only on c and n.

For the proof of Theorem 3 in case of $\lambda = 1$ we use a modification of the lemma given by Fefferman [Ff] as follows:

LEMMA 1: Assume that $V(x) \geqq 0$, measurable on a cube Q in R^n. Suppose that there exists a $c > 0$ such that

$$m(\{x \in Q;\ V(x) \geqq c(\text{diam } Q)^{-2}\}) \geqq c|Q|. \tag{20}$$

Then for $u \in C^1$ we have

$$\int_Q \{|\nabla u(x)|^2 + v(x)|u(x)|^2\}dx \geqq c'(\text{diam } Q)^{-2} \int_Q |u(x)|^2 dx. \tag{21}$$

The constant $c' > 0$ depends only on n and c.

In the Main Lemma of [Ff, p. 146], it is assumed that $V(x)$ is polynomial and $(Av_Q\ V) \geqq (\text{diam } Q)^{-2}$. The proof in [Ff] is still valid under the above hypotheses.

For the case $0 < \lambda < 1$ we need the following lemma:

LEMMA 2: Assume that $V(x) \geqq 0$, measurable on a cube Q in R^n. For a $0 < \lambda < 1$, suppose that there exists a $c > 0$ such that

$$m(\{x \in Q;\ V(x) \geqq c(\text{diam } Q)^{-2\lambda}\}) \geqq c|Q|. \tag{20)'$$

Then for $u \in C^1$ we have

$$\int_{Q\times Q} \{|u(x)-u(y)|^2/|x-y|^{n+2\lambda}\}dxdy \tag{21)'$$

$$+ \int_Q V(x)|u(x)|^2 dx \leqq c'(\text{diam } Q)^{-2\lambda} \int_Q |u(x)|^2 dx.$$

PROOF OF THEOREM 3: Let Q_{x_0} be a cube in R^n_x, that is,

$$Q_{x_0} = \{x \in R^n;\ |x_j - x_{0j}| \leqq \delta/2 \quad \text{for some } x_0 \in R^n,\ \delta > 0. \tag{22}$$

If $\delta = 2^{-2}n^{1/2}(2/R)^{1/2\lambda}$ it follows from (18) that

$$m(\{x \in Q_{x_0} ; V(x) \geq R/2\}) \geq c|Q_{x_0}|. \tag{23}$$

In fact, this is obvious if we consider a box $Q_{x_0} \times \{\xi : |\xi_j| \leq n^{-1/2}(R/2)^{1/2\lambda}\}$ belonging to C. Let $\{Q_j\}$ be a partition of R^n_x such that each Q_j is a translation of Q_{x_0} with $\delta = 2^{-2}n^{1/2}(2/R)^{1/2\lambda}$. Note diam $Q_j = 2^{-1}n(2/R)^{1/2\lambda}$. When $\lambda = 1$, in view of (23) we have (19) by Lemma 1 because $\| |D_x|u \|^2 = \|\nabla u\|^2$. In case of $0 < \lambda < 1$, note that

$$\| |D_x|^\lambda u \|^2$$

$$= c \int_{R^n \times R^n} \{|u(x)-u(y)|^2/|x-y|^{n+2\lambda}\}dxdy, \ u \in C_0^\infty(R^n),$$

where c depends only on λ and n. (See [S1]). Since we have

$$\int_{R^n \times R^n} dxdy \geq \sum_j \int_{Q_j \times Q_j} dxdy,$$

the estimate (19) for the case $0 < \lambda < 1$ also follows from Lemma 2. Q.E.D.

We shall prove (16) by using Theorem 3, that is, we shall check the condition (18) for $a(y,\eta) = |\eta|^{2/(\ell+1)} + g(y)M^2$. For brevity, we write x and ξ instead of y and η. It follows from (13) that for any k there exists a $\tau_k > 0$ such that

$$- \log g(x) = |\log g(x)| \leq 1/(10k|x|^{1/(\ell+1)}) \text{ for } |x| \leq \tau_k. \tag{24}$$

Take a large $M_k > 0$ such that

$$M_k^{-1} \leq \min_{|x| \geq \tau_k} g(x) \quad \text{and} \quad \tau_k \geq (10k \log M_k)^{-1}.$$

When $M \geq M_k$ we have

$$V(x) \geq M \text{ for } |x| \geq \tau_k. \tag{25}$$

Set $R \equiv R(M) = (k \log M)^2$ and note that

$$V(x) = M \exp \{\log M - |\log g(x)|\} \tag{26}$$

$$\geqq M \exp \{\log M - 1/(10k|x|^{1/(\ell+1)})\} \geqq M >> R(M)$$

if $(10R)^{-(\ell+1)/2} \leqq |x| \leqq \tau_k$ and $M \geqq M_k$.

Inequalities (25) and (26) show that for Q_{x_0} defined by (22) with $\delta \geqq (8R)^{-(\ell+1)/2}$ we have

$$m(\{x \in Q_{x_0};\ V(x) \geqq M\}) \geqq |Q_{x_0}|/10. \tag{27}$$

From this we see that (18) holds for $B \in C$ with $\delta \geqq (8R)^{-(\ell+1)/2}$. On account of the term $|\xi|^{2/(\ell+1)}$, (18) is trivial for others $B \in C$.

From now on we shall prove the rest of Proposition 3. Suppose (13) is not fulfilled but P_1 is hypoelliptic in R^4. Then there exists a $\delta > 0$ and a sequence $\{s_k\}_{k=1}^{\infty}$ such that

$$\begin{cases} s_k \to 0 & (k \to \infty) \\ g(s_k)\exp(4\delta^{-1}|s_k|^{-1/(\ell+1)}) \leqq 1. \end{cases} \tag{28}$$

Without loss of generality we may assume $s_k > 0$. Set $\lambda_k = \exp(-\delta^{-1} s_k^{-1/(\ell+1)})$. Then it follows from (28) that

$$\lambda_k^{-4} g(s/(\log \lambda_k^{-\delta})^2 \leqq 1 \text{ for } 0 \leqq s \leqq 1, \tag{29}$$

because g is non-decreasing in R_+. By Theorem 2 (and Proposition 2) there exists a neighbourhood of the origin in $R^3_{x,y,z}$ such that for any $\varepsilon > 0$ the estimate

$$\|(\log \Lambda)u\|^2 \leqq \varepsilon \operatorname{Re}(\tilde{A}u,u) + C_\varepsilon \|u\|^2, \quad u \in C_0^\infty(\omega), \tag{30}$$

holds with a constant C_ε, where $\tilde{A} = P_1 - D_t^2$ and $\Lambda^2 = D_x^2 + D_y^2 + D_z^2 + 1$. Let $\phi(s) \in C_0^\infty(R^1)$ be equal to 1 in a neighbourhood of $s = 0$ and to 0 for $|s| > 1/2$. Note

$$\|(\log\langle D_z\rangle)\phi(\lambda_k^2 D_z - 1)u\| \leqq \|(\log \Lambda)u\|^2, \quad u \in C_0^\infty(\omega).$$

In view of (14), from (30) we have

$$\begin{aligned}
&\|(\log \langle D_z\rangle)\phi(\lambda_k^2 D_z - 1)u\| \\
&\leqq \|D_x u\|^2 + \|x^\ell D_y u\|^2 + (g(y)D_z^2 u, u) \qquad (31)\\
&\qquad + \|zD_z u\|^2 + C_\varepsilon \|u\|^2, \qquad u \in C_0^\infty(\omega).
\end{aligned}$$

Let $\psi(s) \not\equiv 0$ be in C_0^∞ and satisfy supp $\psi \subset [0,1]$. Set $v(\tilde{x},\tilde{y},\tilde{z}) = \phi(\tilde{x})\psi(\tilde{y})\phi(\tilde{z})$ and consider the change of variables

$$\tilde{x} = (\log \lambda_k^{-\delta})x, \quad \tilde{y} = (\log \lambda_k^{-\delta})^{(\ell+1)}y,$$

$$\tilde{z} = \lambda_k^{-2}(\log \lambda_k^{-\delta})^{-1}z.$$

Let $u_0(x,y,z)$ denote the function v after the above change of variables. Then we have supp $u_0 \subset \omega$ if λ_k is small enough. Substitute $\exp(i\lambda_k^{-2}z)u_0$ into (31) and take the above change of variables. Then we have

$$\begin{aligned}
&2 \log \lambda_k^{-1} \|\phi((\log \lambda_k^{-\delta})^{-1}D_{\tilde{z}})v\| \\
&\leqq \varepsilon(\delta \log \lambda_k^{-1}(\|D_{\tilde{x}}v\| + \|\tilde{x}^\ell D_{\tilde{y}}v\| + \|\tilde{z}v\|) \\
&\quad + \lambda_k^{-2} \|g^{1/2}(\tilde{y}/(\log \lambda_k^{-\delta})^{(\ell+1)}v\| + \|\tilde{z}D_{\tilde{z}}v\|) + C_\varepsilon \|v\|,
\end{aligned}$$

because a pseudodifferential operator in $R_{\tilde{z}}$ with a symbol $(\log \lambda_k^4 + ((\log \lambda_k^{-\delta})^{-1} \tilde{\zeta}+1)^2\}^{1/2}\phi((\log \lambda_k^{-\delta})^{-1}\tilde{\zeta})$ is L^2-bounded uniformly with respect to λ_k. Note $\phi((\log \lambda_k^{-\delta})^{-1}D_{\tilde{z}})v$ converges v in L^2 when λ_k tends to 0. Then, there exists a $c_0 > 0$ such that

$$\|\phi((\log \lambda_k^{-\delta})^{-1} D_{\tilde{z}}) v\| \geq c_0$$

if $\lambda_k \leqq \lambda_0$ for a sufficiently small λ_0.

Since it follows from (29) that

$$g^{1/2}(\tilde{y}/(\log \lambda_k^{-\delta})^{(\ell+1)}) \leqq 1 \quad \text{on supp } v,$$

there exists constant c_1 and c_2 independent of ε and C'_ε such that

$$2c_0 \log \lambda_k^{-1} \leqq c_1 \varepsilon \log \lambda_k^{-1} + c_2 \varepsilon + C_\varepsilon,$$

if $\lambda_k \leqq \lambda_0$.

Setting $\varepsilon = c_0/c_1$ we have a contradiction when λ_k tends to 0. Thus we have completed the proof of Proposition 3.

We finally remark the hypoellipticity of $\tilde{A} \equiv P_1 - D_t^2$ in $R^3_{x,y,z}$. We do not know whether the condition (13) is necessary. In view of the aspect in the Introduction, it seems to be necessary for the hypoellipticity of $\tilde{A}$ that the condition (13) with $1/(\ell+1)$ replaced by 1 holds, because if it violates $\tilde{A}$ is hyperbolic with respect to D_x in a certain microlocal sense.

References

[B-G] M.S. Baouendi and C. Goulaouic: Nonanalytic hypoellipticity for some degenerate elliptic operators, Bull. Amer. Math. Soc., 78 (1972), 483-486.

[Fd] V.S. Fediǐ: On a criterion for hypoellipticity, Math. USSR Sb., 14 (1971), 15-45.

[Ff] C. Fefferman: The uncertainty principle. Bull. Amer. Math. Soc., 9 (1983), 129-206.

[F-P] C. Fefferman and D.H. Phong: The uncertainty principle and sharp Garding inequalities, Comm. Pure Appl. Math. 34 (1981), 285-331.

[Hr] L. Hörmander: Hypoelliptic second order differential equations, Acta Math., 119 (1967), 147-171.

[Hs] T. Hoshiro: A property of operators characterized by iteration and a necessary condition for hypoellipticity, J. Math. Kyoto Univ. 27 (1987), 401-416.

[K-S] S. Kusuoka and D. Strook: Applications of the Malliavin calculus, Part II, J. Fac. Sci. Univ. Tokyo Sect. IA, Math., 32 (1985), 1-76.

[Me] G. Métivier: Propriété des itéres et ellipticité, Comm. Partial Differential Equations 3 (1978), 827-876.

[M_1] Y. Morimoto: On the hypoellipticity for infinitely degenerate semi-elliptic operators, J. Math. Soc. Japan 30 (1978), 327-358.

[M_2] Y. Morimoto: Non-hypoellipticity for degenerate elliptic operators, Publ. RIMS Kyoto Univ., 22 (1986), 25-30.

[M_3] Y. Morimoto: On a criterian for hypoellipticity, Proc. Japan Acad., 62 A (1986), 137-140.

[M_3] Y. Morimoto: Hypoellipticity for infinitely degenerate elliptic operators, Osaka J. Math., 24 (1987), 13-35.

[M_5] Y. Morimoto: A criterion for hypoellipticity of second order differential operators, Osaka J. Math., 24 (1987), 651-675.

[M_6] Y. Morimoto: Criteria for hypoellipticity of differential operators, Publ. RIMS Kyoto Univ. 22 (1986), 1129-1154.

[M_7] Y. Morimoto: The uncertainty principle and hypoelliptic operators, Publ. RIMS Kyoto Univ. 23 (1987), 955-964.

[R-S] L. Rothchild and E.M. Stein: Hypoelliptic differential operators and nilpotent groups, Acta Math. 137 (1976), 247-320.

[Sl] L.N. Slobodeskii: Generalized spaces of S.L. Sobolev and their application to boundary value problems for partial differential equations, Uch. Zap. Lenin, Gos. Univ. 197 (1958), 54-112.

Yoshinori Morimoto
Institute of Mathematics,
Yoshida College,
Kyoto University,
606 KYOTO
Japan.

M. NACINOVICH

Tangential Cauchy-Riemann systems

§1. The aim of this paper is to give a general survey of some results and questions about the tangential Cauchy Riemann systems, also in the hope that they will shed some light on directions that some investigations upon general overdetermined systems of linear partial differential equations could take in the future.

Let me recall some basic facts and definitions.

Let X be a complex manifold of dimension n. We denote by J: TX → TX the complex structure of X. Let S be a closed, connected C^∞ real submanifold of X, of real codimension k. We note that

$$H_x S = T_S \cap JT_x S$$

is, for x ∈ S, a complex linear subspace of $T_x X$, of dimension greater or equal to n-k. We say that S is *generic* at x if $\dim_{\mathbb{C}} H_x S = n-k$. Obviously, to be generic, S must be of real dimension greater or equal than n. We say simply that S is generic if it is generic at every point. All smooth real hypersurfaces in X are generic, while a closed complex submanifold not open in X is not generic at any point.

Let A denote the sheaf of germs of complex-valued smooth exterior differential forms on X.

To S we associate the ideal I_S in A generated by the (germs of) functions vanishing on S and by their antiholomorphic differentials. Then we have

$$\bar{\partial}\, I_S \subset I_S$$

and therefore, by passing to the quotient $Q_S = A/I_S$, we obtain a map

$$\bar{\partial}_S : Q_S \to Q_S.$$

In general, Q_S is a sheaf concentrated on S, that has a natural structure of E_S-module (E_S = sheaf of germs of complex-valued, smooth functions on S).

If S is generic, Q_S is a locally free E_S-module. The bigraduation $A = \oplus A^{p,q}$ with respect to the degree of the holomorphic and antiholomorphic differentials induces a bigraduation of Q_S:

$$Q_S = \oplus Q_S^{p,q}, \quad Q_S^{p,q} = A^{p,q}/I_S \cap A^{p,q}.$$

For S generic, $Q_S^{p,q}$ is locally free of rank $\binom{n-k}{q}\cdot\binom{n}{p}$ for $0 \leqq q \leqq n-k$, zero for $q > n-k$. The operator $\bar{\partial}_S$ can be expressed then, in any local trivialization, by systems of first order partial differential operators

$$\bar{\partial}_S : Q_S^{p,q} \to Q_S^{p,q+1}.$$

We have $\bar{\partial}_S^2 = 0$, so that we obtain for every $p = 0,1,\ldots,n$ a complex

$$(*) \quad Q_S^{p,0} \xrightarrow{\bar{\partial}_S} Q_S^{p,q} \xrightarrow{\bar{\partial}_S} Q_S^{p,2} \longrightarrow \ldots \xrightarrow{\bar{\partial}_S} Q_S^{p,n-k} \longrightarrow 0$$

that we call *Tangential Cauchy-Riemann complex.*

This is, for $k \geqq 1$, a very natural example of a non-elliptic complex of linear partial differential operators: for every $x \in S$, all codirections in

$$H_S S^0 = \{\xi \in T_x^*S \mid \langle\xi,\eta\rangle = 0 \quad \forall\, \eta \in H_x S\}$$

are characteristic for the complex (*): indeed the symbols for the operators in (*) vanish on $H_x S^o$. To each point (x,ξ) of the fibre bundle $HS^0 = \bigcup_{x\in S} H_x S^0$ we associate a Hermitean quadratic form on the n-k dimensional complex space $H_x S$

$$H_x S \ni \eta \to L_{x,\xi}(\eta) = \langle\xi,[v,Jv]_x\rangle \text{ where } v \in \Gamma_x(S,HS) \text{ and } v_x = \eta$$

(Levi form).

This is a very important invariant of the Cauchy Riemann structure of S. It can be given an equivalent expression, that is more classical in complex analysis. Denoting by i_S: $S \to X$ the natural imbedding we obtain a map $T^*X \underset{X}{\times} S \xrightarrow{i_S^*} T^*S$. If N_S^*X is the normal bundle to S in X, then we have

$HS^0 = i_S^* \circ J^*N_S^*X$. Given $\xi \in H_xS$, let us fix a real valued, smooth function ϕ on a neighbourhood of x in X, vanishing on S, and such that $i_S^* \circ J^*d\phi(x) = \xi$.

If $v \in \Gamma_x(S,HS)$ is such that $v_x = \eta \in H_xS$, we have

$$L_{x,\xi}(\eta) = \langle i_S^* \circ J^*d\phi(x),[v,Jv]_x\rangle = \frac{1}{2} \langle dJ^*d\phi(x),v_x \wedge Jv_\eta\rangle =$$

$$= \frac{1}{2i} \langle \partial\bar{\partial}\phi(x),\eta\wedge J\eta\rangle = \frac{1}{2i} \langle \partial\bar{\partial}\phi(x),(\eta-J\eta) \wedge (\eta+J\eta)\rangle =$$

$$= \sum_{\alpha,\beta=1}^{n} \frac{\partial^2\phi(x)}{\partial z^\alpha \partial \bar{z}^\beta} \zeta^\alpha \bar{\zeta}^\beta$$

where $\zeta^\alpha = \eta^\alpha - i\eta^{\alpha+n}$, the coordinates having been chosen holomorphic. If locally $S = \{\rho_1 = \ldots = \rho_k = 0\}$ for smooth real valued functions with linearly independent differentials, the genericity assumption can be rewritten as $\partial\rho_1(x) \wedge \ldots \wedge \partial\rho_k(x) \neq 0$ for $x \in S$, while the condition $\eta \in H_xS$ is written by $\sum_{\alpha=1}^{n} \frac{\partial\rho_j}{\partial z^\alpha}(x)\, \zeta^\alpha = 0$ for $j = 1,\ldots,k$.

REMARK: A Cauchy-Riemann structure can be defined on a non-embedded smooth real manifolds, and also one can consider in general a complex of first order linear partial differential operators built up from any set of commuting complex tangent fields. Substituting to the commutator $[v,Jv]$ of our definition above $(\frac{1}{2})$ that of the elements of the complex linear space V generated by the given vector fields with their complex conjugated, and to HS the real annihilator V^0 of V, we obtain also in this case the definition of a Levi form. Problems related to these complexes have been studied, particularly when $\dim_{\mathbb{R}} V^0 = 1$, by several authors, among whom are Trèves, Baouendi, Tartakoff, Stanton, Chang, Tajima. For abstract C.R. structures the problem arises of whether they could be realized by an embedding (at least local) into a complex manifold. The answer is in the affirmative in the case of real analytic manifolds and real-analytic C.R. structures, while an example of L. Nirenberg shows that there are 3-dimensional real manifolds, with a smooth C^∞ Cauchy Riemann structure, that cannot be locally realized as real submanifolds of a complex manifold of dimension two.

This result has been extended by Trèves, to show that the non-embeddable structures are the rule for C^∞ real C.R. structures of codimension one when the Levi form (for some choice of $\xi \in HS^0$) is non degenerate with one only negative eigenvalue, while Kuranishi has shown that manifolds of real dimension ≥ 9 and with definite Levi form are always embeddable. The case of codimension larger than one seems to be completely open up to now.

§2. Let us take up the study of the complex (*) for closed real generic submanifolds of a complex manifold of any codimension. Let us first recall the notion of Whitney function.

Given a locally free $\mathcal{E}$-module E over X, for an open subset U of X and a closed subset K of U we denote by $\mathcal{F}_K^\infty(U,E)$ the subspace of $\Gamma(U,E)$ of sections that vanish with all derivatives at the points of K. Then the space W(K,E) of Whitney sections of E over K is defined by the exact sequence

$$0 \to \mathcal{F}_K^\infty(U,E) \to \mathcal{E}(U) \to W(K,E) \to 0,$$

where the notation W(K,E) is justified by the fact that this space depends only upon the locally closed subset K of X and not on the neighbourhood U. If K is the closure of an open subset of X, then W(K,E) can be identified as the space of sections of E over Ω that are smooth up to the boundary, provided that Ω has the property that every point of $\partial\Omega$ has a neighbourhood V such that $V \cap \Omega$ is connected and $\bar{\Omega} \cap \bar{V}$ is "regular" in the sense of Schwartz (it is sufficient for instance that Ω has the cone-property). If K is a smooth submanifold, then W(K,E) are the sections of E which are formal power series in the normal directions to K, with coefficients that are smooth on K. The importance of Whitney functions for the study of the complex (*) stems from the following

<u>PROPOSITION</u> (Andreotti-Hill [4] and Andreotti-Fredrichs-Nacinovich [2]) (<u>Formal Cauchy Kowalewski theorem for</u> $\bar\partial$): If S is generic, then, for every open set U, we have isomorphisms:

$$H^q(\Gamma(S\cap U, Q_S^{p,*}), \bar\partial_{S_*}) \cong H^q(W(S\cap U, A^{p,*}), \bar\partial_*) \text{ for } 0 \leq p,\ q \leq n.$$

We can reduce the study of the tangential Cauchy Riemann complex to that of the Dolbeault complex on Whitney functions on "wedges", i.e. closures of domains with piece-wise smooth boundaries, by means of a suitable generalization of the Mayer-Vietoris exact sequence of [4] and by the formal Cauchy-Kowalewski theorem.

We say that $\mathcal{U} = \{\Omega_0, \Omega_1, \ldots, \Omega_k\}$ is a *regularly situated* system of closed sets if any pair of sets obtained by intersecting sets of $\mathcal{U}$ is regularly situated in the sense of Łojasiewicz (cf. [29]).

We set $C^s(\mathcal{U},E) = \{$space of alternated chains $f = (f_{i_0 \ldots i_s})$ with $f_{i_0 \ldots i_s} \in W(\Omega_{i_0} \cap \ldots \cap \Omega_{i_s}, E)\}$ and $\delta: C^s(\mathcal{U},E) \to C^{s+1}(\mathcal{U},E)$ is defined by

$$(\delta f)_{i_0 \ldots i_s} = \sum_{h=0}^{s+1} (-1)^h f_{i_0 \ldots \hat{j}_h \ldots j_{s+1}} | \Omega_{j_0} \cap \ldots \cap \Omega_{j_{s+1}}.$$

Set $Z^s(\mathcal{U},E) = \mathrm{Ker}(\delta: C^s(\mathcal{U},E) \to C^{s+1}(\mathcal{U},E))$. Then we have:

PROPOSITION: If $\mathcal{U}$ is regularly situated, then

$$C^0(\mathcal{U},W(E)) \xrightarrow{\delta} C^1(\mathcal{U},W(E)) \xrightarrow{\delta} \ldots \longrightarrow C^k(\mathcal{U},W(E)) \dashrightarrow 0$$

is an exact sequence. Moreover, $Z^0(\mathcal{U},W(E)) \cong W(\Omega_0 \cup \ldots \cup \Omega_k, E)$, $Z^k(\mathcal{U},W(E)) = W(\Omega_0 \cap \ldots \cap \Omega_k, E)$.

In particular, for the Dolbeault complex, we obtain for every $h = 0,1,\ldots,k$ long exact sequences:

$$0 \to H^0(Z^h(\mathcal{U},W(A^{p,*})),\bar\partial_*) \to H^0(C^h(\mathcal{U},W(A^{p,*})), \bar\partial_*)$$

$$\to H^0(Z^{h+1}(\mathcal{U},W(A^{p,*})), \bar\partial_*) \to H^1(Z^h(\mathcal{U},W(A^{p,*})), \bar\partial_*) \to \ldots.$$

$$(**) \qquad \ldots \to H^{q-1}(Z^{h+1}(\mathcal{U},W(A^{p,*})), \bar\partial_*) \to$$

$$H^q(Z^h(\mathcal{U},W(A^{p,*})), \bar\partial_s) \to H^q(C^h(\mathcal{U},W(A^{p,*})), \bar\partial_*) \to$$

$$H^q(Z^{h+1}(\mathcal{U},W(A^{p,*})), \bar\partial_*) \to H^{q+1}(Z^h(\mathcal{U},W(A^{p,*})), \bar\partial_*) \to \ldots$$

where the maps are defined letting $\bar\partial$ operate on each component and noticing that $\bar\partial$ and δ commute. All these maps stay exact also at the germ level. We shall use the notation $H^q_X(\ldots)$ for the local cohomology groups obtained from the H^q's taking the direct limit over the system of all open neighbourhood of the point x in X.

An important instance of this situation is the following: $\Omega_j = \{x \in X | \rho_j(x) \leq 0\}$ for real valued, smooth functions $\rho_0, \rho_1, \ldots, \rho_k$ such that $S = \Omega_0 \cap \Omega_1 \cap \ldots \cap \Omega_k$, $X = \Omega_0 \cup \Omega_1 \cup \ldots \cup \Omega_k$ and, for every $x \in S$, $d\rho_0(x), \ldots, d\rho_k(x)$ are in $N_x S$ the vertices of a K-dimensional simplex containing 0 in its interior. We obtain results about the tangential Cauchy-Riemann equations playing around with tne exact sequence (**) and using some results for the Dolbeault complex on piece-wise smooth domains.

§3. Let $\phi_1, \ldots, \phi_k$ be real valued, smooth functions defined on an open neighbourhood of x_0 in X, with $\phi_1(x_0) = \ldots \phi_k(x_0) = 0$. We assume that $\partial\phi_1(x_0) \wedge \ldots \wedge \partial\phi_k(x_0) \neq 0$. Then we say that

$$\Omega = \{x \in U | \phi_1(x) \leqq 0, \ldots, \phi_k(x) \leqq 0\}$$

has a generic k-edge at x_0.

Let $\Omega = \{\lambda \in \mathbb{R}^k | \lambda_i \geqq 0 \text{ for } i = 1, \ldots, k \text{ and } \lambda_1 + \ldots + \lambda_k = 1\}$ be the standard (k-1)-dimensional symplex in $\mathbb{R}^k$. For $\lambda \in \Lambda$ we set $\phi_{(\lambda)} = \lambda_1\phi_1 + \ldots + \lambda_k\phi_k$, $H_\lambda = \{\eta \in T_{x_0} X | \langle\partial\phi_{(\lambda)}(x_0), \eta\rangle = 0\}$. Then we have (cf. [20], [23])

PROPOSITION: Let $q \geqq 1$. Then we have $H^q_{x_0}(W(\Omega, A^{p,*}), \bar{\partial}_*) = 0$ if, for every $\lambda \in \Lambda$, the Levi form of $\phi_{(\lambda)}$, restricted to $H_{(\lambda)}$, has either at least n-q positive or at least q+k negative eigenvalues. If, for every $\lambda \in \Lambda$, it has at least k negative eigenvalues, then any element of $H^0_{x_0}(W(\Omega, A^{p,*}), \bar{\partial}_*)$ extends to a holomorphic p-form on a neighbourhood of x_0 in X.

Vice versa, we have

PROPOSITION: Let $q \geqq 1$. Assume that, for some $\lambda \in \Lambda$, the Levi form of $\phi_{(\lambda)}$, has q negative and n-q-1 positive eigenvalues on $H_{(\lambda)}$, having q negative eigenvalues on $\bigcap_{\mu\in\Lambda} H_{(\mu)}$. Then $H^q_{x_0}(W(\Omega, A^{p,*}), \bar{\partial}_*)$ is infinite dimensional.

With the same assumption for q = 0 we have obviously that

$$\frac{H^{0}_{x_0}(W(\Omega,A^{p,*}),\bar{\partial}_*)}{H^{0}_{x_0}(W(X,A^{p,*}),\bar{\partial}_*)} \quad \text{is infinite dimensional.}$$

These results have several extensions: one can, for instance, weaken the assumption of genericity of the edge, substitute for the Whitney functions different spaces of functions or distributions in the interior of Ω, consider instead of edges singularities that can be "controlled" by edges, consider identations (i.e. complements of edges), but I think that the two propositions above are sufficient to give a good idea of the argument. The extensions of the propositions above are however important to study the tangential Cauchy Riemann complex for distributions or for hyperfunctions. (cf. [24]).

4. THE SINGULAR SPECTRUM OF A COHOMOLOGY CLASS

Let $\phi_1,\ldots,\phi_k$ be real valued, smooth functions on a neighbourhood U of a point $x_0 \in S$, such that $\partial\phi_1(x) \wedge \ldots \wedge \partial\phi_k(x_0) \neq 0$, and $S \cap U = \{x \in U | \phi_1(x) = \ldots \phi_k(x) = 0\}$. Then we call $\Omega = \{x \in U | \phi_1(x) \leq 0,\ldots,\phi_k(x) \leq 0\}$ a (germ of) wedge at x_0, with edge S. The map that associates, to a $\bar{\partial}$ closed Whitney (p,q)-form f, defined on the intersection of Ω with a neighbourhood of x_0 in X, its restriction to S, induces a natural map

$$H^{q}_{x_0}(W(\Omega,A^{p,*}),\ \bar{\partial}_*) \rightarrow H^{q}_{x_0}(\Gamma(S,Q^{p,*}),\bar{\partial}_{S_*}).$$

The image of a class of cohomology α in the wedge under this map will be called its *boundary value*.

We associate to such a wedge Ω a cone in $N_{x_0} S$ (= annihilator of $T_{x_0} S$ in $T^*_{x_0} X$) by setting $\Lambda_\Omega = \{t_1 d\phi_1(x_0) + \ldots + t_k d\phi_k(x_0) | t_1 \geq 0,\ldots,t_k \geq 0\}$.

We note that $i^*_S \circ J^* : T^*_{x_0} X \rightarrow T^*_{x_0} S$ transforms $N_{x_0} S$ into the "characteristic set" for the C.R. system at x_0, that is the annihilator of $H_{x_0} S$ in $T^*_{x_0} S$. The map $N_{x_0} S \rightarrow (H_{x_0})^0 \subset T^*_{x_0} S$ is a linear isomorphism.

We have the following (cf. [13], [15], [20], [23]).

PROPOSITION: Let $x_0 \in S$ be a point where the Levi form of S is non-degenerate (this means that for some $\xi \in (H_{x_0} S)^0$, L_{ξ,x_0} has all eigenvalues different from zero). Let us fix $q \geq 1$. Then, having fixed $\xi^0 \in H_{x_0} S^0$, a metric on the sphere of $N_{x_0} S$, a real $\varepsilon > 0$, we can find wedges $\Omega_j = \{x \in U | \phi_i^j(x) \leq 0, \ldots, \ldots, \phi_k^j(x) \leq 0\}$, for $j = 1,\ldots,N$, such that

(i) $\cup \Lambda_{\Omega_j} = N_{x_0} S$, interior $\Lambda_{\Omega_j} \cap$ interior $\Lambda_{\Omega_k} = \emptyset$ if $j \neq k$

(ii) $i_S^* J^* \Lambda_{\Omega_1}$ contains ξ^0 as an interior point

(iii) The map $\bigoplus_{j=1}^{N} H^q_{x_0}(W(\Omega_j, A^{p,*}), \bar\partial_*) \to H^q_{x_0}(\Gamma(S,Q^{p,*}), \bar\partial_{S_*})$ given by the sum of boundary values is an isomorphism.

DEFINITION: Let $q \geq 1$ be fixed. The codirection $\xi^0 \in (H_{x_0} S)^0$ is said not to belong to the singular spectrum of $\alpha \in H^q_{x_0}(\Gamma(S,Q^{p,*}), \bar\partial_{S_*})$, if, for some decomposition satisfying i, ii, iii above, the component $\alpha_1 \in H^q_{x_0}(W(\Omega_1, A^{p,*}), \bar\partial_*)$ in the decomposition of α as a sum of boundary values of cohomology classes on wedges, is zero.

We denote by $\sigma_{x_0}(\alpha)$ the singular spectrum of α at x_0. Let us set $\tilde\Sigma^q_{x_0} = \{\xi \in H_{x_0} S^0 | L_{\xi,x_0}$ restricted to H_{x_0} has either at least q+1 negative or at least n-q-k+1 positive eigenvalues$\}$.

Then we define $\Sigma^q_{x_0} = H_{x_0} S^o_* - \tilde\Sigma^q_{x_0}$ = q-*characteristics for the C.R. complex.* Note that $\Sigma^q_{x_0}$ is a closed cone. We have

PROPOSITION: For every $\alpha \in H^q_{x_0}(\Gamma(S,Q^{p,*}), \bar\partial_{S_*})$ we have $\sigma_{x_0}(\alpha) \subset \Sigma^q_{x_0}$. Let $\Sigma^q_{x_0} = \{\xi \in H_{x_0} S^0 | L_{\xi,x_0}$ has exactly q negative and n-q-k positive eigenvalues on $H_{x_0} S\}$.

Then we have

PROPOSITION: For every $\xi^o \in \dot{\Sigma}^p_{x_o}$, we can find an infinite dimensional linear subspace V of $H^q_{x_o}(\Gamma(S,Q^{p,*}_S),\bar{\partial}_{S_*})$ such that, for every $\alpha \in V-\{0\}$, $\sigma_{x_o}(\alpha) \in \xi^o$.

For a cohomology class in $H^q(\Gamma(S \cap U,Q^{p,*}),\bar{\partial}_{S_*})$ defined on an open subset U of S, the singular spectrum is defined as the reunion of the singular spectra of the local cohomology classes that it defines by restrictions.

We say that S is q-*concave* at x_o if, for every $\xi \in H_{x_o}S^o$, the Levi from L_{ξ,x_o} has at least q negative eigenvalues on $H_{x_o}S$. Then we have:

PROPOSITION (Henkin [15], Nacinovich [20]): If S is q-concave at $x_o \in S$, then $H^j_{x_o}(\Gamma(S,Q^{p,*}_S),\bar{\partial}_{S_*})$ is zero for $1 \leq j < q$ and for $j > n-q-k$.

If $q \geq 1$, then every $\alpha \in H^o_{x_o}(\Gamma(S,Q^{p,*}_S),\bar{\partial}_{S_*})$ is the restriction to S of a germ of holomorphic p-form defined on a neighbourhood of x_o in X.

REMARK. In the case of hypersurfaces, more precise results were obtained in [4]: if S is q-concave and at a codirection $\xi \in H_{x_o}S^o$ the Levi form has q negative and p positive eigenvalues, then $H^j_{x_o}(\Gamma(S,Q^{p,*}),\bar{\partial}_{S_*}) = 0$ also for $q < j < p$ and $n-p-1 < j < n-q-1$.

The problem of the extension of Cauchy-Riemann functions and forms has been studied by Hunt and Wells [16], Boggess and Polking [11], and, in a very general setting, by Baouendi, Chang, Trèves [8].

When $q = 0$, the singular spectrum of a Cauchy-Riemann function (or form) can be defined in an analogous way: we have an isomorphism

$$\text{(iii)'} \quad \bigoplus_{j=1}^{N} H^o_{x_o}(W(\Omega_j,A^{p,*}),\bar{\partial}_*)/H^o_{x_o}(W(X,A^{p,*}),\partial_*) \to$$

$$\to H^o_{x_o}(\Gamma(S,Q^{p,*}_S),\bar{\partial}_{S_*})/H^o_{x_o}(W(X,A^{p,*}),\bar{\partial}_*)$$

and an interior point ξ^o of $i^*_S J^* \Lambda_{\Omega_j}$ does not belong to the singular spectrum of $\alpha \in H^o_{x_o}(\Gamma(S,Q^{p,*}),\bar{\partial}_{S_*})$ if the element $\alpha_j \in H^o_{x_o}(W(\Omega_j,A^{p,*}),\bar{\partial}_*)$ of the decomposition extends over x_o.

§5. PROPAGATION OF SINGULARITIES

In [12] Hanges and Trèves proved that extendability of C.R. functions propagates through maximal complex submanifolds contained in S (actually their results apply to hypoanalytic manifolds, that are a generalization of embedded C.R. manifolds).

For the case $q \geqq 1$, we note that the non-vanishing theorem can be further precised by:

PROPOSITION: Given $\xi^0 \in \dot{\Sigma}^q_{x_0}$, we can find an open neighbourhood U of x_0 in X and a class $\alpha \in H^q(\Gamma(S \cap U, Q^{p,*}_S), \bar{\partial}_{S_*})$ with $\sigma_{x_0}(\alpha) = \{t\xi^0 | t > 0\}$ and $\sigma_x(\alpha) = \emptyset$ for $x \in U - \{x_0\}$.

Therefore, we can say that the singularities in $\dot{\Sigma}^q_{x_0}$ do not propagate.

EXAMPLE: Consider in $\mathbb{C}P^4$ the two-codimensional real quadric

$$S = \{z_0\bar{z}_1 + z_1\bar{z}_2 + z_2\bar{z}_3 + z_3\bar{z}_4 + z_4\bar{z}_0 = 0\}.$$

Then S is a generic codimension 2 C.R. submanifold of $\mathbb{C}\,\mathbf{P}^4$, that is 1-concave.

We have $\Sigma^1_{x_0} = \dot{\Sigma}^1_{x_0} = H_{x_0}S^0$, $\Sigma^0_{x_0} = \Sigma^2_{x_0} = \emptyset$ for every $x_0 \in S$, so that all C.R. functions extend, $H^1_{x_0}(\Gamma(S,Q^{p,*}_S),\bar{\partial}_{S_*})$ is infinite dimensional, $H^2_{x_0}(\Gamma(S,Q^{p,*}_S),\bar{\partial}_{S_*}) = 0$.

The singularities of $\alpha \in H^1/\Gamma(S,Q^{p,*}_S),\bar{\partial}_{S_*})$ do not propagate, in accordance with the previous proposition, while through every point of S two distinct complex lines in S can be drawn. This shows that complex submanifolds of S are not by themselves propagators of singularities.

Let us consider a point $\xi^0 \in \Sigma^q_{x_0} - \dot{\Sigma}^q_{x_0}$.

At that point the Levi form degenerates, so that we can find a vector $\eta_0 \in H_{x_0}S - \{0\}$ such that

(***) $L_{x_0,\xi^0}(\eta_0,\eta) = 0 \quad \forall\eta \in H_{x_0}S.$

If S is represented locally by functions $\rho_1,\dots,\rho_k$ as in §1, this condition can be rewritten as

$$\sum_{j,\alpha} \lambda_j\, \partial^2\rho_j/\partial z^\alpha\, \partial\bar{z}^\beta\, (\eta^\alpha - i\eta^{\alpha+n}) = 0 \text{ for } \beta = 1,\dots,n$$

$$\sum_{\alpha} \frac{\partial\rho_j}{\partial z^\alpha} (\eta^\alpha - i\eta^{\alpha+n}) = 0 \text{ for } j = 1,\dots,k.$$

A characteristic solution for this system can be obtained in the following way:

assume that we are given: a real-valued, smooth function defined on an open neighbourhood U of $x_0 \in S$ in X, vanishing on $S \cap U$ and with $d\phi(x) \neq 0$ on U; a holomorphic imbedding of the unit disc D^1 of $\mathbb{C}$, $\gamma: D^1 \to X$, with $\gamma(D^1) \subset S$, $\gamma(0) = x_0$ such that

(a) $i_S^* \circ J^*d\phi(\gamma(\cdot)) \in \Sigma_X^q - \dot{\Sigma}_X^q$

(b) The Levi form of ϕ has n-q positive eigenvalues at all points of U

(c) $\sum_{\alpha,\beta} \frac{\partial^2\phi}{\partial z^\alpha \partial\bar{z}^\beta} (\dot{\gamma}^\alpha(t) - i\dot{\gamma}^{\alpha+n}(t)) = 0$ for $\beta = 1,\dots,n$.

In that case we have:

PROPOSITION: If $q < \frac{n-1}{2}$ and conditions (a), (b), (c) above are fulfilled, then for $\alpha \in H^q(\Gamma(U \cap S, Q_S^{p,*}), \bar{\partial}_*)$, we have:

if $i_S^* \circ J^*d\phi(\gamma(t)) \notin \sigma_{\gamma(t)}(\alpha)$ for $t \in \partial D^1$, then

$$i_S^* \circ J^*d\phi(\gamma(0)) \notin \sigma_{\gamma(0)}(\alpha).$$

REMARK: R. Penrose (cf. for instance [26]) has shown how integration on the fibre allows us to associate, to cohomology groups on some C.R. manifolds, that are fibred in complex manifolds, solutions of hyperbolic equations. His examples are related to Cauchy Riemann structures of codimension one. It should be interesting to consider analogous questions on C.R. submanifolds of higher codimension.

§6. OPEN SUBMANIFOLDS OF C.R. MANIFOLDS

The study of the tangential Cauchy Riemann system on hypersurfaces of a complex manifold originated on one side from the Hans Lewy example ([4]), on the other from the purpose of reducing the problem of solving $\bar{\partial}$-systems with regularity up to the boundary to differential problem on the boundary.

In the same way, our understanding of C.R. structure in higher codimensions shed light on questions related to the boundary cohomology for $\bar{\partial}_S$ of open subsets of a Cauchy Riemann manifold.

Let S be a generic real submanifold of a complex manifold X, of real codimension k. Let Ω be an open submanifold of S and let $x_0 \in \partial\Omega$ in S.

Then, if ϕ is a real valued, smooth function, defined on a neighbourhood U of x_0 in X, with $i_S^* d\phi(x_0) \neq 0$, such that $\Omega \cap U = \{x \in U \cap S | \phi(x) < 0\}$, we consider the form

$$L_{\phi,x_0,\xi}(\eta) = \frac{1}{i} \langle \partial\bar{\partial}\, \phi(x_0), \eta \wedge J\eta \rangle + L_{x_0,\xi}(\eta) \text{ for } \xi \in H_{x_0} S^0,\ \eta \in H_{x_0} S.$$

If, for every $\xi \in H_{x_0} S^0-\{0\}$ it has at least n-k-d positive eigenvalues on $H_{x_0} S$, we say that Ω is d-convex at x_0; if for every $\xi \in H_{x_0} S^0-\{0\}$ it has at least q negative eigenvalues on $H_{x_0} S$, we say that Ω is q-concave at x_0; we say that Ω is strictly q-concave in the codirection $\xi^0 \in H_{x_0}S^0-\{0\}$, if L_{ϕ,x_0,ξ^0} has q negative and n-q-k positive eigenvalues on $H_{x_0} \Omega$. Then we have ([23])

PROPOSITION: If Ω is d-convex at x_0, then $H^s_{x_0} (W(\bar{\Omega},Q_S^{p,*}),\bar{\partial}_{S_*}) = 0$ for $s > d$. If Ω is q-concave at x_0, then $H^s_{x_0} (W(\bar{\Omega},Q_S^{p,*}),\bar{\partial}_{S_*}) = 0$ for $1 \leqq s < q$. If Ω is strictly q-concave at x_0 for some codirection $\xi^0 \in H_{x_0} S^0$, then $H^q_{x_0} (W(\bar{\Omega},Q_S^{p,*}),\bar{\partial}_{S_*})$ is infinite dimensional.

EXAMPLE: Assume that S be a strictly pseudoconcave hypersurface in X. Then the local cohomology groups $H^q_{x_0} (\Gamma(S,Q_S^{p,*}),\bar{\partial}_{S_*})$ are all zero for $1 \leq q < n-1$. But, if we require that for some domain $\Omega \subset S$ at a point $x_0 \in \partial\Omega$ we have $H^q_{x_0} (W(\bar{\Omega},Q_S^{p,*}),\bar{\partial}_{S_*}) = 0$ for all $1 \leqq q \leqq n-2$, we need to require that, writing at x_0 $S = \{\rho = 0\}$, $\Omega = \{x \in S | \phi < 0\}$, one has $\partial\bar{\partial}_S\phi = 0$ at x_0. This is close to requiring that ϕ be the boundary value of a pluriharmonic function on one side of S ([6]), in accordance with some recent results communicated to me by A. Perotti.

References

[1] A. Andreotti, G.A. Fredricks, "Embeddability of Real Analytic Cauchy Riemann Manifolds" Ann. Scuola Norm. Sup. Pisa (4), 6, (1979) pp.285-304.

[2] A. Andreotti, G.A. Fredricks, M. Nacinovich, "On the absence of Poincaré Lemma in tangential Cauchy-Riemann complexes" Ann. Scuola Norm. Sup. Pisa (4), 8, (1981) pp. 365-404.

[3] A. Andreotti, C.D. Hill, "Complex characteristic coordinates and tangential Cauchy Riemann equations" Ann. Scuola Norm. Sup. Pisa (3), 26, (1972) pp. 299-324.

[4] A. Andreotti, C.D. Hill, "E.E. Levi convexity and the Hans Lewy problem I and II" Ann. Scuola Norm. Sup. Pisa (3), 26, (1972) pp.325-363 and pp. 747-806.

[5] A. Andreotti, C.D. Hill, S. Łojasiewica, B. Mackichan, "Complexes of differential operators. The Mayer Vietoris sequence" Invent. Math. 26 (1976) pp. 43-86.

[6] A. Andreotti, M. Nacinovich, "Non characteristic hypersurfaces for complexes of differential operators" Ann. Mat. Pura e Appl. (4), CXXV, (1980) pp. 13-83.

[7] A. Andreotti, F. Norguet, "Problème de Levi et convevité holomorphe pour les classes de cohomologie". Ann. Scuola Norm. Sup. Pisa (3), 20, (1966) pp. 197-241.

[8] M.S. Baouendi, C.H. Chang, F. Trèves, "Microlocal hypo-analicity and extension of C.R. functions" J. Differential Geometry 18 (1983), pp. 331-391.

[9] M.S. Baouendi, F. Trèves, "A property of the functions and distributions annihilated by a locally integrable system of complex vector fields" Ann. of Math. 113 (1981) pp. 387-421.

[10] M.S. Baouendi, F. Trèves, "A local constancy principle for the solutions of certain overdetermined systems of first-order linear partial differential equations". Math. Analysis and Applications Part A; Adv. in Math. suppl. Studies 7A (1981) pp. 245-262.

[11] A. Boggess, J. Polking, "Holomorphic extension of C.R. function" Duke Math. J. 49 (1982) pp. 757-784.

[12] N. Hanges, F. Trèves, "Propagation of holomorphic extendability of C.R. functions" Math. Ann. 263 (1983) pp. 157-179.

[13] G.M. Henkin "Analytic representation for C.R. functions on submanifolds of codimension 2 in $\mathbb{C}^N$". Analytic Functions-Kozubnik, 1979. Lecture Notes in Math. vol. 798, pp. 169-191. Berlin, 1980.

[14] G.M. Henkin, "Resolution of Cauchy Riemann equations on q-concave C.R. manifolds" International Conference on Analytic Functions and their applications in Mathematical Physics. Moskow, 1980.

[15] G.M. Henkin, "Integral representation of differential forms on Cauchy Riemann manifolds and the theory of C.R. functions" (in Russian) Uspiehi Math. Nauk 39 (1984) pp. 39-106.

[16] L.R. Hunt, R.O. Wells Jr., "Extensions of C.R. functions" Amer. J. Math. 98 (1976) pp. 805-820.

[17] H. Jacobowitz, F. Trèves, "Nonrealizable C.R. structures" Invent. Math. 66 (1982) pp. 231-249.

[18] M. Kuranishi, "Strongly pseudoconvex C.R. structures over small balls".

[19] J. Leiterer, "The Penrose transform for bundles non-trivial on the general line". Math. Nachr. 112 (1983) pp. 35-67.

[20] M. Nacinovich, "Poincaré lemma for tangential Cauchy Riemann complexes" Math. Ann. 268 (1984) pp. 449-471.

[21] M. Nacinovich, "On the absence of Poincaré Lemma for some systems of linear partial differential equations" Compositio Math. 44 (1981) pp. 241-303.

[22] M. Nacinovich, "On boundary Hilbert differential complexes" Annales Polonici Math. XLVI (1985) pp. 213-235.

[23] M. Nacinovich, "On strict Levi q-convexity and q-concavity on domains with piecewise smooth boundaries" Preprint n. 183 (1987) Pubbl. Dipartimento di Matematica, Pisa.

[24] M. Nacinovich, G. Valli, "Tangential Cauchy Riemann complexes on distributions" Ann. di Matem. Pura e Appl. (1986).

[25] L. Nirenberg, "On a question of Hans Lewy" Russ. Math. Sur. 29 (1974) pp. 251-262.

[26] R. Penrose, "Physical space-time and nonrealizable CR-structure", Bull. Amer. Math. Soc. 8 (1983) pp. 427-448.

[27] L. Schwartz, "Theorie des Distributions" 2nd Ed. Hermann, Paris, 1966.

[28] S. Tajima, "Analyse Microlocal sue les variétés de Cauchy Riemann et problèmes de prolongement des solutions holomorphes des equations aux derivées partielles" Publ. R.I.M.S. Kyoto Univ. 18 (1982) pp. 911-945.

[29] J.C. Tougeron, "Ideaux de fonctions differentiables" Springer, Berlin, 1972.

[30] R.O. Wells, Jr., "Complex manifolds and mathematical physics" Bull. Amer. Math. Soc. 1 (1979) pp. 296-336.

Mauro Nacinovich
Università di Pisa,
Dipartimento di Matematica,
Via Buonarroti, 2,
56100 - Pisa,
Italy.

T. NISHITANI

Une classe d'opérateurs hyperboliques à caractéristiques multiples

ABSTRACT: A class of hyperbolic operators with multiple characteristics

Let P be a classical pseudo-differential operator of order m in an open set $\Omega \subset \mathbf{R}^{d+1}$ with principal symbol $p(x,\xi) \in C^\infty(T^*\Omega\setminus 0)$. Assume that $p(x,\cdot)$ is hyperbolic with respect to $dt(x)$ near $\hat{x} \in \Omega$, $t(x) \in C^\infty(\Omega)$ with $dt(\hat{x}) \neq 0$. Let $\rho \in T^*_x\Omega\setminus 0$ be a multiple characteristic of p. When ρ is double p is effectively hyperbolic at ρ if and only if the propagation cone $C(p_\rho, dx_0)$ of the localization p_ρ of p at ρ is transversal to KerHess $p(\rho)$ (the Kernel of the Hessian of p at ρ) and p_ρ is strictly hyperbolic in $T_\rho(T^*\Omega)/\text{KerHess } p(\rho)$.

Let Σ be the set of characteristics of order m of p and assume that Σ is a C^∞ manifold near ρ. Then the above conditions naturally extend to: $C(p_\rho, dx_0) \cap T_\rho\Sigma = \{0\}$ and p_ρ is strictly hyperbolic in $T_\rho(T^*\Omega)/T_\rho\Sigma$. We also assume that p_ρ is reducible to a product of polynomials of degree 1 or 2. Under these assumptions and some conditions on lower order terms (see [4]) we shall show that the Cauchy problem for P is well posed in C^∞. We also study the propagation of wave front sets in this context.

§1. INTRODUCTION

Soit P un opérateur pseudo-différentiel classique (o.p.d. en abrégé) sur un ouvert $\Omega \subset \mathbf{R}^{d+1}$ de symbole principal $p(x,\xi) \in C^\infty(T^*\Omega\setminus 0)$. Supposons que $p(x,\cdot)$ est hyperbolique par rappor à $dt(x)$ près de $\hat{x}$ où $\hat{x} \in \Omega$, $t(x) \in C^\infty(\Omega)$ avec $dt(x) \neq 0$. Si $\rho \in T^*_{\hat{x}}\Omega\setminus 0$ est une caractéristique double de p, p est dit effectivement hyperbolique en ρ si l'application Hamiltonienne de p en ρ (qui est le Hessien de p/2 en ρ lu par l'intermédiaire de la 2-forme symplectique sur $T^*\Omega$, voir [3], [4]) possède des valeurs propres réelles non nulles. Notons que si ρ est une caractéristique d'ordre supérieur à 2 l'application Hamiltonienne est toujours nulle.

D'autre part, on peut carácteriser l'hyperbolicité effective plus géométriquement (Lemme 1.1 ci-desous). Cette caractérisation a un sens pour des caractéristiques d'ordre quelconque. Notre but dans cette note est alors d'étudier des relations entre l'hyperbolicité C^∞ et cette *"l'hyperbolicité effective"* pour une classe d'o.p.ds. classiques à caractéristiques multiples.

Soit U une partie ouverte dans $\mathbf{R}^d$. Notons par T^*U le fibré cotangent de U et par $(x',\xi') = (x_1,\ldots,x_d,\xi_1,\ldots,\xi_d)$ des coordonnées naturelles sur T^*U. Soit I un intervalle ouvert dans $\mathbf{R}$ et posons $\Omega = I \times U$. On désigne par $(x,\xi) = (x_0,x',\xi_0,\xi')$ des coordonnées naturelles sur $T^*\Omega$ et

$$D_j = -i\frac{\partial}{\partial x_j},\ j = 0,1,\ldots,d,\ D = (D_0,D'),\ D' = (D_1,\ldots,D_d).$$

Soit

$$P(x,D) = D_0^m + \sum_{j=1}^{m} A_j(x,D')D_0^{m-j},$$

un opérateur différentiel en D_0 d'ordre m ayant pour des coefficients $A_j(x,D')$ d'o.p.ds. classiques d'ordre j définis près de $(\hat{x},\hat{\xi}') \in I \times (T^*U\setminus 0)$. On note par $p(x,\xi)$ le symbole principal de P et on suppose que $p(x,\cdot)$ est hyperbolique par rapport à dx_0 près de $(\hat{x},\hat{\xi}')$, c'est-à-dire que l'équation en ξ_0

$$p(x,\xi_0,\xi') = 0, \tag{1.1}$$

admet m racines réelles pour tout (x,ξ') près de $(\hat{x},\hat{\xi}')$. Soit ℓ une caractéristique multiple de p. Nous désignons par $p_\rho(x,\xi)$ la partie homogène de degré le plus bas dans le développement de Taylor de p en ρ. Alors $p_\rho(x,\xi)$ est un polynôme en (x,ξ) qui est hyperbolique par rapport à $\theta = (0,\ldots,0,1,0,\ldots,0)$, vecteur unité de $\mathbb{R}^{2d+2}$ à ξ_0-composante égale à 1. On note par $\Gamma(p_\rho,dx_0)$ le cône d'hyperbolicité de p_ρ;

$\Gamma(p_\rho,dx_0)$ = la composante de dx_0 dans $\{X \in T_\rho(T^*\Omega);\ p_\rho(X) \neq 0\}$.

Puis on désigne par $C(p_\rho,dx_0)$ le cône de la propagation de p_ρ;

$$C(p_\rho,dx_0) = \{X \in T_\rho(T^*\Omega);\ \sigma(X,Y) \leqq 0 \text{ pour tout } Y \in \Gamma(p_\rho,dx_0)\},$$

où σ est la 2-forme symplectiques naturelle sur $T^*\Omega$.

<u>LEMME 1.1</u> (cf. Lemme 3.2 de [7] et Corollaire 1.4.7 de [3]): Supposons que ρ est une caractéristique double de p. Alors p est effectivement hyperbolique en ρ si et seulement si

$$\Gamma(p_\rho,dx_0) \cap (\text{KerHess } p(\rho))^\sigma \cap T_\rho(T^*\Omega|_{x_0=\hat{x}_0}) \neq \emptyset. \tag{1.2}$$

De plus cette condition est équivalente à la condition;

$$C(p_\rho,dx_0) \cap (\text{KerHess } p(\rho)) = \{0\}, \tag{1.3}$$

où KerHess $p(\rho)$ désigne le noyau du Hessien de p en ρ et $(\text{KerHess } p(\rho))^\sigma$ désigne l'espace véctoriel orthogonal de KerHess $p(\rho)$ par rapport à σ.

<u>REMARQUE 1.1</u>: De (1.2) (ou bien (1.3)), il découle que

$$\begin{array}{l} p_\rho(x,\xi) \text{ est strictement hyperbolique dans} \\ T_\rho(T^*\Omega)/\text{KerHess } p(\rho) \text{ par rapport à } j(\theta), \end{array} \tag{1.4}$$

où j est l'application naturelle de $T_\rho(T^*\Omega)$ sur $T_\rho(T^*\Omega)/\text{KerHess } p(\rho)$ (cf. Introduction de [1]).

§2. RÉSULTATS

Désormais nous nous intéresserons au cas où $\rho = (\hat{x},\hat{\xi})$ est une caractéristique d'ordre m de p. Nous notons par $\Sigma_m(p)$ l'ensemble des caractéristiques d'ordre m de p,

$$\Sigma_m(p) = \{\rho' \in T^*\Omega\setminus 0;\ p(\rho') = dp(\rho') = \ldots = d^{m-1}p(\rho') = 0\},$$

et supposons que

$$\Sigma_m(p) \text{ est une variété } C^\infty \text{ passant par } \rho \text{ près de } \rho. \tag{2.1}$$

Alors la condition (1.3) se généralise naturellement à la condition;

$$C(p_\rho, dx_0) \cap T_\rho(\Sigma_m(p)) = \{0\}. \tag{2.2}$$

De notre hypothèses $p_\rho(x,\xi)$ est bien définie comme un polynôme dans $T_{\Sigma_m(p)}(T^*\Omega)|_\rho$ alors la condition (1.4) se généralise à la condition:

$$\begin{array}{l} p_\rho(x,\xi) \text{ est strictement hyperbolique dans} \\ T_{\Sigma_m(p)}(T^*\Omega)|_\rho \text{ par rapport à } j(\theta), \end{array} \tag{2.3}$$

où j est l'application naturelle de $T_\rho(T^*\Omega)$ sur $T_{\Sigma_m(p)}(T^*\Omega)|_\rho =$ $T_\rho(T^*\Omega)/T_\rho(\Sigma_m(p))$. Nous introduisons une autre hypothèse. On suppose l'une des 2 conditions suivantes;

$$\begin{array}{l} p_\rho(x,\xi) \text{ est réductible en produit des polynômes de degré 1} \\ \text{ou 2.} \end{array} \tag{2.4}$$

$$H^\sigma_{x_0} \cap T_\rho(\Sigma_m(p))^\sigma \text{ est involutif.} \tag{2.5}$$

Ici H_ϕ désigne le champ Hamiltonien de ϕ. La condition (2.5) équivalente à; dans des coordonnées symplectiques convenables, conservant des plans x_0 = const., $\Sigma_m(p)$ est donnée par

$$\Sigma_m(p) = \{\xi_0 = 0,\ b_j(x,\xi') = 0,\ \ j = 1,2,\ldots,k\}$$

avec $b_j(x,\xi')$, homogènes de degré 1 en ξ', telles que

$$(H_{b_i} b_j)(\hat{x},\hat{\xi}') = 0 \text{ pour tout } i,j = 1,2,\dots,k.$$

REMARQUE 2.1: Toutes les conditions (2.1) - (2.5) sont invariantes sous les changements des coordonnées symplectiques homogènes près de ρ conservant les plans initiaux x_0 = const.

D'autre part une condition nécessaire pour le problème de Cauchy pour des opérateurs à caractéristiques d'ordre m soit bien posé dans C^∞ est connue (cf. Théorème 4.1 de [4]). Nous supposons une version microlocale de cette condition. Pour la formuler on note par $P(x,\xi)$ le symbole entier de $P(x,D)$. Donc on a l'expression suivante

$$P(x,\xi) = p(x,\xi) + p_{m-1}(x,\xi) + \dots + p_i(x,\xi) + \dots ,$$

où $p_i(x,\xi)$ est la partie homogène de degré i de $P(x,\xi)$. Alors l'hypothèse sur les termes d'ordre inférieur s'énonce

$$\begin{aligned} &p_{m-j}(x,\xi) \text{ s'annule d'ordre } m-2j \text{ sur } \Sigma_m(p) \text{ près de } \rho \text{ pour} \\ &j = 1,2,\dots,[(m-1)/2], \end{aligned} \tag{2.6}$$

où $[k]$ $(k \in \mathbf{R})$ désigne la partie entière de k. Sous ces hypothèses on a le

THÉORÈME 2.1: Supposons que (2.1) - (2.3), (2.4) (ou bien (2.5)) et (2.6) sont vérifiées. Alors il existe une paramétrix en $(\hat{x}',\hat{\xi}')$ à vitesse finie de propagation du front d'onde de $P(x,D)$.

Plus précisément, si $\tilde{P}(x,D)$ est une extension de $P(x,D)$, alors il existe une constante β et un opérateur G;

$$G: C^s(\tilde{I},H^p) \to C^{s+m-1}(\tilde{I},H^{p-\beta}) \text{ (pour tout } p \in \mathbf{R}), \ \tilde{I} = (\hat{x}-\varepsilon,\hat{x}+\varepsilon) \subset I,$$

vérifiant les conditions suivantes;

$$\begin{cases} Gf = 0 \text{ dans } x_0 < \hat{x}_0 \text{ si } f = 0 \text{ dans } x_0 < \hat{x}_0 \text{ et} \\ \qquad \tilde{P}GH \equiv H \\ \text{modulo un opérateur dans } C^\infty(\tilde{I},S^{-\infty}) + V \text{ pour tout} \\ H = H(x',D') \text{ supporté près de } (\hat{x}',\hat{\xi}'), \end{cases} \tag{2.7}$$

$$\begin{cases} \text{pour tout } M = M(x',D') \text{ (supporté près de } (\hat{x}',\hat{\xi}')), \\ L = L(x',D') \text{ avec } \operatorname{supp}(L) \subset T^*\mathbb{R}^d \setminus (\operatorname{supp}(M)), \text{ on a} \\ \qquad D_0^j LGH \in V, \quad 0 \leqq j \leqq m-1, \end{cases} \tag{2.8}$$

où par V on note l'ensemble des opérateurs R tels qu'on ait avec $\delta = \delta(R) > 0$,

$$\| D_0^k Rf(x_0,\cdot) \|_q^2 \leqq c_{k,p,q} \sum_{j=0}^{k} \int_{\hat{x}_0-\varepsilon}^{t} \| D_0^j f(x_0,\cdot) \|_p^2 dx_0, \quad t \leqq \hat{x}_0 + \delta(R)$$

pour tout $k \in \mathbb{N}$, $p,q \in \mathbb{R}$ et $f \in C^k(\tilde{I},H^p)$ s'annulant dans $x_0 < \hat{x}_0$, et par $H^p = H^p(\mathbb{R}^d)$, $\| \cdot \|_p$ on note l'espace de Sobolev et la norme respectivement.

REMARQUE 2.2: S'il existe une paramétrix de P(x,D) à vitesse finie de propagation du front d'onde en tout $(\hat{x}',\xi')$, $|\xi'| = 1$, alors le problème de Cauchy pour P(x,D) est soluble près de $\hat{x}$ avec des données sur $x_0 = \hat{x}_0$ (voir [8]).

Puis nous traitons le cas où les coefficients $A_j(x,D')$ de P(x,D) sont des o.p.ds. classiques sur Ω. On suppose que $p(x,\cdot)$ est hyperbolique par rapport à dx_0 près de $\hat{x}$, c'est-à-dire que l'équation (1.1) en ξ_0 admet m racines réelles pour tout (x,ξ'), $\xi' \neq 0$, x est près de $\hat{x}$. Soit $\kappa \in T^*_{\hat{x}} \setminus 0$ une caractéristique multiple de p. On note par $m(\kappa)$ sa multiplicité. On désigne par $\Sigma_{m(\kappa)}(p,\kappa)$ la composante de κ dans l'ensemble des caractéristiques d'ordre $m(\kappa)$ de p. Nous notons par $(2.i)_\kappa$ (i = 1,2,...,6) l'hypothèse correspondante à (2.i) où ρ, m et $\Sigma_m(p)$ sont remplacés par κ, $m(\kappa)$ et $\Sigma_{m(\kappa)}(p,\kappa)$. Alors on a

COROLLAIRE 2.1: Supposons que $(2.1)_\kappa$-$(2.3)_\kappa$, $(2.4)_\kappa$ (ou bien $(2.5)_\kappa$)et $(2.6)_\kappa$ sont satisfaites pour toute caractéristique multiple $\kappa \in T^*_{\hat{x}} \setminus 0$. Alors le problème de Cauchy pour P(x,D) est localement soluble dans C^∞ près de $\hat{x}$ avec des données sur $x_0 = \hat{x}_0$.

Ensuite nous donnons un résultat de la propagation des singularités des solutions pour P(x,D). Dans la suite, WF(u) désigne le front d'onde de u.

THÉORÈME 2.2: Supposons que (2.1) - (2.3), (2.4) (ou bien (2.5)) et (2.6) sont vérifiées. Soit ϕ une fonction homogène de degré 0 en ξ à valeur réelle definie près de ρ telle que

$$\phi(\rho) = 0, \quad -H_\phi(\rho) \in \Gamma(p_\rho, dx_0),$$

et soit ω un voisinage conique de ρ suffisament petit. Alors il résulte de

$$\omega \cap \{\phi < 0\} \cap WF(u) = \emptyset, \; \rho \notin WF(Pu),$$

que

$$\rho \notin WF(u),$$

pour des distributions u.

REMARQUE 2.3: Si m = 3, tous résultats ci-dessus sont contenus dans [9].

Ici nous donnons une autre formulation d'hypothèse (2.2) d'où on a une surface séparante.

LEMME 2.1: Supposons (2.1). Alors l'hypothèse (2.2) est équivalente à la condition suivante,

$$\Gamma(p_\rho, dx_0) \cap (T_\rho(\Sigma_m(p)))^\sigma \cap T_\rho(T^*\Omega|_{x_0} = \hat{x}_0) \neq \emptyset. \tag{2.9}$$

§3. EXEMPLES

Dans ce paragraphe nous donnons quelques exemples. Premiers deux exemples montrent que les hypothèses (2.2) et (2.3) sont indispensables en général quand on traite le problème de Cauchy pour P(x,D) sous l'hypothèse (2.6) sur les termes d'ordre inférieur.

EXEMPLE 3.1 ([9]): Considérons l'opérateur suivant dans $\mathbb{R}^2$ avec ρ - (0,0,0,1) $\in T^*\mathbb{R}^2 \setminus 0$,

$$P(x,D) = (D_0 - x_0 D_1)\{(D_0 + x_0 D_1)^2 + aD_1\}, \quad a \neq 0.$$

On voit facilement que cet opérateur vérifie toutes les hypothèses (2.1)-(2.6) sauf (2.3). D'autre part, pour l'opérateur $(D_0 + x_0 D_1)^2 + aD_1$ ($a \neq 0$) le problème de Cauchy n'est pas bien posé dans C^∞ et par conséquent celui de P(x,D) est aussi n'est pas bien posé dans C^∞.

EXEMPLE 3.2 ([9]): Comme deuxième exemple, nous considérons l'opérateur suivant dans $\mathbb{R}^2$ avec $\rho = (0,0,0,1) \in T^*\mathbb{R}^2 \setminus 0$,

$$P(x,D) = (D_0 - ax_1 D_1)(D_0 - bx_1 D_1)(D_0 - cx_1 D_1) + \alpha D_1, \quad \alpha \neq 0,$$

où a, b, c sont les constantes réelles et mutuellement distinctes. Cet opérateur satisfait aux hypothèses (2.1) - (2.6) sauf (2.2). D'après le Théorème 4.1 de [4], le problème de Cauchy pour ce P n'est pas bien posé dans C^∞.

Ensuite nous donnons deux exemples simples qui vérifient (2.1)-(2.3), (2.4) (ou bien (2.5)).

EXEMPLE 3.3: Considérons l'opérateur de symbole principal suivant

$$p(x,\xi) = \prod_{j=1}^{m} (\xi_0 - \phi(x,\xi')\lambda_j(x,\xi')),$$

où $\lambda_j(x,\xi')$, $\phi(x,\xi')$ sont réels homogènes de degré 1 et 0 respectivement définis près de $(\hat{x},\hat{\xi}') \in I \times (T^*U \setminus 0)$. On suppose que $\lambda_j(x,\xi')$ sont mutuellement distincts près de $(\hat{x},\hat{\xi}')$ et $\phi(x,\xi')$ vérifie les conditions

$$\phi(\hat{x},\hat{\xi}') = 0, \quad (\partial/\partial x_0)\phi(\hat{x},\hat{\xi}') \neq 0.$$

La vérification d'hypothèses (2.1) - (2.4) est immédiate.

REMARQUE 3.1: Dans le cas m = 2, le problème de Cauchy pour l'opérateur de ce symbole principal est étudié dans [5] et le Corollaire 2.1 est dû [5]. Des études détaillées pour la propagation des singularités pour cet opérateur sont données, par exemple, dans [2], [6].

EXEMPLE 3.4: Considérons une application

$$\Phi(x,\xi) = (\xi_0,\phi_1(x,\xi'),\dots,\phi_k(x,\xi'))$$

d'un ouvert conique de $T^*\Omega\setminus 0$ dans $\mathbf{R}^{k+1}$ où $\phi_j(x,\xi')$ sont définies près de $(\hat{x},\hat{\xi}') \in I \times (T^*U\setminus 0)$ et homogènes de degré 1 en ξ'. Supposons que avec $\rho = (\hat{x},\hat{\xi}_0,\hat{\xi}')$ que

$$\Phi(\rho) = 0, \quad d\Phi(\rho) \text{ est du rang maximal.}$$

Prenons $q(\zeta)$ un polynôme en $\zeta = (\zeta_0,\zeta_1,\dots,\zeta_k)$ de degré m qui est strictement hyperbolique par rapport à $\theta = (1,0,\dots,0) \in \mathbf{R}^{k+1}$. Notons par $C(q,\theta)$ le cône de la propagation de q (voir [1]), et supposons que

$$(\operatorname{Ker} d\Phi(\rho)) \cap (H_\Phi(\rho)C(q,\theta)) = \{0\}, \tag{3.1}$$

où H_Φ est l'application de $\mathbf{R}^{k+1}$ dans $T_\rho(T^*\Omega)$ définie par

$$\sigma(X,H_\Phi(\rho)Y) = \langle d\Phi(\rho)X,Y\rangle, \text{ pour tout } X \in T_\rho(T^*\Omega)$$

avec le produit scalaire naturel $\langle\cdot,\cdot\rangle$ dans $\mathbf{R}^{k+1}$. La condition (3.1) est équivalente à;

$$\operatorname{Ker}\,(\{\phi_i,\phi_j\}(\rho))_{i,j=0}^{k} \cap C(q,\theta) = \{0\} \text{ avec } \phi_0 = \xi_0.$$

où $\{\phi_i,\phi_j\}$ désigne la paranthèse de Poisson.

Nous supposons l'une des 2 conditions suivantes;

$$q(\xi) \text{ se décompose en produit des polynômes de degré 1 ou 2.} \tag{3.2}$$

$$\{\phi_i,\phi_j\}(\hat{x},\hat{\xi}') = 0 \text{ pour tout } i,j = 1,2,\dots,k. \tag{3.3}$$

Alors $q(\Phi(x,\xi))$ vérifie les conditions (2.1)-(2.3) et (2.4) (ou bien (2.5)). Ici nous remarquons que sous la condition (3.3), l'hypothèse (3.1) se réduit à; il existe n, $1 \leq n \leq k$ telle que

$$(\partial/\partial x_0)\phi_n(\hat{x},\hat{\xi}') \neq 0.$$

§4. L'INÉGALITÉ D'ÉNERGIE

Dans ce paragraphe, nous allons énoncer l'inégalité d'énergie pour $P(x,D)$ lorsque $p(x,\xi)$ vérifie les conditions (1.1)-(1.3) et (1.4). D'abord nous étudions le symbole principal $p(x,\xi)$ de $P(x,D)$. En choisissant des coordonnées symplectiques (x,ξ) conservant x_0 = const., on peut supposer que

$$p(x,\xi) = \xi_0^m + \sum_{j=2}^{m} \hat{a}_j(x,\xi')\xi_0^{m-j},$$

où $\hat{a}_j(x,\xi')$ sont les symboles homogènes de degré j près de $(\hat{x},\hat{\xi}')$. Vu $\Sigma_m(p) \subset \{\xi_0 = 0\}$, $\Sigma_m(p)$ s'exprime;

$$\Sigma_m(p) = \{\xi_0 = b_0(x,\xi) = 0,\ \ b_j(x,\xi') = 0,\ j = 1,2,\ldots,k\}$$

où $db_j(\rho')$ $(\rho' = (\hat{x},\hat{\xi}'))$ sont linéairement indépendantes. A l'aide du Lemme 1.1, il existe $c_j \in \mathbb{R}$ $(1 \leq j \leq k)$ telles que

$$\tilde{f}(x,\xi') = \sum_{j=1}^{k} c_j b_j(x,\xi'), \quad - H_{\tilde{f}}(\rho') \in \Gamma(p_\rho, dx_0).$$

D'après $\Gamma(p_\rho,dx_0) \subset \{\xi_0 > 0\}$, $\tilde{f}(x,\xi')$ s'écrit

$$\tilde{f}(x,\xi') = \tilde{e}(x,\xi')(x_0 - f_1(x',\xi')) = \tilde{e}(x,\xi')f(x,\xi'),\ \tilde{e}(\rho') > 0.$$

Alors on peut prendre des coordonnées symplectiques (x',ξ') avec $(\hat{x}',\hat{\xi}') = (0\ e_2)$ telles qu'on ait

$$df(\rho') = dx_0 - adx_1 - bdx_2 \quad \text{avec des constantes réelles } a,b.$$

Remarquons que avec une constante $c > 0$ on a

$$c \sum_{j=1}^{k} b_j(x,\xi')^2 \geq f(x,\xi')^2\ |\xi'|^2 \quad \text{près de } \rho'.$$

Puis il résulte d'hyperbolicité de $p(x,\xi)$ que

$$p(x,\xi) = b_0(x,\xi)^m + \sum_{\substack{|\alpha|=m \\ \alpha_0 \leq m-2}} \tilde{a}\,(x,\xi')b(x,\xi)^\alpha$$

où $b(x,\xi) = (b_0(x,\xi), b_1(x,\xi'),\ldots,b_k(x,\xi'))$. Nous posons

$$q(\zeta) = \zeta_0^m + \sum_{\substack{|\alpha|=m \\ \alpha_0 \leq m-2}} \tilde{a}_\alpha(\rho')\zeta^\alpha, \quad \zeta = (\zeta_0,\ldots,\zeta_k).$$

Alors il est claire que $p\,(x,\xi)$ est donnée par

$$p_\rho(x,\xi) = q(db(\rho)(x,\xi))$$

d'où on obtient

$$p(x,\xi) = q(b(x,\xi)) + \sum_{\substack{|\alpha|=m \\ \alpha_0 \leq m-2}} a_\alpha(x,\xi')b(x,\xi)^\alpha, \quad a_\alpha(\rho') = 0.$$

D'autre part vu l'hypothèse (1.4), il résulte que $q(\zeta)$ est strictement hyperbolique par rapport à $(1,0,\ldots,0) \in \mathbb{R}^{k+1}$ et

$$q(\zeta) = \prod_{j=1}^{s} q_j(\zeta).$$

Si $q_j(\zeta)$ est de degré 2, $q_j(b(x,\xi))$ s'écrit

$$q_j(b(x,\xi)) = (\xi_0-a_j(b'(x,\xi')))^2 - \sum_{i=1}^{k} (m_{ji}(b'(x,\xi')))^2$$

où

$$a_j(\zeta') = \sum_{s=1}^{k} c_{js}\zeta_s, \quad m_{ji}(\zeta') = \sum_{s=1}^{k} c_{is}^j\zeta_s, \quad b' = (b_1,b_2,\ldots,b_k)$$

avec des matrices inversibles $(c_{is}^j)_{i,s}$.

Maintenant nous définissons les localisations $B_j(x,\xi,\mu)$ des symboles $b_j(x,\xi)$ le long de la surface $f(x,\xi') = 0$. On choisit des fonctions $\chi(s)$, $\chi_0(s) \in C^\infty(\mathbb{R})$ en sorte que

$$\chi(s) = s,\ |s| \leqq 1,\ |\chi(s)| = 2,\ |s| \geqq 2,\ \ 0 \leqq \chi'(s) \leqq 1,$$

$$\chi_0(s) = 1,\ |s| \leqq 1,\ \chi_0(s) = 0,\ \ |s| \geqq 2,\ \ 0 \leqq \chi_0(s) \leqq 1.$$

On pose avec un paramétre $0 < \mu \leqq 1$,

$$y_0 = \mu x_0,\ y_j = \mu\chi(x_j),\ j = 1,2,\ y_j = \mu^{1/2}\chi_0(\mu^{-1/2}x_j)x_j,\ 3 \leqq j \leqq d,$$

$$\eta_0 = \mu^{-1}\xi_0,\ \eta_j = \mu^{-1/2}\chi_0(\mu^{-1/2}\xi_j|\xi'|^{-1})\xi_j,\ 3 \leqq j \leqq d,$$

$$\eta_j = \mu^{-1}\chi_0(\mu^{-1}(\xi_j|\xi'|^{-1} - \delta_{2j}))(\xi_j - \delta_{2j}|\xi'|) + \mu^{-1}\delta_{2j}|\xi'|,\ j = 1,2.$$

Ici δ_{ij} est le symbole de Krönecker. Ensuite on definit $B_j(x,\xi,\mu)$ par

$$B_j(x,\xi,\mu) = b_j^{(0)}(\rho)(\xi_0 - i\theta) + b_j^{(1)}(\rho)\xi_1 + \mu\tilde{b}_j(y,\eta')$$

où

$$\tilde{b}_j(x,\xi) = b_j(x,\xi) - \sum_{i=0}^{1} b_j^{(i)}(\rho)\xi_i,\quad b_j^{(i)}(\rho) = \partial_{\xi_i} b_j(\rho).$$

Aussi on définit $a_\alpha(x,\xi',\mu)$, $\phi(x,\xi',\mu)$ par

$$a_\alpha(x,\xi',\mu) = a_\alpha(y,\eta'),\ \phi(x,\xi',\mu) = \mu^{-1}f(y,\eta').$$

Nous obtenerons l'inégalité d'énergie pour $p(x,D,\mu)$;

$$p(x,\xi,\mu) = q(B(x,\xi,\mu)) + \sum_{\substack{|\alpha|=m \\ \alpha_0 \leqq m-2}} a_\alpha(x,\xi',\mu)B(x,\xi,\mu)^\alpha.$$

Ici notons que si $|x_j| \leqq \mu^{1/2}$, $|\xi_j|\xi'|^{-1} - \delta_{2j}| \leqq \mu$ $(j \geqq 1)$, on a

$$p(x,\xi,\mu) = \mu^m p(\mu x_0, \mu x_1, \mu x_2, \mu^{1/2}\tilde{x}, \mu^{-1}(\xi_0 - i\theta), \mu^{-1}\xi_1, \mu^{-1}\xi_2, \mu^{-1/2}\tilde{\xi})$$

où $\tilde{x} = (x_3,\ldots,x_d)$, $\tilde{\xi} = (\xi_3,\ldots,\xi_d)$.

Pour formuler l'inégalité d'énergie, nous introduisons quelques notations. D'abord prenons des fonctions $\chi_1(s)$, $\chi_2(s) \in C^\infty(\mathbf{R})$ telles que

$$\chi_1(s) = 0,\ s \leqq -1/2,\ \chi_1(s) = 1,\ \ s \geqq -1/4,\ 0 \leqq \chi_1(s) \leqq 1,$$

$$\chi_2(s) = 0,\ s \leqq -1,\ \ \chi_2(s) = 1,\ s \geqq 1,\ \chi_2(s) + \chi_2(-s) = 1.$$

En utilisant ces fonctions, on introduit:

$$\langle\mu\xi'\rangle^2 = 1+\mu^2 \sum_{j=1}^{d} \xi_j^2,\quad \alpha_\varepsilon(x,\xi',\mu) = \chi_2(\varepsilon n^{1/2}\phi(x,\xi',\mu)\langle\mu\xi'\rangle^{1/2}),$$

$$J_\varepsilon(x,\xi',\mu) = \varepsilon(2\chi_1(\varepsilon\phi(x,\xi',\mu)\langle\mu\xi'\rangle^{1/2})-1)\phi(x,\xi',\mu)+\langle\mu\xi'\rangle^{-1/2},$$

$$I_\varepsilon(r)(x,\xi',\mu) = \langle\mu\xi'\rangle^{n\varepsilon^-}\ J_\varepsilon(x,\xi',\mu)^{-n\varepsilon-r},$$

$$m(\phi)(x,\xi',\mu) = (\phi(x,\xi',\mu)^2 + \langle\mu\xi'\rangle^{-1})^{1/2},$$

où $\varepsilon = \pm 1$, $\varepsilon^- = \max(0,-\varepsilon)$, $r \in \mathbf{R}$, $n \in \mathbf{R}^+$.

Dans la suite, par $\|\cdot\|$ on note la norme dans $L^2(\mathbf{R}^d)$.

<u>THÉORÈME 4.1</u>: On a

$$c \sum_{\varepsilon=\pm 1} \int^t \| I_\varepsilon(0)\alpha_\varepsilon p(x,D,\mu)u(x_0,\cdot)\|^2 dx_0 \geqq$$

$$\geqq \sum_{\varepsilon=\pm 1} \sum_{\substack{|\gamma|+j=m \\ j\geqq 1}} \sum_{k=0}^{[j/2]} \sum_{s=j-k}^{2j-2k} \theta^{2j-2k-s} n^s \times$$

$$\times \int^t \| I_\varepsilon(s/2-k)\langle\mu D'\rangle^k \alpha_\varepsilon B(x,D,\mu)^\gamma u(x_0,\cdot)\|^2 dx_0,$$

pour toute $u \in C^\infty(\mathbf{R}^{d+1})$ s'annulant dans $|x'| \geqq 1$ et pour tout $n_0 \leqq n$, $0 < \mu \leqq \mu_0(n)$, $\theta_0(n,\mu) \leqq \theta$.

<u>COROLLAIRE 4.1</u>: On a

$$c \int^t \| m(\phi)^n \langle\mu D'\rangle^n p(x,D,\mu)u(x_0,\cdot)\|^2 dx_0 \geqq$$

$$\geqq \sum_{\substack{|\gamma|+j=m \\ j\geqq 1}} n^{2j} \int^t \| m(\phi)^{-n-j}(\theta^{1/2} n^{-1/2} m(\phi)^{-1/2}+1)^j B(x,D,\mu)^\gamma u(x_0,\cdot)\|^2 dx_0,$$

pour toute $u \in C^\infty(\mathbb{R}^{d+1})$ s'annulant dans $|x'| \geq 1$ et pour tout $n_0 \leq n$, $0 < \mu \leq \mu_0(n)$, $\theta_0(n,\mu) \leq \theta$.

REMARQUE 4.1: Nous remarquons que on a

$$\phi(x,\xi',\mu) = \langle df(\rho'),x\rangle + O(\mu), \quad |x| \leq 1.$$

Alors cela donne

$$cm(\phi) \geq \{1+O(\mu) + \langle\mu\xi'\rangle^{-1/2}\}, \quad si\ x \notin T_{\rho'}\{f(x,\xi') = 0\},\ |x| \leq 1,$$

$$m(\phi) = O(\mu) + \langle\mu\xi'\rangle^{-1/2}, \quad si\ x \in T_{\rho'}\{f(x,\xi') = 0\},\ |x| \leq 1.$$

RÉFÉRENCES

[1] M.F. Atiyah, R. Bott et L. Garding: Lacunas for hyperbolic differential operators with constant coefficients, I Acta Math., 124 (1970), 109-189.

[2] N. Hanges: Parametrix and propagation of singularities for operators with non-involutive characteristics, Indiana Univ. Math. J., 28 (1979), 87-97.

[3] L. Hörmander: The Cauchy problem for differential equations with double characteristics, J. Analyse Math., 32 (1977), 118-196.

[4] V. Ja. Ivrii et V.M. Petkov: Necessary conditions for the Cauchy problem for non strictly hyperbolic equations to be well posed, Russian Math. Surveys, 29 (1974), 1-70.

[5] V. Ja. Ivrii: Sufficient conditions for regular and completely regular hyperbolicity, Trans. Moscow Math. Soc., 33 (1978), 1-65.

[6] V. Ja. Ivrii: Wave fronts of solutions of certain pseudodifferential equations, Trans. Moscow Math. Soc., 39 (1981), 49-86.

[7] T. Nishitani: Microlocal energy estimates for hyperbolic operators with double characteristics, Taniguchi Sym. HERT, Katata, 1984.

[8] T. Nishitani: Système effectivement hyperbolique, Séminaire sur les équations aux dérivées partielles hyperboliques et holomorphes (J. Vaillant) 1984-1985.

[9] T. Nishitani: Une classe d'opérateurs hyperboliques à caractéristiques triple, Séminaire sur les équations aux dérivées partielles hyperboliques et holomorphes (J. Vaillant) 1985-1986, à paraître.

Tatsuo Nishitani
Department of Mathematics
College of General Education,
Osaka University,
JAPAN.

C. PARENTI, E. BERNARDI & A. BOVE

Propagation of C^∞-singularities for a class of hyperbolic operators with double characteristics

0. INTRODUCTION

In this lecture we give an outer estimate of the C^∞-wave front set for the solutions of a class of hyperbolic (pseudo)-differential equations with double characteristics. Detailed proofs will appear in [2].

We consider here a second order pseudodifferential operator for which a suitable factorization exists and suppose that the subprincipal symbol of the operator satisfies a Levi condition (see §1). The Levi condition insures that the Cauchy problem is microlocally well posed (we refer to Ivrii [5] and B. Lascar-Sjöstrand [10] for results when the Levi condition is not satisfied).

Our approach is based on microlocal energy estimates. For some cases where a parametrix has been constructed we refer to B. Lascar-R. Lascar [9], R. Lascar [8], Melrose-Uhlmann [12]. In deriving energy estimates we follow Ivrii [4]. However, since the crucial lemma 2.3 in [4] (which is also the basic step in [5]) is not correct in the generality stated, correct energy estimates are obtained under some geometric restrictions (see §2). From the energy estimates we obtain a microlocal Hölmgren uniqueness result. As a consequence, using the machinery of Wakabayashi [14], [15], we can define the microlocal influence and dependence domains where propagation of the singularities occurs.

The main motivation for this work was to treat non-effectively hyperbolic operators. For the effectively hyperbolic case we refer to the important papers of Melrose [11], N. Iwasaki [6] and Nishitani [13].

1. CLASS OF OPERATORS AND STATEMENT OF THE MAIN RESULT

Denote by (x,ξ) the points in $T^* \mathbb{R}^{n+1}$, $x = (x_0,x')$, $\xi = (\xi_0,\xi')$, $x_0,\xi_0 \in \mathbb{R}$, x', $\xi' \in \mathbb{R}^n$. We put $D = (D_0,D')$, $D' = (D_1,\dots,D_n)$, $D_j = \frac{1}{\sqrt{-1}}\partial_{x_j}$, $j = 0,\dots,n$ and denote by $\sigma = \sum_0^n d\xi_j \wedge dx_j$ the symplectic form on $T^* \mathbb{R}^{n+1}$.

Consider in $\mathbf{R}^{n+1}$ an operator P of the form

$$P(x,D) = -D_0^{\,2} + A_1(x,D')D_0 + A_2(x,D') \tag{1.1}$$

where A_j is a classical properly supported pseudodifferential operator of order j, j = 1,2 depending smoothly on x_0, and denote by $p(x,\xi)$ the principal symbol of P. We suppose that P is hyperbolic with respect to dx_0, i.e.

$$\forall (x,\xi') \in \mathbf{R}^{n+1} \times (\mathbf{R}^n \setminus 0),\ p(x,\xi) = 0 \text{ has only real roots in } \xi_0. \tag{1.2}$$

Define

$$\Sigma = \{(x,\xi) \mid \xi' \neq 0,\ p(x,\xi) = 0\},\ \Sigma_1 = \{(x,\xi) \mid \xi' \neq 0,\ dp(x,\xi) = 0\}. \tag{1.3}$$

We have $\Sigma_1 \subset \Sigma$ and shall suppose that the double characteristic set Σ_1 is non empty. For every $\rho \in \Sigma$ we denote by $p_\rho(\delta z)$, $\delta z \in T_\rho(T^* \mathbf{R}^{n+1})$ the localization polynomial of P at ρ, namely

$$p_\rho(\delta z) = \begin{cases} \sigma(\delta z, H_p(\rho)), & \rho \in \Sigma \setminus \Sigma_1 \\ \sigma(\delta z, F_p(\rho)\delta z), & \rho \in \Sigma_1 , \end{cases} \tag{1.4}$$

where $H_p(\rho) = \sum_0^n {}_j \left(\frac{\partial p}{\partial \xi_j}(\rho)\frac{\partial}{\partial x_j} - \frac{\partial p}{\partial x_j}(\rho)\frac{\partial}{\partial \xi_j}\right)$ is the Hamiltonian vector field of p and $F_p(\rho)$ is the fundamental matrix of p at $\rho \in \Sigma_1$, i.e.

$$F_p(\rho) = \frac{1}{2} J \operatorname{Hess} p(\rho), \quad J = \begin{bmatrix} 0 & I_n \\ -I_n & 0 \end{bmatrix}. \tag{1.5}$$

We know that p_ρ is hyperbolic with respect to $\theta = (\delta x = 0,\ \delta\xi = (1,0,\ldots,0))$ (see Hörmander [3]), and we denote by $\Gamma_\rho \subset T_\rho(T^* \mathbf{R}^{n+1})$ the hyperbolicity cone of p_ρ, i.e. the connected component of $\{\delta z \mid p_\rho(\delta z) \neq 0\}$ containing θ.

It is well known that the spectrum of $F_p(\rho)$, $sp(F_p(\rho))$, is confined on the imaginary axis with the possible exception of two real eigenvalues $\pm\lambda$, $\lambda > 0$ (see Hörmander [3]).

We say that P is effectively hyperbolic (resp. non-effectively hyperbolic) at $\rho \in \Sigma_1$ if $F_p(\rho)$ has real non-zero eigenvalues (resp. $sp\ (F_p(\rho)) \subset i\,\mathbf{R}$).

At non-effectively hyperbolic points a further classification can be given. Precisely, if $sp(F_p(\rho)) \subset i\mathbb{R}$, then either $\ker F_p^2(\rho)$ is symplectic and $\ker F_p^2(\rho) = \ker F_p^3(\rho)$ or $\ker F_p^2(\rho)$ is non-symplectic and $\ker F_p^2(\rho) \neq \ker F_p^3(\rho) \neq \ker F_p^4(\rho) = \ker F_p^5(\rho)$ with 1-dimensional gaps between $\ker F_p^2$, $\ker F_p^3$ and $\ker F_p^4$ (see Hörmander [3]).

We restrict the class of operators (1.1), (1.2) by supposing that P can be factored as follows:

$$P = -\Lambda_-(x,D)\Lambda_+(x,D) + q(x,D') \tag{1.6}$$

where $\Lambda_\pm(x,D) = D_0 - \lambda_\pm(x,D')$ and $\lambda_\pm(x,D')$ (resp. $q(x,D')$) are classical properly supported pseudodifferential operators of order 1 (resp. 2) with real principal symbols $\lambda_\pm(x,\xi')$ (resp. $q(x,\xi')$).

We will assume:

(A.1) $q(x,\xi') \geqq 0 \quad \forall(x,\xi') \in \mathbb{R} \times (T^*\mathbb{R}^n \smallsetminus 0)$.

(A.2) For every conic set $V \subset\subset \mathbb{R} \times (T^*\mathbb{R}^n \smallsetminus 0)$ there exists $C > 0$ s.t.

$$|\{\xi_0 - \lambda_+(x,\xi'), q(x,\xi')\}| \leqq C\, q(x,\xi'), \ \forall(x,\xi') \in V.$$

(A.3) (Levi conditions). Put $\Sigma' = \{(x,\xi') | \xi' \neq 0,\ q(x,\xi') = 0\}$ and denote by $q_1^s(x,\xi') = q_1(x,\xi') - \frac{1}{2i}\sum_1^n{}_j \frac{\partial^2 q}{\partial x_j \partial \xi_j}(x,\xi')$ the subprincipal symbol of $q(x,D')$. We suppose:

a) $\operatorname{Re} q_1^s(\rho) + \operatorname{Tr}^+ F_q(\rho) > 0, \quad \forall \rho \in \Sigma'$,

where $F_q(\rho)$ is the fundamental matrix of $q(x,\xi')$ at ρ and

$$\operatorname{Tr}^+ F_q(\rho) = \sum_{\substack{\mu_j \geqq 0,\ i\mu_j \in sp(F_q(\rho))}} \mu_j$$

b) For every conic set $V \subset\subset \mathbb{R} \times (T^*\mathbb{R}^n \smallsetminus 0)$ there exists $C > 0$ s.t.

$$|\operatorname{Im} q_1^s(x,\xi')| \leqq C\, q(x,\xi')^{1/2}, \ \forall(x,\xi') \in V.$$

(A.4) $F_q^2(\rho)$ has constant rank on Σ'.

REMARK: Since we work microlocally we need only to require (1.6) and conditions (A.1)-(A.4) to hold near a given point of Σ_1.

For $\rho \in \Sigma$ we define

$$C_\rho = \begin{cases} \Gamma_\rho, & \rho \in \Sigma \setminus \Sigma_1 \\ \Gamma_\rho \cap \ker F_q^2(\rho), & \rho \in \Sigma_1 \end{cases} \tag{1.7}$$

and put

$$C_\rho^\sigma = \{\delta z \in T_\rho(T^* \mathbf{R}^{n+1}) | \sigma(\delta z', \delta z) \geq 0, \forall \delta z' \in C_\rho\}. \tag{1.8}$$

We can now define the microlocal influence and dependence domains:

$$K_\rho^\pm = \{\rho' \in \Sigma \mid \exists\, \gamma\ [0,1] \to \Sigma,\ \gamma(t) \text{ Lipschitz-continuous, s.t.} \tag{1.9}$$
$$\gamma(0) = \rho,\ \gamma(1) = \rho',\ \frac{d}{dt}\gamma(t) \in \pm\, C_{\gamma(t)}^\sigma, \text{ a.e. in } t\}.$$

We have the following main result.

THEOREM: Let P be given by (1.6) and satisfy conditions (A.1)-(A.4). Let $u \in \mathcal{D}'(\mathbf{R}^{n+1})$ with $\rho \in \Sigma \setminus WF(Pu)$. Suppose that for some conic neighbourhood $\mathcal{U}$ of ρ and for a choice of the sign + or - we have $WF(u) \cap \mathcal{U} \cap (K_p^\pm \setminus \{\rho\}) = \emptyset$. Then $\rho \notin WF(u)$.

REMARKS:

1. When P is effectively hyperbolic at Σ_1, N. Iwasaki [6] proved the existence of a factorization $p = -(\xi_0 - \lambda_-)(\xi_0 - \lambda_+) + q$ for which conditions (A.1) and (A.2) are satisfied (actually N. Iwasaki proved that $\{\xi_0 - \lambda_+, q\} = c\, q$ for some smooth function $c(x,\xi')$).

2. Suppose that Σ_1 is a smooth submanifold of $T^* \mathbf{R}^{n+1}$ and that

i) $T_\rho \Sigma_1 = \ker F_p(\rho), \quad \forall \rho \in \Sigma_1$

ii) The restriction of σ to $T\Sigma_1$ has constant rank

iii) $Sp(F_p(\rho)) \subset i\mathbf{R}$ and $\ker F_p^2(\rho)$ is symplectic for every $\rho \in \Sigma_1$.

Under the above conditions Ivrii [4] and Hörmander [3] proved the existence of a factorization $p = -(\xi_0 - \lambda_-)(\xi_0 - \lambda_+) + q$ satisfying (A.1) and (A.2) actually one can take $\lambda_- = -\lambda_+$ and q vanishing exactly to second order on Σ').

3. Suppose that conditions i), ii) or Remark 2 are satisfied and replace iii) by iii)' $\ker F_p^2(\rho)$ is non-symplectic for every $\rho \in \Sigma_1$.
(iii)' implies that P is non-effectively hyperbolic at every point of Σ_1).
Under the above conditions there is a smooth vector field θ defined on Σ_1 for which $\ker F_p^4(\rho) = \ker F_p^3(\rho) \oplus [\theta(\rho)]$, $\rho \in \Sigma_1$. Moreover, we can assume that $V = [-F_p^3\theta, -F_p\theta;\theta, F_p^2\theta]$ is symplectic and $\ker F_p^4 = V \oplus L$ (symplectic decomposition) with $F_p^2(L) = (0)$ and $\ker F_p^2 \cap \operatorname{Im} F_p^2 = [-F_p^3, \theta, F_p^2\theta]$.
We can then find a factorization $p = -(\xi_0 - \lambda_-)(\xi_0 - \lambda_+) + q$ satisfying condition (A.1), with $H_{\xi_0-\lambda_+}(\rho) \equiv H_{\Lambda_+}(\rho) // F_p^3(\rho)\theta(\rho)$, $\rho \in \Sigma_1$.

Such a factorization satisfies condition (A.2) if and only if for any function $s(x,\xi)$ for which $H_s(\rho) - F_p^2(\rho)\theta(\rho) \in \ker F_p(\rho)$, $\rho \in \Sigma_1$, we have $p_\rho(H_{\{s,\Lambda_+\}}(\rho)) = 0$, $\rho \in \Sigma_1$.
(For other results concerning this case see Nishitani [13] and Bernardi-Bove [1]). We explicitly remark that in cases 2- and 3-, the factorization of p satisfies condition (A.4). Furthermore, for every conic set $V \subset\subset \mathbb{R} \times (T^*\mathbb{R}^n \setminus 0)$ there exists $C > 0$ s.t. on V we have:

$$(A.2)' \quad |\{\xi_0-\lambda_-(x,\xi'),\xi_0-\lambda_+(x,\xi')\}| \leqq C[q(x,\xi')^{1/2} + |\lambda_+(x,\xi')-\lambda_-(x,\xi')|].$$

Ivrii proved in [5] that under conditions (A.1), (A.2) and (A.2)' the null-bicharacteristics of p starting in $\Sigma \setminus \Sigma_1$ do not have limit points in Σ_1.

4. Concerning Levi conditions we remark that if $\Sigma' = \{(x,\xi') \mid \exists\ \xi_0, (x,\xi_0,\xi') \in \Sigma_1\}$, then condition (A.3)a) is equivalent to

$$\operatorname{Re} p_1^s(\rho) + \operatorname{Tr}^+ F_p(\rho) > 0, \quad \forall\ \rho \in \Sigma_1,$$

p_1^s being the subprincipal symbol of P and $\operatorname{Tr}^+ F_p(\rho)$ being defined as $\operatorname{Tr}^+ F_q(\rho)$ (actually these traces are identical).

Condition (A.3)b) is more involved, but reduces trivially to $\operatorname{Im} q_1^s|_{\Sigma'} = 0$ when Σ' is smooth and q vanishes exactly to second order on Σ'.

Note that

$$\operatorname{Im} p_1^s = 1/2\ \{\xi_0 - \lambda_-, \xi_0 - \lambda_+\} + \operatorname{Im} q_1^s.$$

5. It is easy to see that in the definition of $K^{\pm}$ one can replace C_ρ^σ with $C_\rho^\sigma \cap \{\delta z \in T_\rho(T^* \mathbf{R}^{n+1}) | p_\rho(\delta z) = 0\}$.

If P is non-effectively hyperbolic at $\rho \in \Sigma_1$ and ker $F_p^2(\rho)$ is non symplectic on can show that

$$C_\rho^\sigma \cap \{\delta z | p_\rho(\delta z) = 0\} = [H_{\Lambda_+}(\rho)]^+ = \{t\, H_{\Lambda_+}(\rho) | t \geq 0\}.$$

On the other hand, if P is non-effectively hyperbolic at $\rho \in \Sigma_1$ and ker $F_p^2(\rho)$ = ker $F_q^2(\rho)$ (which amounts to requiring $F_q^2(\rho) H_{\Lambda_-}(\rho) = 0$) then we have

$$C_\rho^\sigma \cap \{\delta z | p_\rho(\delta z) = 0\} = \Gamma_\rho^\sigma \cap \ker F_p(\rho).$$

2. SKETCH OF THE PROOF

From now on we suppose that P is given by (1.4) and conditions (A.1)-(A.4) are satisfied.

Fix any bounded interval $I \subset \mathbf{R}$ and $K \subset\subset \mathbf{R}^n$. For any $u \in C^\infty(I, C_0^\infty(K))$ and $s \in \mathbf{R}$, define the "s-energy" of u as:

$$E_s(u;t) = \|(\Lambda_+ u)(t); H^s(\mathbf{R}^n)\|^2 + \mathrm{Re}(q(t,x',D')\, u(t,\cdot), u(t,\cdot))_s + \alpha \|u(t); H^s(\mathbf{R}^n)\|^2, \quad t \in I, \tag{2.1}$$

where $\alpha > 0$ is any constant (indpendent of u) such that

$$\mathrm{Re}(q(t,x',D')\, u(t,\cdot), u(t,\cdot))_s + \alpha \|u(t); H^s(\mathbf{R}^n)\|^2 \geq \text{const.} \|u(t); H^{s+1/2}(\mathbf{R}^n)\|^2, \quad t \in I \tag{2.1)'$$

((2.1)' is a consequence of Melin's inequality).

Ivrii proved in [4] that there exists $C > 0$ s.t. for every t_0, $t \in I$ the following inequality holds $\forall u \in C^\infty(I; C_0^\infty(K))$:

$$E_s^{1/2}(u;t) \leq C[E_s^{1/2}(u;t_0) + |\int_{t_0}^{t} \|Pu(t'); H^s(\mathbf{R}^n)\|\, dt'|]. \tag{2.2}$$

To microlocalize ineq. (2.2) near Σ_1 we introduce the notion of symbol of space type (see [4], [5]).

A smooth real function $\varphi(x,\xi')$ defined in a conic neighbourhood of a point $\rho_0 \in \Sigma_1$ and positively homogeneous of degree 0 in ξ' is called a symbol of space type at ρ_0 iff $H_\varphi(\rho_0) \in \Gamma_\rho$. This condition can be shown to be equivalent to the existence of a conic neighbourhood V of ρ_0 s.t.

i) $\{\varphi,\Lambda_\pm\} > 0$ in V $(\Lambda_\pm = \xi_0 - \lambda_\pm)$

ii) $\forall\, V' \subset\subset V \;\exists\, C > 1/4$ s.t. (2.3)

$$q\,\{\varphi\,,\Lambda_+\}\;\;\{\varphi\,,\Lambda_-\} \geqq C\;\;\{\varphi,q\}^2 \text{ in } V'.$$

To a symbol of space type φ we associate the symbol

$$q_\varphi = q - 1/4\;\;\{\varphi,q\}^2/\{\varphi,\Lambda_+\}\;\;\{\varphi,\Lambda_-\}). \tag{2.4}$$

It is easily seen that q_φ is locally equivalent to q and that the fundamental matrix of q_φ at Σ' is given by

$$F_{q_\varphi}(\rho)\delta z = F_q(\rho)\;\delta z - k(\rho)\;\sigma(\delta z, F_q(\rho)H_\varphi(\rho))F_q(\rho)H_\varphi(\rho), \tag{2.5}$$

$$k(\rho) = 1/(\{\varphi,\Lambda_+\}(\rho)\{\varphi,\Lambda_-\}(\rho)).$$

In deriving microlocal energy estimates from (2.2) Ivrii [4, Lemma 2.3] claims that $F_{q_\varphi}(\rho)$ has the same spectrum as $F_q(\rho)$. This is not correct since we only have $\mathrm{Tr}^+ F_{q_\varphi}(\rho) \leqq \mathrm{Tr}^+\;F_q(\rho)$; moreover, the following conditions can be shown to be equivalent (see [2]);

1. $\mathrm{Sp}(F_{q_\varphi}(\rho)) = \mathrm{Sp}(F_q(\rho))$
2. $\mathrm{Tr}^+ F_{q_\varphi}(\rho) = \mathrm{Tr}^+ F_q(\rho)$
3. $H_\varphi(\rho) \in \ker F_q^2(\rho)$.

Let now $\varphi(x,\xi')$ be a symbol of space type defined near $\rho_0 \in \Sigma'$ and make the further assumption:

$$\mathrm{Re}\; q_1^s(\rho_0) + \mathrm{Tr}^+\;F_{q_\varphi}(\rho_0) > 0. \tag{2.6}$$

Define, microlocally,

$$\psi(x,\xi') = \begin{cases} e^{-1/\varphi(x,\xi')} & , \text{ where } \varphi > 0 \\ 0 & , \text{ where } \varphi \leq 0 \end{cases} \tag{2.7}$$

Then taking real symbols $h(x,\xi')$, $h'(x,\xi')$ (positively homogeneous of degree 0 in ξ') with conic support close to ρ_0 and satisfying

$$\begin{cases} 0 \leq h \leq 1, \quad h \equiv 1 \text{ near } \rho_0 \\ 0 \leq h' \leq 1, \; h' \equiv 0 \text{ near } \rho_0, \; h' \equiv 1 \text{ near } \{0 < h < 1\}, \end{cases}$$

the following inequality holds for every $u \in C^\infty(I, C_0^\infty(K))$ and any $t_0, t \in I$, $t_0 < t$:

$$E_s^{1/2}(\psi(x,D')h(x,D')u;t) \leq \tag{2.8}$$

$$C[E_s^{1/2}(\psi(x,D')\, h(x,D')u;t_0) + \int_{t_0}^{t} \|\psi(x,D')\, h(x,D')\, Pu(t');H^s(\mathbb{R}^n)\|\, dt'$$

$$+ \sup_{t'\in[t_0,t]} (E_{s-1/2}^{1/2}\,(u;t') + E_s^{1/2}\,(\psi(x,D')\, h'(x,D')u;\, t'))].$$

Using ineq. (2.8) one can prove a microlocal uniqueness result.

<u>LEMMA 1</u>: Let $\gamma(x,\xi)$ be a smooth real function (positively homogeneous of degree 1 in ξ) defined near $\rho_0 \in \Sigma_1$ and satisfying:

1. $\gamma(\rho_0) = 0$, $H_\gamma(\rho_0) \in \Gamma_{\rho_0}$
2. $|(1-\pi_{\rho_0})H_\gamma(\rho_0)| \leq \varepsilon|\pi_{\rho_0}\, H_\gamma(\rho_0)|$ for $\varepsilon > 0$ small,

   π_{ρ_0} being the symplectic projector on $\ker F_q^2(\rho_0)$.

Then for every $u \in \mathcal{D}'(\mathbb{R}^{n+1})$ for which:

i) $\rho_0 \notin WF(Pu)$

ii) $WF(u) \cap \{\gamma > 0\} \cap \mathcal{U} = \emptyset$, for some conic neighbourhood $\mathcal{U}$ of ρ_0,

we have $\rho_0 \notin WF(u)$.

From Lemma 1 we get the crucial result:

<u>LEMMA 2</u>: Let $u \in \mathcal{D}'(\mathbb{R}^{n+1})$ with $\rho_0 = (x^{(0)},\xi^{(0)}) \in \Sigma_1 \cap (WF(u) \setminus WF(Pu))$.

Then for every relatively open cone $C \subset\subset \Gamma_{\rho_0} \cap \ker F_q^2(\rho_0)$ there exists

T > 0 such that:

$$WF(u) \cap (\rho_0 - C^\sigma) \cap \{(x,\xi)|x_0 = x_0^{(0)} - t\} \neq \emptyset, \quad \forall t \in]0,T].$$

REMARK: In proving Lemma 2 we use condition (A.4) and a sweeping-out argument inspired by John [7].

Having Lemmas 1 and 2, the proof of the Theorem is obtained by adapting some arguments from Wakabayashi [14], [15].

References

[1] E. Bernardi - A. Bove: Cauchy problem for a class of hyperbolic operators with double characteristics: Geometric results, preprint.

[2] E. Bernardi - A. Bove - C. Parenti: to appear.

[3] L. Hörmander: The Cauchy problem for differential equations with double characteristics, J. Analyse Math., 32 (1977), 118-196.

[4] V. Ja Ivrii: The well-posedness of the Cauchy problem for non-strictly hyperbolic operators III, The energy integral, Trans. Moscow Math. Soc., 34 (1978), 149-168.

[5] V. Ja Ivrii: Wave fronts of solutions of certain hyperbolic pseudo-differential equations, Trans. Moscow Math. Soc., 39 (1981), 87-119.

[6] N. Iwasaki: The Cauchy problem for effectively hyperbolic equations (A standard type), Publ. RIMS, Kyoto Univ., 20 (1984), 543-584.

[7] F. John: On linear partial differential equations with analytic coefficients, unique continuation of data, Comm. Pure Appl. Math. Math., 2 (1949), 209-254.

[8] R. Lascar: Propagation des singularités ..., Springer Lecture Notes in Math., 856 (1981).

[9] B. Lascar - R. Lascar: Propagation des singularités pour des équations hyperboliques à caractéristiques de multiplicité au plus doubles et singularités masloviennes II, J. Analyse Math., 41 (1982), 1-38.

[10] B. Lascar - J. Sjöstrand: Equation de Schrödinger et propagation pour des O.P.D. à caractéristiques réelles de multiplicité variable II, Comm. P.D.E., 10 (1985), 467-523.

[11] R. Melrose: The Cauchy problem for effectively hyperbolic operators, Hokkaido Math. J., 12 (1983), 371-391.

[12] R. Melrose - G. Uhlmann: Microlocal structure of involutive conical refraction, Duke Math. J., 46 (1979), 571-582.

[13] T. Nishitani: Microlocal energy estimates for hyperbolic operators with double characteristics, Taniguchi Symp. Hert Katata (1984), 235-255.

[14] S. Wakabayashi: Singularities of solutions of the Cauchy problem for symmetric hyperbolic systems, Comm. P.D.E., 9 (1984), 1147-1177.

[15] S. Wakabayashi: Generalized flows and their applications, Advances in Microlocal Analysis (Ed. H.G. Garnir), D. Reidel Publishing Company (1986), 363-384.

E. Bernardi, A. Bove and C. Parenti
Department of Mathematics
University of Bologna,
Piazza di Porta S. Donato, 5,
40127 Bologna,
Italy.

J. PERSSON

The regularization problem for the wave equation with measures as potentials and driving forces

1. INTRODUCTION

Let μ and η be signed Borel measures in $\mathbf{R}^2$. We shall treat the Cauchy problem

$$\partial^2u/\partial t^2 - \partial^2u/\partial x^2 + \mu u = \eta, \; u(x,0) = \partial u/\partial t(x,0) = 0. \tag{1.1}$$

Formally one gets the solution of (1.1) as the solution of

$$u(x,t) = -2^{-1}\int_{D(x,t)} u(r,s)d\mu(r,s) + 2^{-1}\int_{D(x,t)} d\eta(r,s), \; t \geqq 0. \tag{1.2}$$

Here $D(x,t) = D(x,t,t)$ where

$$D(x,t,\varepsilon) = \{(r,s);\; x-t+s \leqq r \leqq x+t-s,\; t-\varepsilon < s \leqq t\},\; \varepsilon > 0. \tag{1.3}$$

According to Persson [2] (1.2) has a solution if

$$\mu(\{(x,t)\}) \neq -2,\; x \in \mathbf{R},\; t > 0. \tag{1.4}$$

If (1.2) represents a physical model then condition (1.4) is very unnatural. It seems more natural to regularize the measures and then take the limit of the solutions of the regularized problems as the solution of (1.1). We prove that the limit always exists, Theorem 1.1 below. Under some extra conditions on μ and η Theorem 1.2 below tells us how one can modify (1.2) so that the modified equation has the limit as its solution.

We regularize μ and η by convolving them with the functions $\phi(\cdot,\cdot,\varepsilon)$, $\varepsilon > 0$, where

$$\phi(x,t,\varepsilon) = \varepsilon^{-2},\; (x,t) \in D(0,0,\varepsilon) \text{ and } \phi(x,t,\varepsilon) = 0 \text{ elsewhere.} \tag{1.5}$$

We let

$$E(x,t,\varepsilon) = \{(r,s);\ x+t-s \leqq r \leqq x-t+s,\ t \leqq s < t+\varepsilon\}. \tag{1.6}$$

Then we define $\mu(x,t,\varepsilon) = \int_{R^2} \phi(x-r,t-s,\varepsilon)d\mu(r,s) = \varepsilon^{-2} \int_{E(x,t,\varepsilon)} d\mu(r,s)$, and $\eta(x,t,\varepsilon) = \varepsilon^{-2} \int_{E(x,t,\varepsilon)} d\eta(r,s)$. In the following we write $d\mu(r,s,\varepsilon) = \mu(r,s,\varepsilon)drds$ and $d\eta(r,s,\varepsilon) = \eta(r,s,\varepsilon)drds$. The regularized problem corresponding to (1.2) is

$$u(x,t,\varepsilon) = -\ 2^{-1} \int_{D(x,t)} u(r,s,\varepsilon)d\mu(r,s,\varepsilon) + 2^{-1} \int_{D(x,t)} d\eta(r,s,\varepsilon). \tag{1.7}$$

It is pointed out in Persson [4] that for certain μ and η $u(x,t,\varepsilon) \nrightarrow u(x,t)$ of (1.2) when ε tends to zero.
The theorems run as follows.

<u>THEOREM 1.1</u>: Let μ and η be signed Borel measures in $\mathbf{R}^2$. Let $u(x,t,\varepsilon)$ be the solution of (1.7). Then $u(x,t,0) = \lim_{\varepsilon\to 0} u(x,t,\varepsilon)$ exists for each (x,t), $t > 0$.

<u>THEOREM 1.2</u>: Let $g(s) = 2s^{-1} (1-\cosh\sqrt{-s}) (\cosh \sqrt{-s})^{-1}$, $s \neq 0$, and let $g(0) = 1$. Let μ and η be signed Borel measures in $\mathbf{R}^2$. Let $((x_j,t_j))_{j=1}^{\infty}$ be a sequence of points in $\mathbf{R}^2$ such that on the characteristic lines through (x_j,t_j) $|\mu| + |\eta|$ has no positive mass outside (x_j,t_j), $j = 1,2,\ldots$. Further let $|\mu|$ have no positive mass on any characteristic line not containing a point (x_j,t_j). Then the conclusion of Theorem 1.1 is true and

$$u(x,t,0) = -2^{-1} \int_{D(x,t)} u(r,s,0)g(\mu(\{(r,s)\}))d\mu(r,s) + \tag{1.8}$$
$$+\ 2^{-1} \int_{D(x,t)} g(\mu(\{(r,s)\}))d\eta(r,s).$$

The proof of Theorem 1.2 is given in two steps. In Section 2 we prove that the solutions of (1.7) are equibounded on compact sets without any restrictions on μ and η, Theorem 2.1. The rest of the proof follows in Section 3. It is a modification of the corresponding proof for ordinary measure differential equations in Persson [3, Theorem 3.1]. It depends heavily on the explicit example given in Persson [4, Section 2].

Theorem 1.1 is proved in Section 4. Although the proof contains explicit computations we have not found any modified version of (1.2) having $u(x,t,0)$ as its solution. Still one may conjecture that such a modification exists. We also believe that the limit exists independent of the choice of regularization at least outside a set of Lebesgue measure zero. In the ordinary differential equation case this question is settled, Persson [3]. Another question is how the results of this note correspond to the physical model.

The equation (1.2) or other forms of the wave equation with measures involved are treated in Persson [1], [2] and [4]. We also refer to the references of these papers.

2. EQUIBOUNDEDNESS OF SOLUTIONS

We prove that the solutions of (1.7) are equibounded on compact sets as a family with parameter ε.

THEOREM 2.1: Let the hypothesis be as in Theorem 1.1. Then the solutions $(u(\cdot,\cdot,\varepsilon))_{0<\varepsilon\leq 1}$ are locally equibounded.

PROOF: We solve (1.7) by successive approximations. Let

$$u_0(x,t,\varepsilon) = 2^{-1}\int_{D(x,t)} d\eta(r,s,\varepsilon), \tag{2.1}$$

and

$$u_{j+1}(x,t,\varepsilon) = -2^{-1}\int_{D(x,t)} u_j(r,s,\varepsilon)d\mu(r,s,\varepsilon),\ j = 0,1,\ldots, \tag{2.2}$$

with $D(x,t)$ defined by (1.3). We shall prove that $u(x,t,\varepsilon) = \sum_{j=0}^{\infty} u_j(x,t,\varepsilon)$ is bounded by a constant independent of ε on each compact set.

We need some notations for sets in $\mathbf{R}^4$ connected with the integration in (2.1) and (2.2) when one also takes into account the integration due to the regularization. On this point we also remark that here and in the following we use Fubini's theorem without referring to it at the individual steps. Let

$$L(x,t,\varepsilon) = \{(r,s,r',s');\ x-t+s \leq r \leq x+t-s,\ 0 < s \leq t, \tag{2.3}$$
$$s \leq s' < s+\varepsilon,\ r+s-s' \leq r' \leq r-s+s'\},$$

and

$$H(x,t,\varepsilon) = \{(r,s,r',s');\ x-t+s' \leq r' \leq x+t-s',\ 0 < s' \leq t, \tag{2.4}$$
$$r'-s'+s \leq r \leq r'+s'-s,\ \max(0,s'-\varepsilon) < s < s'\}.$$

The reader may easily prove the following lemma.

<u>LEMMA 2.2</u>: Let $L(x,t,\varepsilon)$ and $H(x,t,\varepsilon)$ be defined as in (2.3) and (2.4). Then.

$$H(x,t,\varepsilon) \subset L(x,t,\varepsilon) \subset H(x,t+2\varepsilon,\varepsilon). \tag{2.5}$$

Let (x',t') be fixed with $t' > 0$. We shall prove that $(u(x,t,\varepsilon))_{0<\varepsilon\leq 1}$ is equibounded on $D(x',t')$. Let $D(x,t) \subset D(x',t')$. Then Lemma 2.2 gives

$$|u_0(x,t,\varepsilon)| \leq 2^{-1}\int_{D(x,t)} d|\eta|(r,s,\varepsilon) = \tag{2.6}$$
$$= 2^{-1}\varepsilon^{-2}\int_{D(x,t)} \left(\int_{E(r,s,\varepsilon)} d|\eta|(r',s')\right)drds =$$
$$= 2^{-1}\varepsilon^{-2}\int_{L(x,t,\varepsilon)} d|\eta|(r',s')drds \leq$$
$$\leq 2^{-1}\varepsilon^{-2}\int_{H(x,t+2\varepsilon,\varepsilon)} d|\eta|(r',s')drds \leq$$
$$\leq 2^{-1}\varepsilon^{-2}\int_{D(x,t+2\varepsilon)} \left(\int_{D(r',s',\varepsilon)} drds\right)d|\eta|(r',s') =$$
$$= 2^{-1}\int_{D(x,t+2\varepsilon)} d|\eta|(r',s') \leq$$
$$\leq 2^{-1}\int_{D(x',t'+2)} d|\eta|(r',s') = C,\ 0 < \varepsilon \leq 1.$$

In order to be consistent with the definitions we should have changed the variables in the last three steps of (2.6). We have preferred not to do that. We will follow the same line in the rest of the paper.

With C from (2.6) we now assert that

$$|u_j(x,t,\varepsilon)| \leq 2^{-j} C\exp(\int_{D(x,t)} d|\mu|(r,s,\varepsilon)),\ j = 0,1,\ldots, \tag{2.7}$$

$$D(x,t) \subset D(x',t').$$

It is true for j = 0. Let it be true for a certain j. Then (2.2) and (2.7) give

$$|u_{j+1}(x,t,\varepsilon)| \leq 2^{-j-1} \int_{D(x,t)} |u_j(r,s,\varepsilon)| d|\mu|(r',s',\varepsilon) \leq \tag{2.8}$$

$$\leq 2^{-j-1} C \int_{D(x,t)} |\mu|(r,s,\varepsilon)\exp(\int_{D(r,s)} |\mu|(r',s',\varepsilon)dr'ds')drds.$$

Let $y = 2^{-1/2}(x+t)$ and let $z = 2^{-1/2}(x-t)$. In the new variables D(x,t) is transformed into

$$D(y,z) = \{(y',z');\ y' \leq y,\ z' \geq z,\ 0 \leq y'-z' \leq y-z\ . \tag{2.9}$$

In the new variables (2.8) gives

$$|u_{j+1}(x,t,\varepsilon)| \leq \tag{2.10}$$

$$\leq 2^{-j-1} C \int_z^y \int_z^{y'} |\mu|(y',z',\varepsilon)\exp(\int_{z'}^{y'} (\int_{z'}^{y''} |\mu|(y'',z'',\varepsilon)dz'')dy'')dz'dy' \leq$$

$$\leq 2^{-j-1} C \int_z^y \int_z^{y'} |\mu|(y',z',\varepsilon)\exp(\int_{z}^{y'} (\int_{z}^{y''} |\mu|(y'',z'',\varepsilon)dz'')dy'')dz')dy'.$$

Here we have used that $z' \geq z$. We also notice that the integrand of the outer integration in the last member of (2.10) has $\exp\ (\int_z^{y'} (\int_z^{y''} |\mu|(y'',z'',\varepsilon)dz'')dy'')$ as a primitive function with (y,z) fixed. We use this in (2.10) and get (2.7) with j replaced by j+1. Thus (2.7) is true for all j. It follows from (2.7) and (2.6) with $C' = \int_{D(x',t'+2)} d|\mu|(r,s)$ that

$$|u(x,t,\varepsilon)| \leq 2Ce^{C'},\ D(x,t) \subset D(x',t'),\ 0 < \varepsilon \leq 1.$$

Theorem 2.1 is proved.

3. PROOF OF THEOREM 1.2

At first we assume that $\mu(\{(x_j,t_j)\}) = 0$ for all j. It follows from the hypothesis that no characteristic line has a positive $|\mu|$-measure. It is also obvious that there is at most a countable number of characteristic lines with positive $|\eta|$-measure since $|\eta|$ is a Borel measure. The union of these characteristic lines then has $|\mu|$-measure zero. This fact will be used later on.

Let $\varepsilon > 0$. Let $D(x,t) \subset D(x',t')$. We combine (1.7) and (1.2) and get

$$u(x,t,\varepsilon) - u(x,t) = -2^{-1}\int_{D(x,t)} (u(r,s,\varepsilon)-u(r,s))d\mu(r,s) - \tag{3.1}$$

$$- 2^{-1}\int_{D(x,t)} u(r,s,\varepsilon)d(\mu(r,s,\varepsilon)-\mu(r,s)) +$$

$$+ 2^{-1}\int_{D(x,t)} d(\eta(r,s,\varepsilon)-\eta(r,s)).$$

We let $v(x,t,\varepsilon) = u(x,t,\varepsilon) - u(x,t)$. It follows from (3.1) that

$$v(x,t,\varepsilon) = -2^{-1}\int_{D(x,t)} v(r,s,\varepsilon)d\mu(r,s) + g(x,t,\varepsilon), \tag{3.2}$$

where

$$g(x,t,\varepsilon) = -2^{-1}\int_{D(x,t)} u(r,s,\varepsilon)d(\mu(r,s,\varepsilon)-\mu(r,s)) + \tag{3.3}$$

$$+ 2^{-1}\int_{D(x,t)} d(\eta(r,s,\varepsilon)-\eta(r,s)) = A + B.$$

We prove that

$$g(x,t,\varepsilon) \to 0, \quad \varepsilon \to 0, \tag{3.4}$$

and that

$$\int_{D(x,t)} g(r,s,\varepsilon)d\mu(r,s) \to 0, \quad \varepsilon \to 0, \text{ uniformly in } D(x',t'). \tag{3.5}$$

Let t", $0 < t" \leqq t'$ be the maximal t fulfilling $|\mu|(D(x,t)) \leqq 1$,

$(x,t) \in D(x',t')$. Then (3.4), (3.5) and [2, Section 3] show that $v(x,t,\varepsilon) \to 0$, $\varepsilon \to 0$, $0 < t \leqq t''$. If $t'' < t'$ we choose $\bar{t}$ to be the maximal t such that $|\mu|(D(x,t,t-t'')) \leqq 1$, $(x,t) \in D(x',t')$. Then we get

$$v(x,t,\varepsilon) = - 2^{-1} \int_{D(x,t,t-t')} v(r,s,\varepsilon)d\mu(r,s) +$$

$$- 2^{-1} \int_{D(x,t)\setminus D(x,t,t-t')} v(r,s,\varepsilon)d\mu(r,s) + g(x,t,\varepsilon),$$

$$(x,t) \in D(x',t'),\ t' < t \leqq \bar{t}.$$

Here the second term of the right member tends uniformly to zero because of the result for $0 < t \leqq t''$. Combined with (3.4) and (3.5) it proves that $v(x,t,\varepsilon) \to 0$, $\varepsilon \to 0$, $t'' < t \leqq \bar{t}$. After a finite number of steps one has proved that $v(x,t,\varepsilon) \to 0$ in $D(x',t')$.

We notice that $\varepsilon^{-2} \int_{D(r',s',\varepsilon)} drds = 1$. We look at (2.3) and (2.4). From (2.4) and Lemma 2.2 we get

$$B = 2^{-1} \int_{L(x,t,\varepsilon)} \varepsilon^{-2} d\eta(r',s')drds - \tag{3.7}$$

$$- 2^{-1} \int_{D(x,t)} \left(\int_{D(r',s',\varepsilon)} \varepsilon^{-2} drds\right)d\eta(r',s') =$$

$$= 2^{-1} \int_{L(x,t,\varepsilon)\setminus H(x,t,\varepsilon)} \varepsilon^{-2} d\eta(r',s')drds -$$

$$- 2^{-1}\varepsilon^{-2} \int_{0+}^{\varepsilon} \int_{x-t+s'}^{x+t-s'} \left(\int_{s'-\varepsilon}^{0} \int_{r'-s'+s}^{r'+s'-s} drds\right)d\eta(r',s').$$

Let

$$f(r',s',\varepsilon) = \varepsilon^{-2} \int_{\max(0,s'-\varepsilon)}^{s'} \left(\int_{r'-s'+s}^{r'+s'-s} dr\right)ds. \tag{3.8}$$

We see that

$$0 \leqq f(r',s',\varepsilon) \leqq 1,\ f(r',s',\varepsilon) = 1,\ s' \geqq \varepsilon. \tag{3.9}$$

Lemma 2.2, (3.8), (3.9) and (3.4) give

$$|B| \leq 2^{-1} \int_{D(x,t+2\varepsilon)\setminus D(x,t)} f(r',s',\varepsilon)d|\eta|(r',s') + \tag{3.10}$$

$$+ 2^{-1} \int_{0+}^{\varepsilon} \int_{x-t+s}^{x+t-s'} (\varepsilon^2 - s'^2)\varepsilon^{-2} d|\eta|(r',s') \leq$$

$$\leq 2^{-1} \int_{D(x,t+2\varepsilon)\setminus D(x,t)} d|\eta|(r',s') +$$

$$+ \int_{D(x,t)\cap\{(r',s');\ 0 < s' \leq \varepsilon\}} d|\eta|(r',s').$$

It follows from (3.10) that $B \to 0$ when $\varepsilon \to 0$.

The proof that $A \to 0$ when $\varepsilon \to 0$ is more complicated. Now

$$A = - 2^{-1} \int_{L(x,t,\varepsilon)} u(r,s,\varepsilon)\varepsilon^{-2}\, d\mu(r',s')drds + \tag{3.11}$$

$$+ 2^{-1} \int_{D(x,t)} u(r',s',\varepsilon)(\varepsilon^{-2} \int_{D(r',s',\varepsilon)} drds)d\mu(r',s') =$$

$$= -2^{-1} \int_{L(x,t,\varepsilon)\setminus H(x,t,\varepsilon)} \varepsilon^{-2}u(r,s,\varepsilon)d\mu(r',s')drds -$$

$$- 2^{-1} \int_{H(x,t,\varepsilon)} \varepsilon^{-2}(u(r,s,\varepsilon)-u(r',s',\varepsilon))d\mu(r',s')drds -$$

$$- 2^{-1} \int_{D(x,t)} \varepsilon^{-2}u(r',s',\varepsilon)(f(r',s',\varepsilon)-1)d\mu(r',s').$$

Let

$$C = -2^{-1} \int_{H(x,t,\varepsilon)} \varepsilon^{-2}(u(r,s,\varepsilon)-u(r',s',\varepsilon))d\mu(r',s')drds. \tag{3.12}$$

We look at the computations which lead to the proof of $B \to 0$ when $\varepsilon \to 0$. Then Theorem 2.1 inserted in that argument and (3.11) shows that it suffices to show that $C \to 0$ when $\varepsilon \to 0$. Let $(r,s) \in D(r',s',\varepsilon)$, $s \geq 0$, and define

$$E = - u(r',s',\varepsilon) + u(r,s,\varepsilon) = \tag{3.13}$$

$$= 2^{-1} \int_{L(r',s',\varepsilon)\setminus H(r',s',\varepsilon)} \varepsilon^{-2}u(r'',s'',\varepsilon)d\mu(\bar{r},\bar{s})dr''ds'' -$$

$$- 2^{-1} \int_{L(r,s,\varepsilon)\setminus H(r,s,\varepsilon)} \varepsilon^{-2}u(r'',s'',\varepsilon)d\mu(\bar{r},\bar{s})dr''ds'' +$$

$$+ 2^{-1} \int_{H(r',s',\varepsilon)} \varepsilon^{-2} u(r'',s'',\varepsilon) d\mu(\bar{r},\bar{s}) dr''ds'' -$$

$$- 2^{-1} \int_{H(r,s,\varepsilon)} \varepsilon^{-2} u(r'',s'',\varepsilon) d\mu(\bar{r},\bar{s}) dr''ds'' -$$

$$- 2^{-1} \int_{L(r',s',\varepsilon) \setminus H(r',s',\varepsilon)} \varepsilon^{-2} d\eta(\bar{r},\bar{s}) dr''ds'' +$$

$$+ 2^{-1} \int_{L(r,s,\varepsilon) \setminus H(r,s,\varepsilon)} \varepsilon^{-2} d\eta(\bar{r},\bar{s}) dr''ds'' -$$

$$- 2^{-1} \int_{H(r',s',\varepsilon) \setminus H(r,s,\varepsilon)} \varepsilon^{-2} d\eta(\bar{r},\bar{s}) dr''ds''.$$

According to Theorem 2.1 there is a constant M such that $|u(r'',s'',\varepsilon)| \leqq M$, $0 < \varepsilon \leqq 1$, $(r'',s'') \in D(x',t'+2)$. Let $\lambda = M|\mu| + |\eta|$. Then earlier considerations applied to (3.13) shows that

$$|E| \leqq 2^{-1} \int_{D(r',s'+2\varepsilon) \setminus D(r',s')} d\lambda(\bar{r},\bar{s}) + \tag{3.14}$$

$$+ 2^{-1} \int_{D(r',s'+2\varepsilon) \setminus D(r',s'-2\varepsilon)} d\lambda(\bar{r},\bar{s}) +$$

$$+ 2^{-1} \int_{D(r',s') \setminus D(r',s'-2\varepsilon)} d\lambda(\bar{r},\bar{s}).$$

It follows that $|E|$ is majorized by a number independent of (r,s) and that this number tends to zero when $\varepsilon \to 0$ unless (r',s') lies on a characteristic line with positive $|\eta|$-mass. The union of such lines has $|\mu|$-mass zero. It follows that C in (3.12) tends to zero even uniformly on $D(x',t')$. By that we have proved that $A \to 0$ when $\varepsilon \to 0$. Combined with the same result for B proved earlier this gives (3.4). Then (3.4) gives (3.5) at once. By that we have proved Theorem 1.2 for μ without point masses.

Let $(x',t') = (x_1,t_1)$ and let $\mu(\{(x_j,t_j)\}) = 0$, $j = 2,3,\ldots$. Then $|\eta|$ has no positive mass on the characteristic lines through (x',t') outside (x',t'). Let $\mu_1 = a\delta(\cdot-x',\cdot-t')$ and let $\eta_1 = b\delta(\cdot-x',\cdot-t')$ with δ the Dirac measure at $(0,0)$. We have already proved that if $(x,t) \in D(x',t')$, $(x,t) \neq (x',t')$ then $u(x,t,\varepsilon) \to u(x,t) = u(x,t,0)$ when $\varepsilon \to 0$. Let $\mu_0 = \mu - \mu_1$. Then no characteristic line has positive $|\mu_0|$-mass. We see that

$$u(x,t,\varepsilon) = -2^{-1}\int_{D(x,t)} u(r,s,\varepsilon)d\mu_1(r,s,\varepsilon) + \tag{3.15}$$

$$+ 2^{-1}\int_{D(x,t)} d\eta_1(r,s,\varepsilon) + h(x,t,\varepsilon),$$

where

$$h(x,t,\varepsilon) = -2^{-1}\int_{D(x,t)} u(r,s,\varepsilon)d\mu_0(r,s,\varepsilon) +$$

$$+ 2^{-1}\int_{D(x,t)} d(\eta-\eta_1)(r,s,\varepsilon).$$

Let

$$h(x,t,0) = -2^{-1}\int_{D(x,t)} u(r,s,0)d\mu_0(r,s) + \tag{3.17}$$

$$= 2^{-1}\int_{D(x,t)} d(\eta-\eta_1)(r,s).$$

One realizes from what we have already proved that $h(x,t,\varepsilon) \to h(x,t,0)$ and especially that $h(x',t',\varepsilon) \to h(x',t',0)$ when $\varepsilon \to 0$. Let $(x,t) \in D(x',t',\varepsilon)$. Then (3.13) and (3.14) shows that

$$|h(x',t',\varepsilon) - h(x,t,\varepsilon)| \leq 2\int_{D(x',t'+2\varepsilon)\setminus D(x',t'-2\varepsilon)} d\lambda(r,s) \tag{3.18}$$

with $\lambda = M|\mu_0| + |\eta-\eta_1|$. Since λ has no positive mass on the characteristics through (x',t') the right member of (3.18) tends to zero. One then realizes that to given $\delta > 0$ there is an $\varepsilon'' > 0$ such that

$$|h(x',t',0) - h(x,t,\varepsilon)| < \delta\ ,\ 0 < \varepsilon < \varepsilon'',\ (x,t) \in D(x',t',\varepsilon). \tag{3.19}$$

We solve

$$w(x,t,\varepsilon) = 2^{-1}\int_{D(x,t)} w(r,s,\varepsilon)d\mu_1(r,s,\varepsilon) + \tag{3.20}$$

$$+ 2^{-1}\int_{D(x,t)} d\eta_1(r,s,\varepsilon) + h(x',t',0).$$

As in [4, Section 2] we see that $w(x',t',\varepsilon) = w(x',t',0) =$ $b\sum_{j=0}^{\infty}(-a)^j((2(j+1))!)^{-1} + h(x',t',0)\sum_{j=0}^{\infty}(-a)^j((2j)!)^{-1}$. If $a \neq 0$ then

$$w(x',t',0) = b(1 - \cosh\sqrt{-a})/a + h(x',t',0)\cosh\sqrt{-a} \tag{3.21}$$

It now follows from (3.15), (3.19) and the proof of Theorem 2.1 that $w(x',t',\varepsilon) - u(x',t',\varepsilon) = w(x',t',0) - u(x',t',\varepsilon) \to 0$ when $\varepsilon \to 0$. Thus $u(x',t',0) = w(x',t',0)$. Then just as in Persson [4] one easily verifies that (1.8) is fulfilled at $(x,t) = (x',t')$. Then a computation actually performed in Section 4 shows that $u(x,t,0)$ exists for all (x,t). The limit solves (1.8) since the regularized measures tend set wise to the unregularized ones outside (x',t').

Then one uses the same technique and induction to show that the theorem is true also for the case when $\mu(\{(x_j,t_j)\}) = 0$, $j = M+1, M+2, \ldots$, where M is an arbitrary positive integer. We do not write down this step.

We look at the general case. Let $\bar{\mu}_j = \mu(\{(x_j,t_j)\})\delta(\cdot - x_j, \cdot - t_j)$, $j = 1,2,\ldots$. Let $\mu_M = \mu - \sum_{j=M+1}^{\infty} \bar{\mu}_j$. Let $u_M(x,t,\varepsilon)$ and $u_M(x,t,0)$ be defined by

$$u_M(x,t,\varepsilon) = -2^{-1}\int_{D(x,t)} u_M(r,s,\varepsilon)d\mu_M(r,s,\varepsilon) + \\ + 2^{-1}\int_{D(x,t)} d\eta(r,s,\varepsilon), \tag{3.22}$$

and

$$u_M(x,t,0) = -2^{-1}\int_{D(x,t)} u_M(r,s,0)g(\mu_M(\{(r,s)\}))d\mu_M(r,s) + \\ + 2^{-1}\int_{D(x,t)} g(\mu_M(\{(r,s)\}))d\eta(r,s). \tag{3.23}$$

Just as in the proof of Theorem 2.1 one easily shows that $(u_M(x,t,\varepsilon))_{0<\varepsilon\leqq 1, M>0}$ is equibounded on compact sets. Since (1.4) is fulfilled for (3.23) an examination of the existence proof for (3.23), Persson [2, Section 3] shows that the same thing is true for $(u_M(x,t,0))_{M>0}$.

Let $v_M(x,t,\varepsilon) = u_M(x,t,\varepsilon) - u(x,t,\varepsilon)$. Then

$$v_M(x,t,\varepsilon) = -2^{-1}\int_{D(x,t)} v_M(r,s,\varepsilon)d\mu(r,s,\varepsilon) - \\ - 2^{-1}\int_{D(x,t)} u_M(r,s,\varepsilon)d(\mu-\mu_M)(r,s,\varepsilon). \tag{3.24}$$

Here the second term of the right member of (3.24) can be made arbitrarily small in $D(x',t')$ independent of ε by choosing M big. The proof of Theorem 2.1 shows that then $v_M(x,t,\varepsilon)$ can be made uniformly small on $D(x',t')$ for all big M and all ε, $0 < \varepsilon \leq 1$. In an analogous way we find that $u(x,t,0) - u_M(x,t,0) \to 0$ uniformly on $D(x',t')$ when $M \to \infty$. Now

$$|u(x,t,\varepsilon) - u(x,t,0)| \leq |u(x,t,\varepsilon) - u_M(x,t,\varepsilon)| +$$

$$+ |u_M(x,t,\varepsilon) - u_M(x,t,0)| + |u_M(x,t,0) - u(x,t,0)|.$$

First we choose M big such that $u(x,t,\varepsilon) - u_M(x,t,\varepsilon)$ and $u_M(x,t,0) - u(x,t,0)$ both are small independent of ε. Then for this fixed M we choose ε' such that $u_M(x,t,\varepsilon) - u_M(x,t,0)$ is small $0 < \varepsilon < \varepsilon'$, That shows that $u(x,t,\varepsilon) \to u(x,t,0)$ when $\varepsilon \to 0$. Theorem 1.2 is proved.

4. PROOF OF THEOREM 1.1

We reduce the proof to a proof of a sequence of special cases. This procedure has already been used in the proof of Theorem 1.2 and in the proof of a corresponding theorem for ordinary differential equations in Persson [3]. The complication here lies in the interplay between masses of η and μ along the characteristics. This is excluded by the hypothesis of Theorem 1.2. Although the proof contains very concrete computations we have not been able to find a modified integral equation having the limit of the solutions of the regularized problems as its solution.

Our efforts to show directly that the solutions of the regularized problems form a Cauchy sequence have not been successful. Instead we have chosen to modify the signed measures μ and η so that the restriction of the modified signed measures along a characteristic with positive $(|\mu|+|\eta|)$ measure are unchanged at points with point masses but in between are signed measures with polynomial desnsities with respect to the Lebesgue measure along the characteristic line. The signed measures are modified only along such characteristics.

One then proves that the limit of the solutions of the regularized modified problem exists and ultimately that the limit of the solutions of the regularized unmodified problem also exists.

We start by a general remark. One realizes that there is a sequence of different characteristic lines $(l_k)_{k=1}^{\infty}$ such that $(|\mu|+|\eta|)(l) = 0$ if l is a characteristic line and $l \neq l_k$, $k = 1,2,\ldots$.

In the first step we assume that $(|\mu|+|\eta|)(l) = 0$ for all characteristic lines l. This is a special case of the hypothesis of Theorem 1.2. We notice that Theorem 1.2 shows that the limit is continuous in this case.

In the next step we assume that there is a characteristic line l_1 such that if l is another characteristic line then $(|\mu|+|\eta|)(l) = 0$. We notice that this excludes point masses on l_1. Let μ_1 and η_1 be signed measures such that $(|\mu_1|+|\eta_1|)$ has no mass outside l_1 and such that $\mu_1(A) = \mu(A)$ and $\eta_1(A) = \eta(A)$, $A \subset l$, A Borel set. Let $\mu_0 = \mu - \mu_1$ and $\eta_0 = \eta - \eta_1$.

As in Section 2 we let $y = 2^{-1/2}(x+t)$ and $z = 2^{-1/2}(x-t)$. We refer to Section 2 for the notation.

In the new coordinates we get

$$u(y,z,\varepsilon) = -2^{-1}\int_{D(y,z)} u(y',z',\varepsilon)d\mu(y',z',\varepsilon) + 2^{-1}\int_{D(y,z)} d\eta(y',z',\varepsilon). \tag{4.1}$$

We assume that l_1 is the line $y = c$, where c is a constant. We define $u_0(y,z,\varepsilon)$ by

$$u_0(y,z,\varepsilon) = -2^{-1}\int_{D(y,z)} u_0(y',z',\varepsilon)d\mu_0(y',z',\varepsilon) + 2^{-1}\int_{D(y,z)} d\eta_0(y',z',\varepsilon). \tag{4.2}$$

If $z \geqq c$ or $y < c$ it follows from the proof of Theorem 1.2 that for small ε

$$u(y,z,\varepsilon) = u_0(y,z,\varepsilon) \to u_0(y,z,0),\quad \varepsilon \to 0.$$

In other words $u(y,z,0)$ exists in these cases.

Let $y \geqq c$ and let $z < c$. Let

$$v(y,z,\varepsilon) = u(y,z,\varepsilon) - u_0(y,z,\varepsilon).$$

It follows from (4.1) and (4.2) that

$$v(y,z,\varepsilon) = -2^{-1}\int_{D(y,z)} v(y',z',\varepsilon)d\mu_1(y',z',\varepsilon) + \tag{4.3}$$

$$+ 2^{-1}\int_{D(y,z)} d\eta_1(y',z',\varepsilon) - 2^{-1}\int_{D(y,z)} u_0(y',z',\varepsilon)d\mu_1(y',z',\varepsilon) -$$

$$- 2^{-1}\int_{D(y,z)} v(y',z',\varepsilon)d\mu_0(y',z',\varepsilon).$$

We notice that the family $(u_0(y,z,\varepsilon))_{0<\varepsilon<1}$ is uniformly equicontinuous on compact sets. To get a fixed compact set we choose $(\bar{y},\bar{z}) = (c,\bar{z})$ with $\bar{z} < c$. Then we look at (y,z) with $D(y,z) \subset D(\bar{y},\bar{z})$, $(\bar{y},\bar{z})$ fixed. For simplicity we also let $\varepsilon' = 2^{-1/2}\varepsilon$.

We rewrite (4.3) as

$$v(y,z,\varepsilon) = -2^{-1}\int_{D(y,z)} v(y',z',\varepsilon)d\mu_1(y',z',\varepsilon) + \tag{4.4}$$

$$+ 2^{-1}\int_{D(y,z)} d\eta_1(y',z',\varepsilon) - 2^{-1}\int_{D(y,z)} u_0(c,z',0)d\mu_1(y',z',\varepsilon) +$$

$$+ g(y,z,\varepsilon),$$

with

$$g(y,z,\varepsilon) = -2^{-1}\int_{D(y,z)} v(y',z',\varepsilon)d\mu_0(y',z',\varepsilon) - \tag{4.5}$$

$$- 2^{-1}\int_{D(y,z)} (u_0(y',z',\varepsilon)-u_0(c',z',0))d\mu_1(y',z',\varepsilon).$$

For $y \leq c - \varepsilon'$ $v(y,z,\varepsilon) = 0$ and $|\mu_0|(D(\bar{y},\bar{z}) \cap \{(y',z');\ c-\varepsilon' \leq y' \leq c\}$ tends to zero when $\varepsilon \to 0$. Let M be all (y',z') in $D(\bar{y},\bar{z})$ with $c-\varepsilon' \leq y' \leq c$. Then one realizes that

$$\sup_{(y',z')\in M} |u_0(y',z',\varepsilon)-u_0(c,z',0)| \to 0,\quad \varepsilon \to 0.$$

All this means that $g(y,z,\varepsilon) \to 0$ uniformly on $D(\bar{y},\bar{z})$. It follows from (4.4) and the proof of Theorem 2.1 that it suffices to prove that $\lim_{\varepsilon\to 0} v(y,z,\varepsilon)$

exists when $g = 0$. This we assume from now on. Still we have not found any direct method to do this. We approximate the signed measures μ_1 and η_1 by signed meausres $\mu_1(\cdot;\alpha)$ and $\eta_1(\cdot;\alpha)$ given by polynomial densities with respect to the Lebesgue measure along $y = c$. They are both zero outside $y = c$. The densities are chosen such that $\mu_1(A;\alpha) \to \mu_1(A)$ and $\eta_1(A;\alpha) \to \eta_1(A)$ when $\alpha \to 0$, for each Borel set $A \subset D(\bar{y},\bar{z})$. This we can achieve by regularizing the signed measures along $y = c$ and then use Weierstrass approximation theorem on the continuous densities.

We have

$$d\mu_1(c,z;\alpha) = (\sum_{j=0}^{m(\alpha)} a_j(\alpha))c-z)^j)dz \tag{4.6}$$

$$d\eta_1(c,z;\alpha) = (\sum_{j=0}^{n(\alpha)} b_j(\alpha)(c-z)^j)dz \tag{4.7}$$

$$h(z;\alpha) = \sum_{j=0}^{k(\alpha)} d_j(\alpha)(c-z)^j \tag{4.8}$$

where $h(z;\alpha) \to u(c,z,0)$, $\alpha \to 0$, uniformly on $\bar{z} \leq z \leq c$. We define $v(y,z;\alpha,\varepsilon)$ by letting

$$v(y,z;\alpha,\varepsilon) = -2^{-1}\int_{D(y,z)} v(y',z';\alpha,\varepsilon)d\mu_1(y',z';\alpha,\varepsilon) - \tag{4.9}$$

$$-2^{-1}\int_{D(y,z)} h(z';\alpha)d\mu_1(y',z';\alpha,\varepsilon) +$$

$$+2^{-1}\int_{D(y,z)} d\eta_1(y',z';\alpha,\varepsilon).$$

It follows from the proof of Theorem 2.1 that $v(y,z;\alpha,\varepsilon) - v(y,z,\varepsilon) \to 0$ equiuniformly on compact sets when $\alpha \to 0$. Thus it suffices to show that $\lim_{\varepsilon\to 0} v(y,z;\alpha,\varepsilon) = v(y,z;\alpha,0)$ exists.

We start by treating the case $\eta_1 = 0$ and $h(z;\alpha) = (c-z)^k$ and $\mu_1(c,z;\alpha) = (c-z)^m$, $k \geq 0$, $m \geq 0$. At first we calculate $\mu_1(y,z;\alpha,\varepsilon)$.

$$\mu_1(y,z;\alpha,\varepsilon) = \int_{R^2} \phi(y-y', z-z',\varepsilon)d\mu_1(y',z';\alpha) =$$

$$= \varepsilon^{-2}\int_{E(y,z,\varepsilon)} d\mu_1(y',z';\alpha) = \varepsilon^{-2}\int_{c-y+z-\varepsilon'}^{z} (c-z')^m dz' =$$

$$= \varepsilon^{-2}((y-z+\varepsilon')^{m+1} - (c-z)^{m+1})/(m+1) =$$

$$= \varepsilon^{-2}(m+1)^{-1}((y-c+\varepsilon') + (c-z))^{m+1} - (c-z)^{m+1}) =$$

$$= \varepsilon^{-2}(m+1)^{-1} \sum_{i=1}^{m+1} \binom{m+1}{j}(y-c+\varepsilon')^{j}(c-z)^{m+1-j} =$$

$$= \varepsilon^{-2}(y-c+\varepsilon')((c-z)^{m} + O(\varepsilon)), \quad c-\varepsilon' \leqq y \leqq c.$$

It means that we commit an error of order $O(\varepsilon)$ if we in the following let

$$\mu_1(y,z;\alpha,\varepsilon) = \varepsilon^{-2}(y-c+\varepsilon')(c-z)^{m}, \quad c-\varepsilon' \leqq y \leqq c. \tag{4.11}$$

This we do from now on. In the following we also let

$$\int_{D(y,z)} = \int_{c-\varepsilon'}^{y} \int_{z}^{y} \tag{4.12}$$

By this we commit an error of order $o(1)$ when $\varepsilon \to 0$, uniformly in α since μ_1 has no point masses.

Let $c-\varepsilon' \leqq y \leqq c$, $D(y,z) \subseteq D(\bar{y},\bar{z})$. We now solve (4.9) by successive approximations committing an error of order $o(1)$ when $\varepsilon \to 0$ using (4.11) and (4.12). Let

$$w_1(y,z,\varepsilon) = 2^{-1} \int_{D(y,z)} (c-z)^{k} d\mu_1(y',z';\alpha,\varepsilon) =$$

$$= -2^{-1}\,\varepsilon^{-2} \int_{c-\varepsilon'}^{y} \int_{z}^{y} (c-z')^{k+m}(y'-c+\varepsilon')dy'dz' =$$

$$= -2^{-1}\,\varepsilon^{-2}(k+m+1)^{-1} \int_{c-\varepsilon'}^{y} ((c-z)^{k+m+1}-(c-y\ )^{k+m+1})(y'-c+\varepsilon')dy' =$$

$$= -\ (k+m+1)^{-1}2^{-2}\varepsilon^{-2}(y-c+\varepsilon')^{2}((c-z)^{k+m+1}-\varepsilon'^{k+m+1}H(y))$$

where $|H(y)| < 1$, $c - \varepsilon' \leqq y \leqq c$. We omit the last term of the last factor committing an error of order less than

$$2^{-1}(k+m+1)^{-1}\varepsilon'^{k+m+1}C, \quad C \text{ constant},$$

where C is taken from the proof of Theorem 2.1. Then we get

$$w_2(y,z,\varepsilon) = 2^{-1} \int_{c-\varepsilon'}^{y} \int_{z}^{y} (k+m+1)^{-1} 2^{-2\varepsilon-4}(y'-c+\varepsilon')^3(c-z')^{k+2m+1}dz'dy' =$$

$$= 2^{-2}(4!!)^{-1}\varepsilon^{-4}(k+m+1)^{-1}(k+2m)^{-1}(y-c+\varepsilon')^4$$

$$((c-z)^{k+2m+2} + \varepsilon'^{k+2m+2}H_2(y))$$

where $|H_2(y)| < 1$, $c - \varepsilon' \leq y \leq c$. We omit the last term of the last factor committing an error or order

$$(4!!)^{-1}(k+m+1)^{-1}(k+2m+2)^{-1}\varepsilon'^{k+2m+2}C.$$

Continuing in the same way we get

$$w_j(y,z,\varepsilon) = (-1)^j 2^{-2j}(j!)^{-1}\varepsilon^{-2j} \prod_{l=1}^{j} (k+lm+l)^{-1}(y-c+\varepsilon')^{2j} \times \tag{4.13}$$

$$\times ((c-z)^{k+jm+j} + \varepsilon'^{k+jm+j}H_j(y)),$$

where $|H_j(y)| \leq 1$, $c - \varepsilon' \leq y \leq c$. Then we let $H_j(y) = 0$ before we determine w_{j+1}. Then we commit an error of order

$$(2j!!)^{-1} \prod_{l=1}^{j} (k+lm+l)^{-1}\varepsilon'^{k+jm+j}C$$

at the jth step. It is obvious that the total error is $O(\varepsilon)$. It follows that

$$v(y,z;\alpha,\varepsilon) = \sum_{j=1}^{\infty} (-1)^j 2^{-2j}(j!)^{-1}\varepsilon^{-2j} \prod_{l=1}^{j} (k+lm+l)^{-1} \times \tag{4.16}$$

$$\times (y-c+\varepsilon')^{2j}(c-z)^{k+jm+j} + O(\varepsilon)$$

It is obvious from (4.16) that $v(y,z;\alpha,\varepsilon)$ is continuous in ε, $0 \leq \varepsilon \leq 1$. So $v(y,z;\alpha,0)$, $y \leq c$ exists in this case. Especially we notice that the convergence is uniform on $y = c$.

In the general case we must use the full polynomial in (4.6). Let $a_j(\alpha) = a_j$. Then

$$\mu_1(y,z;\alpha,\varepsilon) = \varepsilon^{-2}(y-c+\varepsilon') \sum_{j=0}^{m} a_j(c-z)^j + O(\varepsilon).$$

As before we omit the error term. We get

$$w_1(y,z,\varepsilon) = - 2^{-2}\varepsilon^{-2}(y-c+\varepsilon')^2 \times$$

$$\times \sum_{l=0}^{m} a_l((k+l+1)^{-1}((c-z)^{k+l+1} + \varepsilon'^{k+l+1}H_{1l}(y)).$$

Here $|H_{1l}(y)| \leq 1$. Let $a = \sum_{l=0}^{m} |a_l|$. Then we let $H_{1l}(y) = 0$, $0 \leq l \leq m$. By this we commit an error of order

$$2^{-1}a(k+1)^{-1}\varepsilon'^{k+1}C, \; \varepsilon' < 1, \tag{4.18}$$

in this step. Then

$$w_1(y,z,\varepsilon) = - 2^{-2}\varepsilon^{-2}(y-c+\varepsilon')^2 p_1(z). \tag{4.19}$$

Here $p_1(z)$ is a polynomial majorized by $a((c-z) + (c-z)^{k+m+1})$. Then with the same procedure we get

$$w_2(y,z,\varepsilon) = 2^{-2}(4!!)^{-1}\varepsilon^{-4}(y-c+\varepsilon')^4 p_2(z).$$

Here $p_2(z)$ is majorized by

$$a^2((c-z)^2+(c-z)^{m+2}+(c-z)^{m+k+2}+(c-z)^{2m+k+2})2^{-1} \leq$$

$$\leq 4a^2((c-z)^2+(c-z)^{2m+k+2})2^{-1}.$$

The error is majorized by

$$4a^2\varepsilon'^2 2^{-1}C, \; \varepsilon' \leq 1,$$

at this step. In general we get

$$w_j(y,z,\varepsilon) = 2^{-2j}(j!)^{-1}\varepsilon^{-2j}(y-c+\varepsilon')^{2j}p_j(z)$$

where $p_j(z)$ is majorized by

$$2^{2j}a^j((c-z)^j + (c-z)^{k+jm+j})(j!)^{-1}. \qquad (4.20)$$

The error committed at the jth step is majorzied by

$$2^{j+1}a^j\varepsilon'^j(j!)^{-2}C. \qquad (4.21)$$

As before we now realize that $v(y,z;\alpha,0)$ exists and that $v(c,z;\alpha,0)$ is continuous due to the uniform convergence on $y = c$.

For (y,z), $z < c$, $D(y,z) \supset D(c,z)$ we let

$$u(y,z,\varepsilon) = u(c,z,\varepsilon) - 2^{-1}\int_{D(y,z)\setminus D(c,z)} u(y',z',\varepsilon)d\ (y',z',\varepsilon) + $$
$$+ 2^{-1}\int_{D(y,z)\setminus D(c,z)} d\eta(y',z',\varepsilon). \qquad (4.22)$$

We want to prove that $(u(y,z,\varepsilon))_{0<\varepsilon<1}$ is a Cauchy sequence when $\varepsilon \to 0$. Let $\varepsilon'' > 0$ and let $v(y,z,\varepsilon,\varepsilon'') = u(y,z,\varepsilon) - u(y,z,\varepsilon'')$. We get

$$v(y,z,\varepsilon,\varepsilon'') = -2^{-1}\int_{D(y,z)\setminus D(c,z)} v(y',z',\varepsilon,\varepsilon'')d\mu(y',z',\varepsilon) + \qquad (4.23)$$
$$+ g(y,z,\varepsilon,\varepsilon'').$$

Here

$$g(y,z,\varepsilon,\varepsilon'') = u(c,z,\varepsilon) - u(c,z,\varepsilon'') +$$
$$+ 2^{-1}\int_{D(y,z)\setminus D(c,z)} u(y',z',\varepsilon'')d(\mu(y',z',\varepsilon'')-\mu(y',z',\varepsilon)) +$$
$$+ 2^{-1}\int_{D(y,z)\setminus D(c,z)} d(\eta(y',z',\varepsilon)-\mu(y',z',\varepsilon'')).$$

It follows from Theorem 2.1, the properties of μ and η in $y > c$ and the fact that $u(c,z,\varepsilon) \to u(c,z,0)$ uniformly on compact sets of $y = c$ that $g(y,z,\varepsilon,\varepsilon'') \to 0$ uniformly on compact subsets of $y \geq c$. We then modify μ and η by letting them zero $y < c+\varepsilon'$ and unchanged elsewhere. Then we commit an error of order $o(1)$ when ε and ε'' tend to zero. This is true uniformly on compact set in the value of $v(y,z,\varepsilon,\varepsilon'')$. If we then change $D(y,z)\setminus D(c,z)$ to $D(y,z)$ in (4.23) nothing is changed. Then it is clear from the proof of Theorem 2.1 that $v(y,z,\varepsilon,\varepsilon'')$ tends to zero when ε and ε'' tend to zero, uniformly on compact sets of $y \geqq c$. By that we have proved the theorem when there is only one characteristic line l with $(|\mu|+|\eta|)(l) \neq 0$.

Let the theorem be true when there are no point masses in $|\mu|+|\eta|$ and there are precisely q characteristic lines l with $(|\mu|+|\eta|)(l) \neq 0$. First we assume that the lines are parallel. Let $y = c$ be such a characteristic line l such that $(|\mu|+|\eta|)(l) \neq 0$ and such that if l' is given by $y = c'$, $c' > c$, implies $(|\mu|+|\eta|)(l') = 0$. Let μ_1 and η_1 be defined by $|\mu_1|(\mathbb{R}\setminus l) = 0$, $|\eta_1|(\mathbb{R}\setminus l) = 0$, $\mu_1(A) = \mu(A)$, $\eta_1(A) = \eta(A)$, $A \subset l$, A Borel set. Let $\mu_0 = \mu - \mu_1$ and $\eta_0 = \eta - \eta_1$. Let u_0 be the solution of (4.2) with μ replaced by μ_0 and η by η_0. By induction over the number of the characteristic parallel lines mentioned above and the computations we have already done one realizes that $u(y,z,\varepsilon) = u_0(y,z,\varepsilon) \to u(y,z,0)$ when $\varepsilon \to 0$, $y < c$, $u(c,z,\varepsilon) \to u(c,z,0)$ and finally that $u(y,z,\varepsilon) \to u(y,z,0)$, $\varepsilon \to 0$, uniformly on compact subsets of $y \geqq c$.

Let the line l_1 be defined by $z = d$ be such that $(|\mu|+|\eta|)(l_1') = 0$ when l_1' is given by $z = d'$, $d' < d$. Let l_2 be given by $y = c$ and be such that $(|\mu|+|\eta|)(l_2') = 0$, when l_2' is defined by $y = c'$, $c' > c$. At first we assume that $(|\mu|+|\eta|)(l) = 0$, $l \neq l_1$, $l \neq l_2$, l characteristic. One sees that for $D(y,z) \subset D(c,d)$, $(y,z) \neq (c,d)$, $u(y,z,\varepsilon) \to u(y,z,0)$, $\varepsilon \to 0$, just as before. We now construct μ_j and η_j from $\mu_j(\mathbb{R}\setminus l_j) = 0$, $\eta_j(\mathbb{R}\setminus l_j) = 0$, $\mu_j(A) = \mu(A)$, $\mu_j(A) = \mu(A)$, $A \subset l_j$ A Borel set, $j = 1,2$. Let $\mu_0 = \mu - \mu_1 - \mu_2$ and $\eta_0 = \eta - \eta_1 - \eta_2$. We let these μ_0 and η_0 into (4.2). We see that

$$u(y,z,\varepsilon) = -2^{-1}\int_{D(y,z)} u(y',z',\varepsilon)d\mu_1(y',z',\varepsilon) -$$

$$-2^{-1}\int_{D(y,z)} u(y',z',\varepsilon)d\mu_2(y',z',\varepsilon) + 2^{-1}\int_{D(y,z)} d\eta_1(y',z',\varepsilon) +$$

$$+ 2^{-1}\int_{D(y,z)} d\eta_2(y',z',\varepsilon) - 2^{-1}\int_{D(y,z)} u(y',z',\varepsilon)d\mu_0(y',z',\varepsilon).$$

Let $v(y,z,\varepsilon) = u(y,z,\varepsilon) - u_0(y,z,\varepsilon)$. Then

$$v(y,z,\varepsilon) = -2^{-1} \int_{D(y,z)} v(y',z',\varepsilon)d\mu_1(y',z',\varepsilon) -$$
$$- 2^{-1} \int_{D(y,z)} v(y',z',\varepsilon)d\mu_2(y',z',\varepsilon) -$$
$$- 2^{-1} \int_{D(y,z)} u_0(y',z',\varepsilon)d\mu_1(y',z',\varepsilon) -$$
$$- 2^{-1} \int_{D(y,z)} u_0(y',z',\varepsilon)d\mu_2(y',z',\varepsilon) + 2^{-1} \int_{D(y,z)} d\eta_1(y',z',\varepsilon) +$$
$$+ 2^{-1} \int_{D(y,z)} d\eta_2(y',z',\varepsilon) - 2^{-1} \int_{D(y,c)} v(y',z',\varepsilon)d\mu_0(y',z',\varepsilon).$$

Let $K_j \subset \mathbb{R}$, be compact $j = 1,2$. We notice that when supremum is taken over $y' \in K_1$, $0 \leqq z'-d \leqq \varepsilon'$,

$$\sup |u_0(y',z',\varepsilon)-u_0(y',d,\varepsilon)| \to 0, \ \varepsilon \to 0,$$

and that when supremum is taken over $z' \in K_2$, $- \varepsilon < y-c \leqq 0$,

$$\sup |u_0(y',z',\varepsilon)-u_0(c,z',\varepsilon)| \to 0, \ \varepsilon \to 0.$$

We commit an error of order $o(1)$, $\varepsilon \to 0$, uniformly on $D(c,d)$ when we let

$$v(y,z,\varepsilon) = - 2^{-1} \int_{D(y,z)} v(y',z',\varepsilon)d\mu_1(y',z',\varepsilon) - \tag{4.24}$$
$$- 2^{-1} \int_{D(y,z)} v(y',z',\varepsilon)d\mu_2(y',z',\varepsilon) -$$
$$- 2^{-1} \int_{D(y,z)} u_0(y',d,\varepsilon)d\mu_1(y',z',\varepsilon) -$$
$$-2^{-1} \int_{D(y,z)} u_0(c,z',\varepsilon)d\mu_2(y',z',\varepsilon) +$$
$$+ 2^{-1} \int_{D(y,z)} d\eta_1(y',z',\varepsilon) + 2^{-1} \int_{D(y,z)} d\eta_2(y',z',\varepsilon).$$

Let

$$v^+(y,z,\varepsilon) = - 2^{-1} \int_{D(y,z)} v^+(y',z',\varepsilon)d\mu_2(y',z',\varepsilon) - \tag{4.25}$$
$$- 2^{-1} \int_{D(y,z)} u_0(c,z',\varepsilon)d\mu_2(y',z',\varepsilon) + 2^{-1} \int_{D(y,z)} d\eta_2(y',z',\varepsilon),$$

and let

$$v^-(y,z,\varepsilon) = -2^{-1}\int_{D(y,z)} v^-(y',z',\varepsilon)d\mu_1(y',z',\varepsilon) - \tag{4.26}$$
$$-2^{-1}\int_{D(y,z)} u_0(y',d,\varepsilon)d\mu_1(y',z',\varepsilon) + 2^{-1}\int_{D(y,z)} d\eta_1(y',z',\varepsilon).$$

It is obvious from what we have already proved that

$$v(y,z,\varepsilon) = v^+(y,z,\varepsilon) + v^-(y,z,\varepsilon) + o(1),\ \varepsilon \to 0, \tag{4.27}$$

uniformly on $D(c,d)$. It follows that $v(c,d,0) = v^+(c,d,0) + v^-(c,d,0)$. In other words $u(c,d,0)$ exists. It also follows from what we have already proved that $u(y,z,0)$ exists for $z < y < c$ and for $d < z < y$.

Let

$$u_1(y,z,\varepsilon) = -2^{-1}\int_{D(y,z)} u_1(y',z',\varepsilon)d(\mu_0+\mu_1)(y',z',\varepsilon) \tag{4.28}$$
$$+ 2^{-1}\int_{D(y,z)} d(\eta_0+\eta_1)(y',z',\varepsilon).$$

Let $v = u - u_1$. Then

$$v(y,z,\varepsilon) = -2^{-1}\int_{D(y,z)\setminus D(y,d)} (v(y',z',\varepsilon)-u_1(y',z',\varepsilon))d\mu_2(y',z',\varepsilon) +$$
$$+ 2^{-1}\int_{D(y,z)\setminus D(y,d)} d\eta_2(y',z',\varepsilon) + v(y,d,\varepsilon),\ y \leqq c. \tag{4.29}$$

It follows from (4.27) and (4.16) that $v(y,d,\varepsilon)$ can be approximated by an entire function in $(y-c+\varepsilon')^2$ in $c - \varepsilon' \leqq y \leqq c$, since $v^-(y,d,\varepsilon)$ can be approximated by a constant. Then we just repeat the procedure when $d' \leqq z \leqq d$, $c - \varepsilon' \leqq y \leqq c$, d' a fixed constant. Instead of approximating with polynomials in $(c-z)$ we now use polynomials in $(d-z)$. We notice that the integration in (4.29) corresponds to

$$\int_{c-\varepsilon'}^{y} \int_{z}^{d} .$$

Continuing in this way we arrive at showing the existence of the limit along l_2 of the regularized approximating solutions. Then we prove the same thing along l_1. Then the limit will also exist for the original solutions of the regularized problems. The convergence is uniform on compact subsets of l_2, $y \geqq c$, and on compact subsets of l_1, $z \leqq d$. If $D(y,z) \supset D(c,d)$ then

$$u(y,z,\varepsilon) = -2^{-1}\int_c^y\int_z^d u(y',z',\varepsilon)d\mu_0(y',z',\varepsilon) + 2^{-1}\int_c^y\int_z^d d\eta_0(y',z',\varepsilon) -$$

$$- u(y,d,\varepsilon) - u(c,z,\varepsilon) + u(c,d,\varepsilon).$$

It is now obvious that $u(y,z,\varepsilon) \to u(y,z,0)$, $\varepsilon \to 0$, everywhere.

Let the theorem be true when there are precisely q characteristic lines l with $(|\mu|+|\eta|)(l) \neq 0$ and no point masses in $|\mu| + |\eta|$. A generalization of the argument above then shows that the theorem must be true also when there are q+1 characteristic lines l with $(|\mu|+|\eta|)(l) \neq 0$.

Let there be precisely two characteristic lines l_1, $z = d$, and l_2, $y = c$, such that $(|\mu|+|\eta|)(\{(c,d)\}) \neq 0$ and such that $(|\mu|+|\eta|)(l) = 0$ for all other characteristic lines l. Let $\delta = \delta(\cdot,(c,d))$ be the Dirac measure at (c,d). Let $\mu' = \mu(\{(c,d)\})\delta$ and let $\eta' = \eta(\{(c,d)\})\delta$. Let μ_j and η_j be defined by $|\mu_j|(\mathbb{R}\setminus l_j) = 0$, $|\eta_j|(\mathbb{R}\setminus l_j) = 0$, $\mu_j(A) = (\mu-\mu()(A)$, $\eta_j(A) = (\eta-\eta')(A)$, $A \subset l_j$, A Borel set, $j = 1,2$. Let $\mu_0 = \mu - \mu' - \mu_1 - \mu_2$ and $\eta_0 = \eta - \eta' - \eta_1 - {}_2$. With this μ_0 and η_0 (4.2) defines u_0 such that $u_0(y,z,\varepsilon) \to u(y,z,0)$, $\varepsilon \to 0$, uniformly on compact sets. Let $v(y,z,\varepsilon) = u(y,z,\varepsilon) - u_0(y,z,\varepsilon)$. With an error of o(1), $\varepsilon \to 0$, uniformly on compact sets, we get

$$v(y,z,\varepsilon) = -2^{-1}\int_{D(y,z)} v(y',z',\varepsilon)d(\mu' + \mu_1 + \mu_2)(y',z',\varepsilon) - \quad (4.30)$$

$$-2^{-1}\int_{D(y,z)} u_0(y',z',\varepsilon)d(\mu'+\mu_1+\mu_2)(y',z',\varepsilon) +$$

$$+2^{-1}\int_{D(y,z)} d(\eta'+\eta_1+\eta_2)(y',z',\varepsilon),\ D(y,z) \subseteq D(c,d).$$

Let $0 < 4\varepsilon < \varepsilon''$. We modify μ_1,μ_2, η_1 and η_2 such that

$$|\mu_1|([c-\varepsilon'',c] \times \{d\}) = 0,\ |\eta_1|([c-\varepsilon'',c] \times \{d\}) = 0, \quad (4.31)$$

$$|\mu_2|(\{c\}\times[d,d+\varepsilon'']) = 0,\quad |\eta_2|(\{c\}\times[d,d+\varepsilon'']) = 0,$$

keeping them unchanged elsewhere. By this we commit an error of order o(1) in $v(y,z,\varepsilon)$ when $\varepsilon'' \to 0$.

With the notation used earlier we define w by

$$w(y,z,\varepsilon) = v(y,z,\varepsilon) - v^{+}(y,z,\varepsilon) - v^{-}(y,z,\varepsilon).$$

We get with an error of order $o(1)$, $\varepsilon'' \to 0$,

$$w(y,z,\varepsilon) = -2^{-1}\int_{D(y,z)} w(y',z',\varepsilon)d\mu'(y',z',\varepsilon) - \tag{4.32}$$

$$- 2^{-1}\int_{D(y,z)} (u_0(c,d,0)+v^{+}(y'z',\varepsilon)+v^{-}(y',z',\varepsilon))d\mu'(y',z',\varepsilon) +$$

$$+ 2^{-1}\int_{D(y,z)} d\eta'(y',z',\varepsilon).$$

It follows from the linearity and the computations already made in Persson [4] and the symmetry that we may let $u_0(c,d,0) = 0$, $\eta' = 0$, $v^{-} = 0$. One also realizes that one can use the approximation expressed in (4.16) and then a further approximation of $v^{+}(y,z,\varepsilon)$ in $D(c,d,\varepsilon)$ by a polynomial in $((y-c+\varepsilon')/\varepsilon')^2$. Thus we may assume that $v^{+}(y,z,\varepsilon) = ((y-c+\varepsilon')/\varepsilon')^{2k}$, $c-\varepsilon' \leqq y \leqq c$, for some integer $k \geqq 0$.

We let $\mu' = \delta$ for simplicity. We get

$$w_1(y,z,\varepsilon) = - 2^{-1}\int_{D(y,z)} (y-c+\varepsilon')^{2k}\varepsilon'^{-2k}d\mu'(y',z',\varepsilon) =$$

$$= - \varepsilon'^{-2(k+1)}\int_{c-\varepsilon'}^{y}\int_{z}^{y'-c+\varepsilon'+d} (y'-c+\varepsilon')^{2k}dz'dy' =$$

$$= - \varepsilon'^{-2(k=1)}\int_{c-\varepsilon'}^{y} (y'-c+\varepsilon'+d-z)(y'-c+\varepsilon')^{2k}dy' =$$

$$= - \varepsilon'^{-2(k+1)}[(2k+2)^{-1}(y-c+\varepsilon')^{2k+2}+(d-z)(2k+1)^{-1}(y-c+\varepsilon')^{2k+1}].$$

Since we shall solve (4.32) by successive approximations we let

$$w_2(y,z,\varepsilon) = - \varepsilon'^{-2(k+2)} \times$$

$$\times\int_{c-\varepsilon'}^{y}\int_{z}^{y'-c+d+\varepsilon'} ((2k+2)^{-1}(y'-c+\varepsilon')^{2k+2}+(2k+1)^{-1}(d-z')(y'-c+\varepsilon')^{2k+1})dz'dy'=$$

$$= \varepsilon'^{-2(k+2)}[((2k+2)\ (2k+4)^{-1}\ (y-c+\varepsilon')^{2k+4} +$$

$$+ ((2k+2)(2k+3))^{-1}(d-z)(y-c+\varepsilon')^{2k+3} +$$

$$+ ((2(2k+1)(2k+2))^{-1}(d-z)^2(y-c+\varepsilon')^{2k+2} +$$

$$+ ((2(2k+1)(2k+4)^{-1}(y-c+\varepsilon')^{2k+4}].$$

We continue to define the functions w_j recursively. It is then obvious that w_j is an entire function in $(y-c+\varepsilon')/\varepsilon'$ and $(d-z)/\varepsilon'$ for (y,z) in $D(c,d,\varepsilon)$ modulo an error of order $o(1)$, $\varepsilon'' \to 0$. It follows that $w(c,d,\varepsilon)$ is independent of ε modulo an error or order $o(1)$, $\varepsilon'' \to 0$. By that we have proved that $u(y,z,0)$ exists in $D(c,d)$. The other computations made earlier apply to all points outside $D(c,d)$.

Then we assume that there are a finite number of characteristic lines l with $(|\mu|+|\eta|)(l) \neq 0$. An induction argument shows that the theorem is true also in this case.

Then the approximation of the general case with a problem as above as in the proof of Theorem 1.2 shows that the theorem is true.

References

[1] J. Persson, Second order linear ordinary differential equations with measures as coefficients, Matematiche 36 (1981) 151-171.

[2] J. Persson, The wave equation with measures as potentials and related topics, Rend. Sem. Mat. Univ. Politec. Torino, Fascicolo speciale settembre 1983, 207-219.

[3] J. Persson, Fundamental theorems for linear measure differential equations, to appear in Math. Scand.

[4] J. Persson, Wave equations with measures as coefficients, in "Hyperbolic equations. Proceedings of the conference on hyperbolic equations and related topics, University of Padova 1985, ed. F. Colombini and M.K.V. Murthy, Longman, Harlow, 1987, pp. 130-140.

Jan Persson
Matematiska Institutionen
Lunds Universitet
Box 118, S-22100 Lund,
Sweden.

L. RODINO & A. CORLI

Gevrey solvability for hyperbolic operators with constant multiplicity

INTRODUCTION

Let $P(x,D) = \sum_{|\alpha|\leq m} c_\alpha(x)D^\alpha$ be a linear partial differential operator with analytic coefficients in a neighbourhood of a point $x_0 \in R^n$; if there exists at least a multi-index α, $|\alpha| = m$, such that $c_\alpha(x_0) \neq 0$, as a consequence of the Cauchy-Kovalevsky theorem we can deduce that for every analytic function f, defined in a neighbourhood of x_0, there exist another neighbourhood V of x_0 and a solution u, analytic in V, of the differential equation Pu = f.

This local solvability result is not valid when we require f to be only an indefinitely differentiable function, even maintaining the analiticity of the coefficients. In fact in 1957 Lewy, [13], proved that the equation

$$D_1u + iD_2u + i(x_1 + ix_2)D_3u = f$$

has no distribution-solutions u in a neighbourhood of the origin in R^3 (actually, in any open subset $\Omega \neq \emptyset$ of R^3), for suitable functions $f \in C^\infty$. Such an example was the starting point for the study of the problem to characterize differential equations which don't admit solutions; among the first works we quote Hörmander [8] and Mizohata [16]. In particular, Mizohata proved that the operator, in R^2,

$$M = D_1 + ix_1^hD_2$$

is not solvable at the points of the x_2-axis, if h is an odd integer.

In 1970, after some preliminary works, Nirenberg and Treves [17] showed a general necessary and sufficient condition for the local solvability of pseudo-differential operators of principal type (see also Egorov [4]). We briefly recall this condition.

Let Ω be an open set in R^n and $P(x,D) = \sum_{|\alpha|\leq m} c_\alpha(x)D^\alpha$ a differential operator with coefficients $c_\alpha \in C^\infty(\Omega)$.

DEFINITION 1: The operator P is said to be locally solvable at $x_0 \in \Omega$ if there exists a neighbourhood U of x_0, contained in Ω, such that for every $f \in C^\infty(\Omega)$ there is a solution $u \in \mathcal{D}'(\Omega)$ of the equation $Pu = f$ in U.

Suppose now that the coefficients c_α of P are analytic in Ω and, denoting by $p_m(x,\xi) = \sum_{|\alpha|=m} c_\alpha(x)\xi^\alpha$ the principal symbol of P, assume that P is of principal type in the following sense:

$d_\xi \mathrm{Re}\, p_m$ doesn't vanish on the characteristic manifold

$$\Sigma = \{(x,\xi) \in T^*\Omega\setminus 0;\ p_m(x,\xi) = 0\} \text{ of } P;$$

here $\mathrm{Re}\, p_m$ indicates the real part of p_m, $\mathrm{Im}\, p_m$ will be the imaginary one. Under these assumptions we can state the result of Nirenberg and Treves.

THEOREM [17]: The operator P is locally solvable at the point $x_0 \in \Omega$ if and only if the following condition is satisfied:

for all $(x,\xi) \in \Sigma$, x in a small neighbourhood of x_0, $\xi \neq 0$, $\mathrm{Im}\, p_m$ doesn't change sign at (x,ξ) along the bicharacteristic strip of $\mathrm{Re}\, p_m$ through (x,ξ). (NT)

As a consequence of this theorem we have that strictly hyperbolic or elliptic operators are everywhere locally solvable. As regards the multiple characteristic case no general result, similar to the former one, is known, even if the case in which the characteristics are double has been studied by several authors, for example Cardoso-Treves [1], Wenston [24], [25], Menikoff [15], Popivanov [20], Egorov [5].

We can at this point supply a motivation to the problem we are going to treat. As we have recalled above, we have always local solvability where the datum f is analytic and the principal symbol doesn't degenerate, although the neighbourhood where the solution is defined may depend on f. One may then ask if, given an unsolvable operator P, is it possible to find solutions u of the equation Pu = f, for f in subspaces of C^∞ containing the space A of analytic functions.

The subspaces we consider here are the Gevrey function classes $G^{(s)}$, defined in the following standard way. Let Ω be an open subset of R^n, K a compact in Ω, h and s real numbers, $h > 0$, $s \geq 1$; we denote by $G^{(s),h}(K)$ the space of all the functions $\phi \in C^\infty(K)$ for which

$$\sup_{x \in K} |D^\alpha \phi(x)| \leq C h^{|\alpha|} \alpha!^s \qquad |\alpha| = 0,1,2,\ldots$$

for some positive constant C. Moreover we set $G_0^{(s),h}(K) = G^{(s),h}(K) \cap C_0^\infty(K)$ and define then

$$G^{(s)}(\Omega) = \lim_{\substack{\leftarrow \\ K \subset\subset \Omega}} \lim_{\substack{\rightarrow \\ h \to \infty}} G^{(s),h}(K) \qquad G_0^{(s)}(\Omega) = \lim_{\substack{\rightarrow \\ K \subset \Omega}} \lim_{\substack{\rightarrow \\ h \to \infty}} G_0^{(s),h}(K).$$

Observe that $G^{(1)}(\Omega) = A(\Omega)$ though $\bigcap_{s>1} G^{(s)}(\Omega) \supsetneq A(\Omega)$. The elements of strong dual spaces $G^{(s)'}(\Omega)$ and $G_0^{(s)'}(\Omega)$ of $G^{(s)}(\Omega)$ and $G_0^{(s)}(\Omega)$, respectively, are called ultradistributions (see Komatsu [12]).

Let P be a differential operator with analytic coefficients in Ω, s a real number, $s \geq 1$; it is then natural to set the following

DEFINITION 2: The operator P is said to be s-locally solvable at $x_0 \in \Omega$ if there exists a neighbourhood V of x_0, contained in Ω, such that for every $f \in G^{(s)}(\Omega)$ there is a solution $u \in G_0^{(s)'}(\Omega)$ of the equation $Pu = f$ in V.

Since $G^{(s)}(\Omega) \subset C^\infty(\Omega)$ and then $G_0^{(s)'}(\Omega) \supset \mathcal{D}'(\Omega)$, if an operator is locally solvable (in C^∞) then it is also s-locally solvable for $1 < s < \infty$. Furthermore, since $G^{(s_1)}(\Omega) \subset G^{(s_2)}(\Omega)$ if $s_1 < s_2$, the s_2-local solvability implies the s_1-local solvability for every $s_1 < s_2$. Therefore the set of the real numbers s such that one has s-local solvability is an interval, eventually empty, with 1 as left ending point (if non-empty).

We shall show here some theorems concerning non-solvability results in Gevrey classes. More precisely, in the first section we prove that the (NT) condition remains necessary (and sufficient) for the s-local solvability if $s > 1$ (cf. Gramchev [6] for the sufficient condition, when P is a Gevrey pseudo-differential operator and $s > 2$). In the second section we consider first the double characteristic operator

$$P_\lambda = D_1^2 + x_1^2 D_2^2 + \lambda D_2,$$

which is not locally solvable at the origin if $\lambda = \pm 1, \pm 3, \pm 5, \ldots$, as showed by Grušin in [7]. It is proved then that P_λ is not even s-locally solvable there, if $1 < s < \infty$. The results till here reported are entirely due to the second author.

Finally, we study a class of operators with double characteristics which result in being microlocally s-solvable for $1 < s < 2$, but not s-locally solvable when $s > 2$; for these operators, then, the interval of s-local solvability is not empty as in the former cases.

§1. OPERATORS OF PRINCIPAL TYPE

The main result of this section is the following:

THEOREM 1.1: Let Ω be an open subset of R^n, P a linear partial differential operator of principal type with analytic coefficients in Ω, x_0 a point of Ω. Then the condition (NT) is necessary (and sufficient) for the s-local solvability at x_0, for every $s > 1$.

In other words, when an operator of principal type is not locally solvable in C^∞, then it remains unsolvable in every Gevrey class $G^{(s)}$, $s > 1$.

It will be enough to prove Theorem 1.1 for the Mizohata operator M in R^n, since every principal type operator can be reduced to this form after a closure argument and a canonical transformation (see Hörmander [9], Chapter 26). We need however a microlocal version of Definition 2, in such a way it is invariant under conjugation of elliptic Fourier integral operators with analytic phase and amplitude functions. The action of such operators on ultradistributions and Gevrey wave front sets has been studied, for example, by Taniguchi [22] and Liess-Rodino [14].

DEFINITION 1.2: The operator P is said to be s-solvable at $(x_0,\xi_0) \in T^*\Omega \setminus 0$ if for every $f \in G^{(s)}(\Omega)$ there exist an $\varepsilon > 0$ and $u \in G^{(s)'}(\Omega)$ such that $(x_0,\xi_0) \notin WF_{s-\varepsilon}(f-Pu)$.

Obviously, if P is s-locally solvable at x_0, then it is s-solvable at (x_0,ξ) for every $\xi \neq 0$. Therefore for the proof of Theorem 1.1 we shall need to show that there exist functions $f \in G^{(s)}(\Omega)$ such that for every $u \in G_0^{(s)'}(\Omega)$ and $\varepsilon > 0$ one has $(0,\theta_2) \in WF_{s-\varepsilon}(f-Mu)$, where $\theta_2 = (0,1,0,\dots,0)$.

Let then be $0 < \rho \leqq 1$, h an odd integer; we are going to study the pseudo-differential operator, in R^n,

$$Q_\rho = D_1 + ix_1^h|D_2|^\rho.$$

This operator is not locally solvable at the origin: this is essentially contained in Nirenberg-Treves [17]; see also Treves [23]. We remark that Q_1 is microlocally equivalent to M at $(0,\theta_2)$ and so Theorem 1.1 will follow from

PROPOSITION 1.3: Let $s > 1/\rho$; then there exist functions $f \in G_0^{(s)}(R^n)$ such that for every $\varepsilon > 0$ and $u \in G^{(s)'}(R^n)$ it results $(0,\theta_2) \in WF_{s-\varepsilon}(f-Q_\rho u)$.

PROOF: We shall consider, for simplicity, the case $n = 2$; then $\theta_2 = (0,1)$. Define the operator Π_ρ as follows:

$$\Pi_\rho f(x_1,x_2) =$$

$$= \iint_{\xi_2>0} \exp[i\omega(x_1,x_2,y_1,y_2,\xi_2)]\xi_2^{\rho/(h+1)} f(y_1,y_2)dy_1dy_2d\xi_2 \qquad (1.1)$$

where

$$\omega(x_1,x_2,y_1,y_2,\xi_2) = \xi_2(x_2-y_2) + i\xi_2^\rho(x_1^{h+1} + y_1^{h+1})/(h+1).$$

This operator can be thought of as a Fourier integral operator with non-homogeneous complex phase function ω; from (1.1) we see that $\Pi_\rho Q_\rho f = 0$, for every $f \in C_0^\infty(R^2)$. It is not difficult to show that, if $s > 1/\rho$, Π_ρ maps continuously $G_0^{(s)}(R^2)$ in $G^{(s)}(R^2)$ and $G^{(s)'}(R^2)$ in $G_0^{(s)'}(R^2)$; moreover, again when $s > 1/\rho$, Π_ρ is s-microlocaly, that is, if $f \in G_0^{(s)'}(R^2)$ and $(x_0,\xi_0) \in T^*R^2\setminus 0$ is such that $(x_0,\xi_0) \notin WF_s(f)$, then $(x_0,\xi_0) \notin WF_s(\Pi_\rho f)$. The distribution kernel $\Pi_\rho(x,y)$ of Π_ρ is therefore of class $G^{(s)}$, $s > 1/\rho$, outside the diagonal Δ of T^*R^2, though $\Pi_\rho(x,y)$ is not regular there: in fact

$\{((x,y),(\xi,\eta)) \in T^*\Delta\setminus 0;\ x_1 = y_1 = 0,\ x_2 = y_2, \xi_1=\eta_1=0, \xi_2 = -\eta_2\} \subset WF(\Pi_\rho(x,y))$.

This means that we can choose a function $f \in G_0^{(s)}(R^2)$, $s > 1/\rho$, such that $(0,\theta_2) \in WF_{s-\varepsilon}(f)$ and $(0,\theta_2) \in WF_{s-\varepsilon}(\Pi_\rho f)$ for every $\varepsilon > 0$. Then, if the conclusion of Proposition 1.3 didn't hold true, there might exist u and a sufficiently small δ for which $(0,\theta_2) \notin WF_{s-\delta}(f-Q_\rho u)$. In this case, since Π_ρ is s-microlocal if $s > 1/\rho$, it would follow $(0,\theta_2) \notin WF_{s-\delta}(\Pi_\rho(f-Q_\rho u))$ and so $(0,\theta_2) \notin WF_{s-\delta}(\Pi_\rho f)$, contradicting the choice of f. This proves Proposition 1.3 and then Theorem 1.1.

In particular we have that the Lewy and the Mizohata operators are not s-locally solvable at the origin if $s > 1$.

We go on in the study of the operator Q_ρ, showing that it is globally solvable in the class $G^{(s)}$ if $1 < s < 1/\rho$. More precisely:

Theorem 1.4: Let $1 < s < 1/\rho$; then for every f in $G_0^{(s)}(R^n)$ (or in $G^{(s)'}(R^n)$) there exists a solution u in $G^{(s)}(R^n)$ (resp. in $G_0^{(s)'}(R^n)$) of the equation $Q_\rho u = f$ in R^n.

PROOF: As before, we give the proof when $n = 2$. Let $A_\rho^\pm$ be the pseudo-differential operator of infinite order defined by

$$A_\rho^\pm f(x_1,x_2)=(2\pi)^{-2}\iint \exp[i(x_1\xi_1+x_2\xi_2)\pm x_1^{h+1}|\xi_2|^\rho/(h+1)]\hat{f}(\xi_1,\xi_2)d\xi_1 d\xi_2,$$

where $\hat{f}$ denotes the Fourier transform of f. About these operators and the related calculus we refer to Zanghirati [25], Cattabriga-Zanghirati [2], Rodino-Zanghirati [21].

By the definition of $A_\rho^\pm$ it is easy to see that it maps continuously $G_0^{(s)}(R^2)$ (or $G^{(s)'}(R^2)$) in $G^{(s)}(R^2)$ (resp. $G_0^{(s)'}(R^2)$), whenever $1 < s < 1/\rho$. Moreover the operators A_ρ^+, A_ρ^-, Q_ρ can be extended to suitable subspaces of $G^{(s)}(R^2)$ or $G_0^{(s)'}(R^2)$, so that it makes sense to consider their compositions and one verifies that

$$\begin{aligned} A_\rho^+ A_\rho^- &= \text{identity} \\ Q_\rho A_\rho^+ &= A_\rho^+ D_1. \end{aligned} \tag{1.2}$$

We set now

$$\tilde{f}(x_1,x_2) = i \int_0^{x_1} A_\rho^- f(y_1,x_2)dy_1$$

for f in $G_0^{(s)}(R^2)$ or in $G^{(s)'}(R^2)$; a solution of the equation $Q_\rho u = f$ with the required regularity is then $u = A_\rho^+\tilde{f}$. This proves Theorem 1.4.

Since A_ρ^+ and A_ρ^- are s-microlocal for $1 < s < 1/\rho$, from (1.2) it follows that $Q_\rho u = D_1 v = f$ implies $WF_s(u) = WF_s(v)$; therefore, in this case, the operators Q_ρ and D_1 are equivalent as regards the study of the propagation of Gevrey singularities of the solutions.

In order to complete the study of the operator Q_ρ we need to take into account the case $s = 1/\rho$. When s is so, from [18] we deduce that, however fixed a point $x_0 \in R^2$ and a function $f \in G^{1/\rho}(R^2)$, there exists a neighbourhood V of x_0, depending on f, where there is a solution u of class $G^{1/\rho}$ of the equation $Q_\rho u = f$.

Finally we remark that, when the exponent h which appears in Q_ρ is even, there is global solvability for every $s > 1$, in the sense that for every $f \in G_0^{(s)}(R^2)$ there exists a solution $u \in G^{(s)}(R^2)$ of the equation $Q_\rho u = f$ in R^2, for $s > 1$. The same solvability result holds true for the operator

$$Q_\rho^* = D_1 - ix_1^h|D_2|^\rho, \tag{1.3}$$

for every positive integer h, as we may deduce from [19].

§2. OPERATORS WITH MULTIPLE CHARACTERISTICS

We begin this section by considering the operator in R^2

$$P_\lambda = D_1^2 + x_1^2D_2^2 + \lambda D_2.$$

In [7] Grušin has proved that P_λ is not locally solvable at the origin, in the C^∞ category, if

$$\lambda = \pm 1, \pm 3, \pm 5, \ldots \tag{2.1}$$

THEOREM 2.1: The operator P_λ is not s-solvable at the origin if λ assumes the values (2.1), for $1 < s < \infty$.

PROOF: We use here the method of concatenations, introduced by Treves in [23]. We first show that when $\lambda = -1, -3, -5, \ldots$ the operator P_λ is not s-solvable at $(0,\theta_2)$, $s > 1$. In fact we observe that

$$P_{-1} = (D_1 + ix_1D_2)(D_1 - ix_1D_2), \tag{2.2}$$

where the Mizohata operator M, with h = 1, appears as first factor in the right hand side. But M is not s-solvable at $(0,\theta_2)$, if $s > 1$ (Proposition 1.3 with $\rho = 1$), and then we have the same result for P_{-1}.

As regards P_{-3}, we can write

$$(D_1 - ix_1D_2)P_{-3} = (D_1 + ix_1D_2)\tilde{P},$$

where $\tilde{P}$ is a second order operator. The right hand side of this equality is surely s-unsolvable at $(0,\theta_2)$, owing to the presence of the Mizohata operator M as left term. The first factor in the left hand side is instead s-solvable at the same point, since it is microlocally equivalent to Q_1^* in (1.3), with h = 1, and the solutions u of $Q_1^*u = f$, $f \in G_0^{(s)}$, are in $G^{(s)}$. This proves that P_{-3} cannot be s-solvable. Arguing in the same way one easily proves that P_λ is not s-solvable at $(0,\theta_2)$ when λ assumes the negative values in (2.1), if $s > 1$. In the case that λ is positive, a similar method yields to the same non-solvability result at $(0,-\theta_2)$.

For all the operators so far considered, non-local solvability in C^∞ was equivalent to non-local solvability in $G^{(s)}$, for $1 < s < \infty$. The last part of these notes intends to show that this is not always the case, that is, there are operators which are unsolvable in C^∞ but become solvable in some $G^{(s)}$ classes, if s is sufficiently near to one.

More precisely we study operators with double characteristics, written under the form

$$Q(x,D) = P_m^2(x,D) + P_{2m-1}(x,D), \tag{2.3}$$

where $P_m = p_m$ is homogeneous of order m, real, of principal type, and P_{2m-1}

is of order 2m-1; we shall denote by p_{2m-1} its principal symbol.

Necessary (and sufficient) conditions for the local solvability of operators of this type has been given by several authors, as mentioned in the introduction. Wenston in particular, [25], shows that if the coefficients of Q are of class C^∞ in an open subset $\Omega \subset R^n$, then there is local solvability at $x_0 \in \Omega$ if the following conditions are satisfied for every x in a small neighbourhood of x_0 and $\xi \neq 0$:

$$p_m(x,\xi) = \mathrm{Im}\ p_{2m-1}(x,\xi) = 0 \text{ imply } \mathrm{Re}\ p_{2m-1}(x,\xi) \neq 0; \tag{2.4}$$

Im p_{2m-1} has constant sign near the null bicharacteristic strip of p_m through (x,ξ). (2.5)

On the contrary, [24], we have not local solvability at the points $x_0 \in \Omega$ for which there exist $\xi_0 \neq 0$ such that

$$p_m(x_0,\xi_0) = \mathrm{Im}\ p_{2m-1}(x_0,\xi_0) = 0,\ \mathrm{Re}\ p_{2m-1}(x_0,\xi_0) \neq 0; \tag{2.6}$$

Im p_{2m-1} changes sign at (x_0,ξ_0) along the null bicharacteristic strip of p_m through (x_0,ξ_0). (2.7)

This last result is obtained by constructing a suitable "approximate solution" u of the homogeneous equations Qu = 0 and then contradicting an a priori inequality which is a necessary condition for the local solvability (see Hörmander [8]).

On the other hand, when the operator Q has analytic coefficients in Ω, then it is everywhere s-solvable in a microlocal sense for $1 < s < 2$ (see Rodino-Zanghirati [21]). Furthermore, when p_m is strictly hyperbolic with respect to some directions, then the corresponding Cauchy problem is weakly locally $G^{(2)}$ correct in the sense given by Ivrii [10], [11]; hence, however fixed $x_0 \in \Omega$ and $f \in G^{(2)}(\Omega)$, there exists a neighbourhood of the point x_0, depending on f, where solutions u of class $G^{(2)}$ of the equation Qu = f are defined.

We can now consider the case $s > 2$.

THEOREM 2.2: Let us assume that the coefficients of the operator Q are analytic and (2.6), (2.7) are satisfied; then, if $s > 2$, Q is not s-locally solvable at x_0.

The proof of this result is done along the lines of [10], and since it is rather long it will be here only sketched; for further details we refer to [3].

Let us introduce then a sequence of Banach spaces, different from that used in the Introduction, to define Gevrey classes.

Let K be a compact subset of Ω, η a positive real number; we denote by $G^{(s)}(K,\eta)$ the space of all the functions $\phi \in C^\infty(K)$ for which the norm

$$|\phi;K,s,\eta| = \sum_\alpha |D^\alpha\phi;\ L^2(K)|\ \frac{\eta^{|\alpha|}}{\alpha!^s}$$

is finite. Set $G_0^{(s)}(K,\eta) = G^{(s)}(K,\eta) \cap C_0^\infty(K)$; it results then

$$G^{(s)}(\Omega) = \lim_{\substack{\leftarrow \\ K\subset\Omega}}\ \lim_{\substack{\rightarrow \\ \eta\to 0}}\ G^{(s)}(K,\eta),\quad G_0^{(s)}(\Omega) = \lim_{\substack{\rightarrow \\ K\subset\Omega}}\ \lim_{\substack{\rightarrow \\ \eta\to 0}}\ G_0^{(s)}(K,\eta).$$

We need first of all an inequality which is a necessary condition for the s-local solvability; its proof is a consequence of the Baire category theorem.

LEMMA 2.3: Let P be a differential operator which is s-locally solvable at x_0; then there exists a neighbourhood U of x_0 such that for every $\eta > 0$ and every compact subset K of U there is a positive constant C such that

$$(\sup_K |f|)^2 \leq C|f;K,s,\eta|\cdot|{}^tPf;K,s,\eta| \tag{2.8}$$

for all the functions f in $G_0^{(s)}(K)$ for which the right hand side makes sense. Here ${}^tP$ indicates the formal adjoint operator of P.

In order to prove Theorem 2.2 we need then an "approximate solution" w, defined in a neighbourhood of x_0, of the equation ${}^tQw = 0$; we can suppose $x_0 = 0$. Furthermore, since ${}^tQ$ satisfies (2.6), (2.7), we use, for simplicity, Q instead of ${}^tQ$. The "approximate solution" w has to make the right hand side of (2.8) sufficiently small, but leaving bounded from below the left hand side. We shall look for w under the form $v \exp(i\lambda^2 u + i\lambda\psi)$, where λ is a real positive parameter and u the real analytic solution of the characteristic equation $p_m(x,\text{grad } u) = 0$ satisfying $u(0) = 0$, $\text{grad } u(0) = \xi_0$; v and ψ are functions to be determined.

With this choice of u it is easy to see that, set $p_m^{(j)}(x,\xi) = \frac{\partial}{\partial \xi_j} p_m(x,\xi)$,

$$\exp(-i\lambda^2 u - i\lambda\psi)Q(x,D)[v \exp(i\lambda^2 u + i\lambda\psi)] =$$

$$= \lambda^{4m-2} v[p_{2m-1}(x,\text{grad } u) + (\sum_j p_m^{(j)}(x,\text{grad } u) \frac{\partial\psi}{\partial x_j})^2] + \qquad (2.9)$$

$$+ \lambda^{4m-3}[\tilde{Q}(x,D) + \sum_{k=1}^{4m-3} \lambda^{-k} Q_k(x,D)]v,$$

where $\tilde{Q}$ and Q_k are differential operators with analytic coefficients; $\tilde{Q}$ is of first order, with $d_\xi \tilde{q}_1(0,\xi_0) \neq 0$.

The assumptions we have done on p_{2m-1} allow us now to find an analytic function ψ vanishing the coefficient of λ^{4m-2} in (2.9); moreover $\psi(0) = 0$ and

$$\text{Im}\psi(x) \geq c|x|^{\kappa+1} \qquad (2.10)$$

in a neighbourhood of the origin. Here κ is the (odd) order of the zero of Im p_{2m-1} at $(0,\xi_0)$ (see (2.7)).

Since $d_\xi \tilde{q}_1(0,\xi_0) \neq 0$ we can solve the Cauchy problems

$$\tilde{Q}(x,D)v_\lambda^{(k)} = - g_\lambda^{(k)} \qquad k = 0,1,2,\ldots$$

$$v_\lambda^{(0)} = 1 \quad \text{on } x_1 = 0$$

$$v_\lambda^{(k)} = 0 \quad \text{on } x_1 = 0, \qquad k \geq 1,$$

where $g_\lambda^{(0)} = 0$ and, if $k \geq 1$, $g_\lambda^{(k)} = \sum \lambda^{-j} Q_j(x,D) v_\lambda^{(k-j)}$; the sum is performed for j, $1 \leq j \leq \min(k,4m-3)$.

Define now

$$V_\lambda^{(N)} = \sum_{k=0}^{N} v_\lambda^{(k)}.$$

The core of this method of proof is the estimate of the Gevrey norms of $V_\lambda^{(N)}$. Set

$$Q_\lambda = \tilde{Q} + \sum_{k=1}^{4m-3} \lambda^{-k} Q_k;$$

arguing as in Ivrii [10] one shows that:

for every sufficiently small compact neighbourhood K of the origin and $\eta > 0$, there exist positive constants C and M such that, assigned to the parameter λ the value N exp(L), N, L sufficiently large, it results

$$\sum_{|\alpha|\leq 2m} |D^\alpha V_\lambda^{(N)};K,s,\eta| \leq C \exp(MN) \tag{2.11}$$

$$|Q_\lambda V_\lambda;K,s,\eta| \leq C \exp(-LN+MN). \tag{2.12}$$

Let now K be a compact set as above, and $\chi \in G_0^{(s)}(K)$ a cut-off function identically equal to one in a neighbourhood of the origin; the "approximate solution" we were looking for will be then $u_\lambda = \chi V_\lambda^{(N)} \exp(i\lambda^2 u + i\lambda\psi)$. For such a function, given now as in the following the value N exp(L) to the parameter λ, it results

$$|u_\lambda;K,s,\eta| \leq C \exp(2MN) \tag{2.13}$$

if N is sufficiently large and η sufficiently small.

As regards Qu_λ we need the following estimate:

$$|\exp(i\lambda^2 u + i\lambda\psi);K,s,\eta| \leq C \exp(a\lambda + d(\lambda^2\eta)^{1/s})$$

where C and d are positive constants and $a = \sup_K (-\mathrm{Im}\psi)$.

Hence, taking into account (2.11), (2.12), we have that for every $\varepsilon > 0$

$$|Qu_\lambda;K,s,\eta-\varepsilon| \leq C_\varepsilon |Q_\lambda V_\lambda^{(N)};K,s,\eta| \cdot |\exp(i\lambda^2 u + i\lambda\psi);K,s,\eta| +$$

$$+ C_\varepsilon \lambda^{2m} \sum_{|\alpha|\leq 2m} |D^\alpha V_\lambda^{(N)};K,s,\eta| \cdot |\exp(i\lambda^2 u + i\lambda\psi);K \cap \mathrm{supp}(\mathrm{grad}\chi),s,\eta|$$

$$\leq C'_\varepsilon \exp(-LN + MN + d(\lambda^2\eta)^{1/s}) + C'_\varepsilon \exp(MN - c_1\lambda + d(\lambda^2\eta)^{1/s}),$$

where $c_1 = \sup_{K \cap \mathrm{supp}(\mathrm{grad}\chi)} (\mathrm{Im}\psi) > 0$ because of (2.10). If we choose then L sufficiently large we have, if $s > 2$,

$$|Qu_\lambda; K, s, \eta-\varepsilon| \leq C''_\varepsilon \exp(-3MN). \tag{2.14}$$

But $\sup_K |u_\lambda| \geq 1$, since $v_\lambda^{(0)}(0) = 1$; then we deduce from (2.13) and (2.14) that the inequality (2.8) cannot hold if f is replaced by u and N is sufficiently large. This proves Theorem 2.2.

References

[1] Cardoso, F., Treves, F., A necessary condition of local solvability for pseudo-differential equations with double characteristics, Ann. Inst. Fourier Grenoble, 24 (1974), 225-292.

[2] Cattabriga, L., Zanghirati, L., Fourier integral operators of infinite order any Gevrey spaces. Applications to the Cauchy problem for hyperbolic operators, Proceedings NATO ASI "Advances in microlocal analysis", 41-71, Reidel Publ. Comp. (1986).

[3] Corli, A., On local solvability in Gevrey classes of linear partial differential operators with multiple characteristics, Preprint.

[4] Egorov, Ya. V., On necessary conditions for solvability of pseudo-differential operators of principal type, Trudy Moskov. Mat. Obsc., 24 (1971), 29-41; Trans. Moscow Math. Soc., 24 (1971), 29-42.

[5] Egorov, Ya. V., On solvability conditions for equations with double characteristics; Dokl. Akad. Nauk SSSR, 234 (1977); Soviet Math. Dokl., 18 (1977), 632-635.

[6] Gramchev, T.V., Hypoellipticity and solvability in ultra-distribution spaces for principal-type pseudo-differential operators, Comptes rendus Academie bulgare des Sciences, 38 (1985), 1295-1298.

[7] Grušin, V.V., On a class of elliptic pseudo differential operators degenerate on a submanifold, Mat. Sb., 84 (1977), 163-195; Math. USSR Sb., 13 (1971), 155-185.

[8] Hörmander, L., Differential equations without solutions, Math. Ann., 140 (1960), 169-173.

[9] Hörmander, L., The analysis of linear partial differential operators IV, Springer Verlag, Berlin 1985.

[10] Ivrii, V. Ya., Conditions for correctness in Gevrey classes of the Cauchy problem for weakly hyperbolic equations, Sib. Mat. Zh., 17 (1976), 547-563; Sib. Math. J., 17 (1976), 422-435.

[11] Ivrii, V. Ya., Cauchy problem conditions for hyperbolic operators with characteristics of variable multiplicity for Gevrey classes, Sib. Mat. Zh., 17 (1976), 1256-1270; Sib. Math. J., 17 (1976), 921-931.

[12] Komatsu, H., Ultradistributions, I: Structure theorems and a characterization, J. Fac. Sci. Univ. Tokyo, Sect I A, 20 (1973), 25-105.

[13] Lewy, H., An example of a smooth linear partial differential equation without solution, Ann. of Math., 66 (1957), 155-158.

[14] Liess, O., Rodino, L., Fourier integral operators and inhomogeneous Gevrey classes, to appear in Annali Mat. Pura Appl.

[15] Menikoff, A., On hypoelliptic operators with double characteristics, Ann. Sc. Norm. Sup. Pisa, Ser. IV, 4 (1977), 689-724.

[16] Mizohata, S., Solutions nulles et solutions non analytiques, J. Math. Kyoto Univ., 1 (1962), 271-302.

[17] Nirenberg, L., Treves, F., On local solvability of linear partial differential equations, I: Necessary conditions; II: Sufficient conditions, Comm. Pure Appl. Math., 23 (1970), 1-38 and 459-509.

[18] Ohya, Y., Le problème de Cauchy pour les equations hyperboliques à characteristique multiple; J. Math. Soc. Japan, 16 (1964), 268-286.

[19] Parenti, C., Rodino, L., Parametrices for a class of pseudo differential operators, Ann. Mat. Pura Appl., 125 (1980), 221-278.

[20] Popivanov, P.R., Local solvability of pseudodifferential operators with characteristics of second multiplicity, Mat. Sb., 100 (142) (1976), 217-241; Math. USSR Sb., 29 (1976), 193-216.

[21] Rodino, L., Zanghirati, L., Pseudo differential operators with multiple characteristics and Gevrey singularities, Comm. Partial Differential Equations, 11 (7) (1986), 673-711.

[22] Taniguchi, K., Fourier integral operators in Gevrey class on $\mathbb{R}^n$ and the fundamental solution for a hyperbolic operator, Publ. RIMS, Kyoto Univ., 20 (1984), 491-542.

[23] Treves, F., Concatenations of second-order evolution equations applied to local solvability and hypoellipticity, Comm. Pure Appl. Math., 26 (1973), 201-250.

[24] Wenston, P.R., A necessary condition for the local solvability of the operator $P_m^2(x,D) + P_{2m-1}(x,D)$, J. Diff. Equ., 25 (1977), 90-95.

[25] Wenston, P.R., A sufficient condition for the local solvability of a linear partial differential operator with double characteristics, J. Diff. Equ., 29 (1978), 374-387.

[26] Zanghirati, L., Pseudodifferential operators of infinite order and Gevrey classes, Ann. Univ. Ferrara, Sez. VII, 31 (1985), 197-219.

Luigi Rodino
Università di Torino,
Dipartimento di Matematica,
Via Carlo Alberto, 10,
I-10123 Torino,
Italy

Andrea Corli
Università di Ferrara,
Dipartimento di Matematica,
Via Machiavelli, 35
I-44100 Ferrara,
Italy.

X. SAINT RAYMOND

Extensions of Holmgren's uniqueness theorem

Let (t,y) denote the coordinates in $\mathbb{R} \times \mathbb{R}^{n-1}$. We are interested in the following question, which has an obvious "physical" meaning: for two solutions u_1 and u_2 of the same partial differential equation such that $u_1 = u_2$ if $t < 0$, is it true that this equality still holds in a whole neighbourhood of a point $(0,y_0)$? We do not treat nonlinear equations for which very few results are known (first order equations); for linear equations, the property we want to characterize can be written (by setting $u = u_1 - u_2$)

$$Pu = u_{|t<0} = 0 \Rightarrow u = 0 \text{ near } (0,y_0).$$

Here, $P(x,D) = \sum_{|\alpha|\leq m} a_\alpha(x)D^\alpha$ is a linear differential operator with complex valued coefficients. This problem is what Zuily called "Propagation of the zeroes" in his lectures on the uniqueness in the Cauchy problem here at Pisa. There are two kinds of known results on this subject:

1. Characterization of the uniqueness property by the regularity of the coefficients for equations of a fixed type: elliptic equations with Lipschitz coefficients (Hörmander, Pliś), $\Delta + V$ with $V \in L^p$ (Jerison & Kenig), $\partial_t^2 - a(t)\partial_x^2 + b(x,t)$ with Hölder nonnegative a and Gevrey b (Colombini, Jannelli & Spagnolo). The previous references are given only as examples since I know almost nothing on this kind of results.

2. Characterization of the uniqueness property by geometrical conditions (structure or convexity conditions) when the coefficients are assumed to be very regular. To be able to write these geometrical conditions, we set $(t,y) = x$ and $t = \varphi(x)$; in fact, these conditions will not depend on the function φ but only on the hypersurface S defined by the equation $\varphi(x) = 0$. The principal symbol of P will be denoted by $p(x,\xi) = \sum_{|\alpha|=m} a_\alpha(x)\xi^\alpha$. For the regularity of the coefficients, we still have to choose between the following two possibilities:

a). If the coefficients are assumed to be C^∞, the uniqueness property cannot be characterized by purely geometrical conditions: there exist an operator P and a function a flat on S such that the uniqueness property holds for P, but does not hold for P + a. In fact one can then characterize the following stable uniqueness property: the uniqueness property for P + a for any perturbation $a \in C^\infty$. For such results, we refer to the already quoted lectures by C. Zuily.

b). If the coefficients are assumed to be analytic, the main result is the well-known Holmgren's uniqueness theorem: <u>if</u> S <u>is noncharacteristic at</u> x_0 <u>(that is</u> $p(x_0,d\varphi(x_0)) \neq 0$), <u>the uniqueness property holds at</u> x_0: $\forall u \in \mathcal{D}'(\mathbb{R}^n)$, $Pu = u_{|\varphi<0} = 0 \Rightarrow u = 0$ <u>near</u> x_0. In this talk, we want to describe uniqueness results for the characteristic problem that can be obtained from this Holmgren's theorem.

From now on, P is an analytic linear partial differential operator and φ is an analytic real valued function such that $\varphi(x_0) = 0$ and $d\varphi(x_0) \neq 0$. Our problem can be considered as a study of the geometry of the supports of the solutions of the equation $Pu = 0$; for this, there exists a theory of microlocal propagation that we now describe.

1. <u>MICROLOCAL PROPAGATION OF THE SUPPORT</u> (due to Bony & Sjöstrand)

The description of this theory of microlocal propagation is very similar to the corresponding propagation of singularities in the hyperbolic problem. Indeed, to understand the propagation of singularities, it was necessary to introduce a microlocal (that is contained in $T^*\mathbb{R}^n$) object over singsupp u, called WF u, for which good propagation theorems were known, and the corresponding results for singsupp u were obtained by projection:

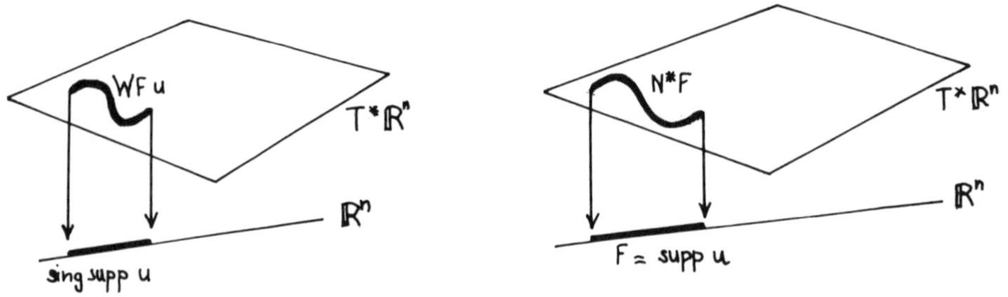

Here, the idea is similar: we introduce over $F = \text{supp } u$ a microlocal object N^*F for which it is possible to prove a good propagation theorem and which will give uniqueness results by projection.

Microlocal object over the support (due to J.-M. Bony): let F be the support of u; the point $(x,\xi) \in T^*\mathbb{R}^n \setminus 0$ is said to be in N^*F if $x \in F$ and F is contained in an analytic half-space whose boundary contains x and is normal to ξ at x:

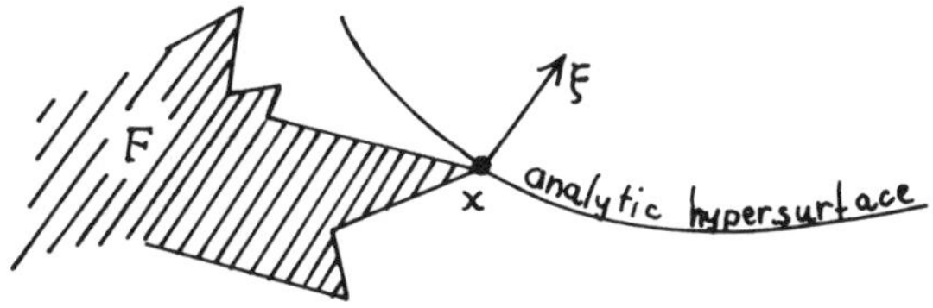

When F is a regular manifold, N^*F is the conormal bundle over F; but here F is the support of an unknown distribution, and that is why we need this definition which makes sense for any closed set F, even if it is very irregular. The projection of N^*F on $\mathbb{R}^n$ can be shown to be dense in ∂F.

PROPAGATION THEOREM (due to J. Sjöstrand): the bicharacteristic curve of any real valued C^∞ function p vanishing on N^*F remains locally in N^*F.

SKETCH OF THE PROOF: $(x_0,\xi_0) \in N^*F \iff$ there exists a real φ_0 such that $x_0 \in F$, $d\varphi_0(x_0) = \xi_0$ and $F \subset \{x;\varphi_0(x) \geq \varphi_0(x_0)\}$. Similarly, we have to prove that there exists a function $\varphi(x,t)$ such that for any fixed t, $x(t) \in F$, $d_x\varphi(x(t),t) = \xi(t)$ and $F \subset \{x;\varphi(x,t) \geq \varphi(x(t),t)\}$ (here, $x(t),\xi(t)$ denotes the bicharacteristic curve of p). By differentiating these conditions it can be seen that $\varphi(x,t)$ must be the solution of the following Cauchy problem:

$$\begin{cases} \dfrac{\partial\varphi}{\partial t}(x,t) + p(x, \dfrac{\partial\varphi}{\partial x}(x,t)) = 0 \\ \varphi(x,0) = \varphi_0(x); \end{cases}$$

however, this solution will not work if φ_0 was not chosen sufficiently convex; finally, the good function $\varphi(x,t)$ will be obtained by taking

$\varphi_0(x) + A|x-x_0|^2$ with a large constant A instead of $\varphi_0(x)$ as initial data.

2. BONY & SJÖSTRAND'S UNIQUENESS THEOREM

DEFINITIONS: Let us denote by I(p) the ideal of analytic functions generated by Rep, Imp and all their iterated Poisson brackets.

(We give this definition for the sake of simplicity; indeed, one can define a larger ideal with the same properties which gives a better uniqueness theorem).

Then, the oriented hypersurface $S = \{x;\varphi(x) = 0\}$ is said to be pseudo-concave at x_0 (with respect to p) if for any function $q \in I(p)$, $q(x_0,d\varphi(x_0))=0$ and the bicharacteristic curve of q starting at this point remains in the future, that is in $\{x;\varphi(x) \geqq 0\}$.

THEOREM (due to Bony & Sjöstrand): if S is not pseudo-concave at x_0, the uniqueness property holds at x_0: $\forall u \in \mathcal{D}'(R^n)$, $Pu = u_{|\varphi<0} = 0 \Rightarrow u = 0$ near x_0.

EXAMPLES: (i) If p is of real principal type, I(p) is simply the ideal generated by p, and S is pseudo-concave at x_0 if and only if it is characteristic at x_0 and the bicharacteristic curve of p starting at $(x_0,d\varphi(x_0))$ remains in the future. Therefore, the theorem claims that the uniqueness property holds if S is noncharacteristic at x_0 or if the bicharacteristic curve meets the past with any order of contact (even odd or infinite).

(ii) If $p = \xi_1 + ix_1\xi_2$ and:

(a) $\varphi = x_2$ in $\mathbb{R}^2$, then $q(x_0,d\varphi(x_0)) \neq 0$ with $q = \{Rep, Imp\}$;

(b) $\varphi = x_3 + x_1^2 - \exp(-1/x_2^2)$, then the bicharacteristic curve of q starting at $(x_0,d\varphi(x_0))$ meets the past with an infinite order of contact, and in these two cases, this is sufficient to imply the uniqueness property.

PROOF OF THE THEOREM: let us set F = supp u.

Pu = 0 implies that p vanishes on N*F according to Holmgren's uniqueness theorem, and therefore this is true separately for Rep and Im p. If $(x_1,\xi_1) \in$ N*F and γ is the bicharacteristic curve starting at (x_1,ξ_1) of a function q vanishing on N*F, then $\gamma \subset$ N*F according to the propagation theorem, and if r is another function vanishing on N*F, we can write $\{q,r\}(x_1,\xi_1) = 0$ since this quantity is equal to the derivative of r along γ where r vanishes; therefore, the function $\{q,r\}$ itself vanishes on N*F. Since this is true for Re p and Im p, this proves that any $q \in I(p)$ vanishes on N*F .

$u_{|\varphi<0} = 0$ implies that $x_0 \in F$ if and only if $(x_0,d\varphi(x_0)) \in N^*F$ (by definition of N*F). Then, by contradiction: if $x_0 \in F$, $(x_0,d\varphi(x_0)) \in N^*F$ and any $q \in I(p)$ vanishes at this point, and its bicharacteristic curve remains in N*F whose projection is contained in $F \subset \{x;\varphi(x) \geqq 0\}$; this means that S is pseudo-concave at x_0, which is false by assumption. This ends the proof of the theorem.

QUESTION: When S is pseudo-concave at x_0, what can be said? More precisely, can we always construct a null solution with support based on the bicharacteristic curves of the elements of I(p)? Actually, the answer is "no": we can sometimes prove the uniqueness property even when S is pseudo-concave at x_0. We now describe these new results.

3. FURTHER RESULTS

To be clear, we first treat the following example: $p = \xi_1^2 - \xi_2^2 - \xi_3^2$ and $\varphi = x_1 + x_2 + x_1x_3$. Since this operator is of real principal type, we only have to consider p and its bicharacteristic curve: $d\varphi(0) = dx_1 + dx_2$, $p(d\varphi(0)) = 0$, $\xi(t) = (1,1,0)$, $x(t) = (2t,-2t,0)$ and $\varphi(x(t)) = 0$; thus, S is pseudo-concave at $x_0 = 0$ (more precisely, the bicharacteristic curve remains in S).

However, we can prove the uniqueness property in the following way: $d\varphi(x(t)) = dx_1 + dx_2 + 2tdx_3$ and then $p(d\varphi(x(t))) = -4t^2$; therefore, for $t \neq 0$, S is noncharacteristic at x(t) and according to Holmgren's uniqueness theorem, $x(t) \notin \text{supp } u$; this implies that $(x(t),\xi(t)) \notin N^*F$, and using Sjöstrand's propagation theorem, that $(x_0,d\varphi(x_0)) \notin N^*F$, and this means that $x_0 \notin \text{supp } u$, hence the uniqueness property.

This example is essentially due to Persson who studied this question in great detail for operators with constant coefficients. For operators of real principal type, similar results were obtained by Tintarev.

The idea of the previous proof can be used to obtain new results: if $(x(t),\xi(t))$ now denotes the bicharacteristic curve of a $q \in I(p)$, and if

$r(x(t),d\varphi(x(t))) \neq 0$ for another $r \in I(p)$, we can get the same conclusion by using Bony & Sjöstrand's uniqueness theorem instead of Holmgren's; moreover, if the bicharacteristic curve does not remain in S, the proof will still work if we can perform a deformation of the initial surface which remains noncharacteristic with respect to r. When measuring what is necessary to construct such a deformation, we get the following definition.

DEFINITION: if S is pseudo-concave at x_0, it is said to be non-twisted at x_0 if for any $q \in I(p)$ such that $q_\xi(x_0,d\varphi(x_0)) \neq 0$,

(i) $q(x,d\varphi(x)) = O(\varphi(x)^{1/2})$ on $\bigcup_{r\in I(p)}$ bichar. curve of r;

(ii) $q(x,d\varphi(x)) = O(\varphi(x)/|x-x_0|+ \varphi(x)^{1/2}|d\varphi\wedge\xi dx|)$ on the bicharacteristic curve of q.

THEOREM: if S is pseudo-concave and twisted at x_0, the uniqueness property holds at x_0: $\forall u \in \mathcal{D}'(\mathbb{R}^n)$, $Pu = u_{|\varphi<0} = 0 \Rightarrow u = 0$ near x_0.

COMPLEMENT: there are nonuniqueness theorems which show that, in rather generic situations, the uniqueness property fails to hold when S is pseudo-concave and non-twisted at x_0. In particular, when m = 1 (first-order equations) or n = 2 (equations in two variables), it can be easily seen that any pseudo-concave surface is non-twisted and then the nonuniqueness theorems give the following result.

THEOREM: if m = 1 or n = 2 and if $p_\xi(x_0,d\varphi(x_0)) \neq 0$, the uniqueness property holds at x_0 if and only if the initial surface S is not pseudo-concave at x_0.

This result, for m = 1, is due to Zachmanoglou.

FINAL REMARK: in all these uniqueness theorems, we do not use, actually, the analyticity of the coefficients, but only what gives Holmgren's uniqueness theorem: the uniqueness property across any noncharacteristic hypersurface. All the previous results will remain valid for any class of operators for which the uniqueness property holds across any noncharacteristic hypersurface; this is the case for locally solvable first-order equations according to a result of Strauss & Treves.

Main references

Bony J.-M., Une extension du théorème de Holmgren sur l'unicité du problème de Cauchy, C.R. Acad. Sc. Paris 268, 1103-1106 (1969).

Persson, J., The Cauchy problem at simply characteristic points and P-convexity, Ann. di Mat. pura ed appl. 102, 117-140 (1979).

Saint Raymond X., Autour du théorème de Holmgren sur l'unicité de Cauchy, J. of diff. geom. 20, 121-135 (1984).

Saint Raymond X., Contribution to the study of the uniqueness in the analytic linear Cauchy problem, to appear in Comm. in P.D.E.

Sjöstrand J., Singularités analytiques microlocales, Chapitre 8: Théorème d'unicité de Holmgren et extensions, Astérisque 95, 52-60 (1982).

Tintarev K., On complementary geometrical conditions of uniqueness and non-uniqueness for operators with simple characteristics, Comm. in P.D.E. 11, 989-1008 (1986).

Zachmanoglou E.C., Propagation of zeroes and uniqueness in the Cauchy problem for first order partial differential equations, Arch. Rat. Mech. Anal. 38, 178-188 (1970).

Xavier Saint Raymond
Université de Paris-Sud
Mathematiques,
Batiment 425,
F 91405 Orsay
France.

J. SJÖSTRAND & B. HELFFER

Semiclassical analysis for Harper's equation

0. INTRODUCTION

In this paper we are interested in the Spectrum of Harper's operator in $l^2(\mathbf{Z})$, given by $H_{\theta,\lambda,h}u(n) = (u(n+1) + u(n-1))/2 + \lambda \cos(hn + \theta)u(n)$, which appears naturally for instance in the study of the Schrödinger operator on $\mathbf{R}^2_{x,y}$ with a potential which is periodic in x and in y and with a magnetic field, having the same periods. Here $\lambda > 0$, $h \in \mathbf{R}$. If $h/2\pi$ is rational, we can use the Floquet theory to reduce the study of the spectrum of $H_\theta = H_{\theta,\lambda,h}$ to that of a finite matrix depending on two parameters. The quantity that we are interested in in this case is the union of the spectra of H_θ when θ varies in $\mathbf{R}$. This set is then a finite union of closed intervals ("bands"). When $h/2\pi$ is irrational, it is easy to show that the spectrum of H_θ *as a set* is independent of θ. We then have a discrete Schrödinger operator with a quasi-periodic potential and there is a vast mathematical and physical literature about such operators. Concerning Harper's operator, we can in particular, mention the work of Bellissard and Simon [3], which shows that for a dense set (and in fact for a countable intersection of open dense sets) in the set of parameters (λ,h), the spectrum of H is not dense in any non-trivial interval. See also [10]. On the other hand, the case $\lambda = 1$ seems to play an important role as a transition between the cases $\lambda < 1$ and $\lambda > 1$. In this case, there seem to be no rigourous results, but it has been conjectured in this case that for $h/2\pi$ irrational, the spectrum is a Cantor set of measure 0. See Azbel [2], Aubry, André [1], Sokoloff [11], Wilkinson [12], Hofstadter [9]. The two last works, indicated to one of us by J. Bellissard, have been for us a particularly important source of inspiration. Hofstadter's results are entirely numerical and indicate very clearly the Cantor structure of the Spectrum.

Wilkinson's work (see also [13, 14]) is based on a WKB analysis with infinitely many "potential" wells in the space $T^*\mathbf{R}$. These wells interact by tunnel effect and Wilkinson indicates how, by analyzing these interactions, one obtains a new Harper's operator with $\lambda = 1$, but with a new h. To get the complete structure of the spectrum, it would be enough in principle, to repeat this procedure indefinitely. The sequence of h's that one obtains is given by the continued fraction expansion of $h/2\pi$, and Wilkinson indicates how his procedure works, if all the h's are sufficiently small. His work shows a remarkable intuition, in view of the fact that his mathematical arguments are rather vague. Using the techniques of [6,7], extended to the case of infinitely many wells by U. Carlsson [4], we have managed to give a rigourous presentation of Wilkinson's (and Azbel's) ideas, but at each step in the iteration procedure there is a small part of the spectrum which requires a different analysis (which also seems quite within reach) and which has to be excluded. This prevents us at the present stage of our investigations to make a complete analysis of the spectrum.

1. REDUCTION TO A SEMICLASSICAL PROBLEM AND SPECTRAL ANALYSIS mod $\mathcal{O}(h^\infty)$

A good part of our analysis works also in the case $\lambda \neq 1$, but the case $\lambda = 1$ is the one which is the best suited for our methods, and we only treat that case in the following. If $Sp(H_\theta)$ denotes the spectrum of H_θ, then it is easy to see that $Sp(H_\theta) = Sp(H_{\theta+h})$, and if we assume that $h > 0$, then $U_\theta Sp(H_\theta) = Sp(H')$, where H' is the operator in $L^2(\mathbf{Z} \times [0,h[)$, defined by

$$(H'u)(k,\theta) = H_\theta u(.,\theta)(k). \qquad (1.1)$$

The substitution $x = \theta + hk$, shows that H' can be identified with the operator

$$P = \cos(hD) + \cos x = (\tau_h + \tau_{-h})/2 + \cos x, \qquad (1.2)$$

acting in $L^2(\mathbf{R})$. Here $D = -id/dx$, and $\tau_h = e^{-ihD}$ is the operator of translation by h. Our problem is then to study $Sp(P) \subset [-2,2]$. We can view P as an h-pseudodifferential operator with symbol $p(x,\xi) = \cos(\xi)+\cos(x)$. For $0 < E < 2$, we have $p^{-1}(E) = \cup U_\alpha$, $\alpha \in \mathbf{Z}^2$, where U_α is a closed curve around $2\pi\alpha$. If $E = 2$, we have the same relation with $U_\alpha = \{2\pi\alpha\}$. For $-2 \leq E < 0$, one has an analogous situation. For $E = 0$, $p^{-1}(0)$ is a union of lines and the study of the spectrum near 0 will require different techniques. If $\tau = \tau_{2\pi}$ and σ denotes the operator of multiplication by $e^{2\pi ix/h}$, then P commutes with τ and σ and hence also with the operators $T_\alpha = \tau^{\alpha_1}\sigma^{\alpha_2}$, for $\alpha = (\alpha_1,\alpha_2) \in \mathbf{Z}^2$. The difficulty in the problem comes from the fact that τ and σ do not always commute. Let $2\pi/h = k + h'/2\pi$, with $k \in \mathbf{Z}$. Then $\tau\sigma = e^{ih'}\sigma\tau$, and

$$T_\alpha T_\beta = e^{i\omega(\alpha,\beta)h'} T_\beta T_\alpha, \tag{1.3}$$

where ω denotes the standard symplectic form on $\mathbf{R}^2$.

P also commutes with the unitary Fourier transformation,

$$F_h u(\xi) = (2\pi h)^{-1/2} \int e^{-ix\xi/h} u(x)dx. \tag{1.4}$$

We now fix $\varepsilon_0 > 0$, and we are interested in the spectrum of P in $[\varepsilon_0, 2]$, for $h > 0$ sufficiently small. The first problem is then to determine the spectrum modulo $O(h^\infty)$, but already here one tries to imitate the approach of Helffer-Sjöstrand [6,7] for the Schrödinger equation with potential wells, extended to the case of infinitely many potential wells by U. Carlsson [4]. Let $0 \leq \theta_0(x,\xi) \in C^\infty(\mathbf{R}^2)$ be of support in $|x| + |\xi| < \pi$, and sufficiently large so that $p-\theta_0-\varepsilon_0<0$ in this square. In the following we identify a symbol $a(x,\xi,h) \in S^m = \{a \in C^\infty(\mathbf{R}^2);\ \partial_x^j \partial_\xi^k a = O(h^{-m})\}$ with its h-Weyl quantization,

$$au(x) = (2\pi)^{-1} \iint e^{i(x-y)\theta/h} a((x+y)/2, \theta, h)u(y)dyd\theta, \tag{1.5}$$

uniformly bounded on L^2 if $m = 0$. Let $\theta_\alpha = T_\alpha \theta_0 T_\alpha^{-1}$, with symbol $\theta_\alpha(x,\xi) = \theta_0(x-2\pi\alpha_1, \xi-2\alpha_2)$, and put

$$P_0 = P - \Sigma_{\alpha\neq 0}\, \theta_\alpha. \tag{1.6}$$

For the corresponding symbol; p_0, we have $p_0^{-1}(E) = U_0 = U_0(E)$, if $\varepsilon_0 \leq E \leq 2$. Combining a result of Helffer-Robert [5] for $\varepsilon_0 \leq E \leq 2-Ch$ with a study of the eigenvalues of P_0 in $[2-Ch,2]$ analogous to the one of the small eigenvalues for the semiclassical Schrödinger operator with a non-degenerate point well, we obtain,

<u>PROPOSITION 1.1</u>: For $h > 0$ sufficiently small $Sp(P_0) \cap [\varepsilon_0, 2] = \cup_{0\leq j\leq N(h)} \{\mu_j\}$, where each μ_j is a simple eigenvalue, and $h/C \leq \mu_j - \mu_{j+1} \leq Ch$ for some sufficiently large constant $C > 0$.

One can indeed determine the μ_j modulo $\mathcal{O}(h^\infty)$, but this proposition suffices in order to describe the essential phenomena in the following. One can also show that the eigenfunction of P_0 corresponding to μ_j, is microlocally concentrated to any neighbourhood of $U_0(\mu_j)$. We now choose one of the eigenvalues; $\mu(h) = \mu_{j(h)}(h)$. All the estimates below will be uniform with respect to all such possible choices. Let $\varphi_0 \in L^2$ be a corresponding normalized eigenvector, and put $\varphi_\alpha = T_\alpha\varphi_0$, $P_\alpha = T_\alpha P_0 T_\alpha^{-1} = P-\Sigma_{\beta\neq\alpha}\theta_\beta$. Then, $P_\alpha\varphi_\alpha = \mu\varphi_\alpha$. Combining the techniques of [6], [7], [4], with some pseudo-differential arguments, one gets,

THEOREM 1.2: For every $N > 0$, there exists $C_N > 0$ such that $Sp(P) \cap [\varepsilon_0,2] \subset \cup_{0\leq j\leq N(h)}[\mu_j-C_Nh^N,\mu_j + C_Nh^N]$. If we choose , φ_0, φ_α as above, and if $F \subset L^2$ denotes the spectral subspace associated to $Sp(P) \cap [\mu-C_Nh^N,\mu+C_Nh^N]$, and Π_F is the orthogonal projection onto F, then the $v_\alpha = \Pi_F\varphi_\alpha$, $\alpha \in \mathbf{Z}^2$ form a Hilbert basis in F in the sense that every vector $f \in F$ is of the form $f = \Sigma f_\alpha v_\alpha$, with $\|f\|^2 \sim \Sigma|f_\alpha|^2$. The matrix $V = ((v_\alpha|v_\beta))$ is of the form $1 + \mathcal{O}(h^\infty)$ in $L(l^2,l^2)$, and in the orthonormalized basis, $u = v\,V^{-1/2}$, the matrix of $P|_F$ is of the form $\mu I+ W$, where $W = \mathcal{O}(h^\infty)$.

2. WEIGHTED INEQUALITIES AND THE INTERACTION MATRIX

It turns out that the interactions hiding in W are exponentially small and cannot be studied by using standard C^∞ pseudodifferential methods. The first step is then as in [6], [7] to develop weighted inequalities. If we consider an operator,

$$Q = (1-\cos(hD)) + V(x), \tag{2.1}$$

where $V \in C(\mathbf{R}) \cap L^\infty(\mathbf{R})$ is real, we first remark that $(1-\cos(hD)) = 2\sin(hD/2)^2$. Hence, if $\varphi \in C^{1,1}(\mathbf{R})$ with φ', $\varphi'' \in L^\infty(\mathbf{R})$, we find,

$$(e^{\varphi/h}(1-\cos(hD))u|u_\varphi) = 2(Ru_\varphi|R^*u_\varphi),\ u \in C_0^\infty, \tag{2.2}$$

where $u_\varphi = e^{\varphi/h}u$, and

$$R = e^{\varphi/h}\circ\sin(hD/2)\circ e^{-\varphi/h} = A + iB, \tag{2.3}$$

$$A = ch(\varphi'/2)\sin(hD/2),\quad B = sh(\varphi'/2)\cos(hD/2).$$

Here A, B are selfadjoint mod.$\mathcal{O}(h)$ and by imitating the procedure in [6], we find that there exists $C > 0$, such that if $z \in \mathbf{C}$ and $0 \leq F_{\pm} \in L^{\infty}$ satisfy

$$V\text{-Re}z\text{-}Ch\text{-}2(\text{sh}(\varphi'/2)) = F_+^2 - F_-^2,$$

then,

$$2\,\|\text{ch}(\varphi'/2)\sin(hD/2)u_{\varphi}\|^2 + (1/4)\|(F_+ + F_-)u_{\varphi}\|^2 \leq \tag{2.4}$$

$$\|(F_+ + F_-)^{-1}e^{\varphi/h}(P-z)u\|^2 + \|F_-u_{\varphi}\|^2,$$

for $u \in C^{\infty}(\mathbf{R})$, $u_{\varphi} = e^{\varphi/h}u$.

We apply this inequality to $Q = 2-P = (1-\cos(hD)) + (1-\cos x)$ with z close to $\nu(h) = 2 - \mu(h)$. Let $U_j = \pi_x(U_{(j,k)})$, $U_{(j,k)} = U_{(j,k)}(\mu(h))$. The weighted inequalities then work very nicely outside $\cup U_j$; if we fix $a > 0$ sufficiently small, then for $|z-\mu(h)| = ah$, we can represent $(P-z)^{-1}$ by a Neumann type series making use of the operators $(P_j-z)^{-1}$, $j \in \mathbf{Z}$, with $P_j = P-\Sigma_{k\neq j}\theta_k$ where $\theta_k = \tau_{2\pi k}\theta_0$, and $0 \leq \theta_0 \in C_0^{\infty}$ is of support close to U_0 and with the property that $1-\cos x-\nu(h)+\theta_0 > 0$ near U_0. One then gets the following results:

If $\delta > 0$ and if $2(\text{sh}(\varphi'/2))^2_- \leq (1-\cos x-\nu(h)-\delta)_+$, then for $0 < h \leq h_{\delta}$ sufficiently small, $(P-z)^{-1}$ is bounded and of norm $\mathcal{O}(h^{-1})$: $L^2_{\varphi} \to L^2_{\varphi}$. (2.5)

Here $L^2_{\varphi} = \{u; e^{-\varphi/h}u \in L^2\}$.

If $\delta > 0$ and if φ_1, φ_2 satisfy $2(\text{sh}(\varphi_k'/2))^2 \leq (1-\cos x-\nu(h)-\delta)_+$, and $\varphi_1 = \varphi_2$ on $\cup U_j$, then for $0 < h \leq h_{\delta}$ sufficiently small, Π_F is bounded and of norm $\mathcal{O}(1)$: $L^2_{\varphi_1} \to L^2_{\varphi_2}$. (2.6)

Let (x_0, x_0) be the unique point in U_0 with $x_0 \geq 0$. Then there exist (x_h, ξ_h) tending to (x_0, x_0), a linear combination; g_0 of the functions m, $F_h m$, $F_h^2 m$, $F_h^3 m$, with coefficients $\mathcal{O}_{\varepsilon}(e^{\varepsilon/h})$ for every $\varepsilon > 0$, where

$m(x) = \exp(i(x-x_h)\xi_h - (x-x_h)^2/2/h$, such that,

$$(g_0|\varphi_0) = 1 + O(h^\infty), \tag{2.7}$$

$$F_h g_0 = \omega g_0, \ |\omega| = 1. \tag{2.8}$$

We then redefine v_0 as $\Pi_F g_0$ and we put $v_\alpha = T_\alpha v_0 = \Pi_F T_\alpha g_0$. One can then assume that $\|v_0\| = 1$, and if we use (2.6), we find,

$$v_0 = \underline{O}(e^{-f(x)/h}),\ f(x) = \min(k\nu_0+D(2\pi k,|x|),(k+1)\nu_0+D(2\pi(k+1),|x|),$$
$$\text{for } 2\pi k \leqq |x| \leqq 2\pi(k+1). \tag{2.9}$$

Here $0 < \nu_0 \leqq S_0 = D(0,2\pi)$, and the first relation means by definition, that for every $\varepsilon > 0$, we have $\|e^{(1-\varepsilon)f/h}v_0\| = O_\varepsilon(e^{\varepsilon/h})$, $h \to 0$. Moreover, we have put $D(x,y) = |\Phi(x)-\Phi(y)|$, where $\Phi(x)$ is an increasing solution of $2(\mathrm{sh}\Phi'/2)^2 = (1-\cos x-\nu(h))_+$. Since P commutes with F_h, we get the same estimate for $F_h v_0$. We then get for $\alpha \neq \beta$: $(v_\alpha|v_\beta) = O_\varepsilon(1)e^{-(1-\varepsilon)\nu_0|\alpha-\beta|_\infty/h}$, $\varepsilon>0$.

Again, the v_α generate F, and if we pass to the (new) orthonormalised basis; u_α, we find that $F_h u_0 = \omega u_0$, $u_\alpha = T_\alpha u_0$. The matrix of $P|_F$ is then of the form,

$$((Pu_\beta|u_\alpha)) = \mu I + W, \tag{2.10}$$

with a new $\mu = \mu(h)$, which differs by $O(h^\infty)$ from the old $\mu(h)$. Here $W = (w_{\alpha,\beta})$, with $w_{\alpha,\alpha} = 0$, and

$$w_{\alpha,\beta} = O_\varepsilon(1)e^{-(1-\varepsilon)\nu_0|\alpha-\beta|/h}. \tag{2.11}$$

On the other hand,

$$w_{\alpha,\beta} = e^{-ih'\beta_2(\beta_1-\alpha_1)}f(\alpha-\beta). \tag{2.12}$$

At first sight, (2.11) seems to be insufficient, but using various tricks, one can first show that $w_{\alpha,\beta} = O(e^{-(S_0+\varepsilon_1)/h})$, for some $\varepsilon_1 > 0$, if $|\alpha-\beta|_\infty \geqq 2$, and for $|\alpha-\beta|_\infty = 1$, one finds for instance in the case $\alpha_1 = 1$, $\beta = 0$:

$$w_{\alpha,0} = -(u_0|[P,\chi]u_\alpha) + \mathcal{O}(e^{-(S_0+\varepsilon_1)/h}),$$

with $\chi = 1_{]-\infty,t]}$, and t close to π. One then establishes that u_0 and u_α have WKB expressions in a complex neighbourhood of π, which permits to show that $|w_{\alpha,0}| = \mathcal{O}(e^{-(S_0+\varepsilon_1)/h})$, if $|\alpha| \geq 2$, and

$$C_\varepsilon^{-1} e^{-(1+\varepsilon)S_0/h} \leq |w_{\alpha,0}| \leq C_\varepsilon e^{-(1-\varepsilon)S_0/h}, \text{ for every } \varepsilon > 0, \text{ if } |\alpha| = 1.$$

3. ITERATION

The problem is now to study the spectrum of W. (2.8) shows that W is a convolution operator in the variables α_1, and if $h' = 0$ it is a convolution in all the variables. In the last case, one obtains $Sp(W) = \{Gf(\theta);\ \theta \in \mathbf{T}^2\}$, where $Gf(\theta) = \Sigma f(\alpha)e^{i\alpha\theta} = 2|f(1,0)|(\cos(\theta_1) + \cos(\theta_2)) + \mathcal{O}(e^{-(S_0+\varepsilon_1)/h})$. The spectrum is then a band of width $\sim |f(1,0)|$. Suppose now that $h' > 0$. Conjugating W by the unitary matrix; $\mathrm{diag}(e^{ih'\alpha_1\alpha_2})$, we obtain a new matrix (with the same spectrum) that we shall also denote by W, which is given by,

$$w_{\alpha,\beta} = e^{-ih'\alpha_1(\alpha_2-\beta_2)}f(\alpha-\beta), \tag{3.1}$$

and which is a convolution in the α_2-variables. By inverse Fourier transform in these variables one then gets the unitarily equivalent operator $W':L^2(\mathbf{Z} \times S^1) \to L^2(\mathbf{Z} \times S^1)$, of the form,

$$W'u(\alpha_1,\theta) = (K_\theta u(.,\theta))(\alpha_1), \tag{3.2}$$

where K_θ is given by the kernel,

$$k(\alpha_1,\beta_1,\theta) = g(\alpha_1-\beta_1,\theta-h'\alpha_1), \tag{3.3}$$

where,

$$g(\alpha_1,\theta) = \Sigma_{\alpha_2} f(\alpha_1,\alpha_2)e^{i\theta\alpha_2}. \tag{3.4}$$

The spectrum of W' is equal to that of its restriction to $\mathbf{Z} \times [0,h'[$, and by the substitution; $x = -(\theta-h'\alpha_1)$ we can identify this restriction of W' with the operator $Q:L^2(\mathbf{R}) \to L^2(\mathbf{R})$, given by

$$Q = \Sigma_{k \in \mathbf{Z}} g(k,-x)\tau_{kh'} = \Sigma_{k \in \mathbf{Z}} g(k,-x)e^{-ikh'D}. \tag{3.5}$$

Using (3.4), we find that Q is the h'-Weyl quantization of the symbol,

$$Q(x,\xi) = \Sigma\Sigma\, f(j,k)e^{-ikjh'/2}e^{i(kx+j\xi)}. \tag{3.6}$$

Using that $F_h u_0 = \omega u_0$, $|\omega| = 1$, one also verifies that Q commutes with $F_{h'}$, so that on the symbol level,

$$Q\circ\kappa = Q, \text{ where } \kappa(x,\xi) = (\xi,-x). \tag{3.7}$$

Using (3.6) and the results of Section 2 we get,

$$Q(x,\xi) = 2|f(1,0)|Q_0(x,\xi) + R(x,\xi)), \tag{3.8}$$

where R is holomorphic and $= \mathcal{O}(e^{-1/C_0 h})$ in $|\mathrm{Im}(x,\xi)| \leq 1/C_0 h$, for some sufficiently large $C_0 > 0$, and

$$Q_0(x,\xi) = \cos(\xi) + \cos(x). \tag{3.9}$$

The h'-quantization of Q_0 is then $Q_0 = \cos(h'D) + \cos(x)$. Here we observe that the symbol Q is 2π-periodic in x and in ξ. The critical points of Q coincide with those of Q_0 (thanks to (3.7)) and Q has precisely 3 critical values: $m_Q = Q(\pi,\pi) = -2 + \mathcal{O}(e^{-1/C_0 h})$, $M_Q = Q(0,0) = 2 + \mathcal{O}(e^{-1/C_0 h})$, $c_Q = Q(\pi,0) = Q(0,\pi) = \mathcal{O}(e^{-1/C_0 h})$. The energy surfaces, Q = E in the cases $E \in \{m_Q\}$, $]m_Q,c_Q[$, $\{c_Q\}$, $]c_Q,M_Q[$, $\{M_Q\}$, have the same structure as those of $Q_0 = E'$ in the corresponding cases (obtained, by replacing m_Q, c_Q, M_Q by -2,0,2).

If $\mu' \in]c_Q,M_Q]$, we can define an action $S_1(\mu')$ as $\mathrm{Im}\int_{[x_0,2\pi-x_0]}\xi(x)dx$, where $[-x_0,x_0]$ is the projection of U_0'; the component of $Q = \mu'$ naturally associated to (0,0), and $\xi(x)$ is the unqiue complex solution of $Q(x,\xi(x)) = \mu'$ with $(x_0,\xi(x_0)) \in U_0'$, $\mathrm{Im}\xi(x) \geq 0$. One can then proceed as for the operator P, but with a few modifications, in particular for the weighted L^2-estimates, and we obtain the following result, (where we recall that $\varepsilon_0 > 0$ has been fixed earlier in the lecture):

THEOREM 3.1: There exists $h_0 > 0$, such that if $0 < h \leq h_0$, then we have: For every $\varepsilon_0' > 0$, there exists $h_0' > 0$, such that for $0 < h' \leq h_0'$, the set $Sp(Q) \cap [c_Q + \varepsilon_0', +\infty[$ is contained in $[c_Q + \varepsilon_0', M_Q + C_0 h']$. and more precisely in a union of bands: $\cup_{0 \leq j \leq N} [\mu_j - 4b_j e^{S_1(\mu_j)/h'}, \mu_j + 4b_j e^{-S(\mu_j)/h'}]$, where $\mu_j - \mu_{j+1} \in [h'/C_0, h'C_0]$, $|b_j|$, $|b_j^{-1}| \leq C_\varepsilon e^{\varepsilon/h'}$, for every $\varepsilon > 0$, and $|M_Q - \mu_0| \leq C_0 h'$. The spectrum of $Q - \mu_j$ in the j:th band is equal to that of the h"-quantification of $Q_j' = b_j e^{-S_1(\mu_j)/h'}(Q_0 + R_j')$, where R_j' is 2π-periodic in x, ξ, holomorphic and of modulus $\leq e^{-1/C_0 h'}$, in $|\mathrm{Im}(x,\xi)| \leq 1/C_0 h'$. Here, C_0, C_ε do not depend on h, h', but depend on ε_0'. Moreover, $2\pi/h' = k + h''/2\pi$, $k \in \mathbb{N}$, and we assume that $0 < h'' < 2\pi$. (if $h'' = 0$, we get a band exactly as for $h' = 0$ in the beginning of this section.)

We can then apply the same theorem to $Q_0 + R_j'$ and so on. Since the same analysis can be applied below $c_Q - \varepsilon_Q'$, we then obtain

THEOREM 3.2: We fix $\varepsilon_0 > 0$. Then there exists $C_0 > 0$, such that if $h/2\pi \in]0, 1[\setminus \mathbb{Q}$ and $h/2\pi$ has the infinite continued fraciton expansion $1/(q_1 + 1/(q_2 + 1/(q_3 + \ldots +)))$, with $q_j \in \mathbb{N}$, $q_j \geq C_0$, then we have:

The smallest closed interval containing $Sp(P)$ is of the form $J = [-2 + O(1/q_1), 2 - O(1/q_1)]$. $Sp(P) \cap J \subset \cup_{N_- \leq j \leq N_+} J_j$, where J_j are closed intervals of length $\neq 0$, with $\partial J_j \subset Sp(P)$. J_{j+1} is to the right of J_j at a distance $\sim 1/q_1$. J_0 is of length $2\varepsilon_0 + O(1/q_1)$, containing 0 at a distance $O(1/q_1)$ from its centre. The other bands are of length $e^{-C(j)q_1}$, with $C(j) \sim 1$. For $j \neq 0$, let κ_j be the increasing affine function, which transforms J_j into $[-2,2]$. Then we have $\kappa_j(J_j \cap Sp(P)) \subset \cup_k J_{j,k}$, where the $J_{j,k}$ have the same properties etc. Here "$a \sim b$" means that a/b and b/a are bounded by a constant which only depends on ε_0.

References

[1] S. Aubry, C. André, Proc. Israel Phys. Soc., ed. C.G. Kuper 3 (Adam Hilger, Bristol, 1979), 133-.

[2] M. Ya. Azbel, Energy spectrum of a conduction electron in a magnetic field, Zh. Eksp. Teor. Fiz. 46, (1964), 939-, Sov. Phys. JETP19, (1964), 634-.

[3] J. Bellisard, B. Simon, Cantor spectrum for the almost Mathieu equation, J. Funct. Anal., 48 (3), 408-419.

[4] U. Carlsson, Manuscript in preparation.

[5] B. Helffer, D. Robert, Puits de potentiel généralisés et asymptotique semi-classique, Ann. de l'IHP, 41 (3)(1984), 291-331.

[6] B. Helffer, J. Sjöstrand, Multiple wells in the semi-classical limit I. Comm. in PDE, 9(4) (1984), 337-408.

[7] B. Helffer, J. Sjöstrand, Puits mulptiples., II, Intéraction moléculaire, symétries, perturbation, Ann. de l'IHP, 42 (2) (1985), 127-212.

[8] B. Helffer, J. Sjöstrand, Effet tunnel pour l'équation de Schrödinger avec champ magnétique. Préprint de l'Ecole Polytechnique, Dec. 1986.

[9] D.R. Hofstadter, Energy levels and wave functions of Bloch electrons in rational and irrational magnetic fields, Phys. Rev. B, 14(6), 15 Sept. 1976.

[10] B. Simon, Almost periodic Schrödinger operators, A review, Adv. Appl. Math. 3, (1982), 463-490.

[11] J. Sokoloff, Unusual band structure, wave functions and electrical conductance in crystals with incommensurate periodic potentials, Physics reports (Review section of Physics letters), 126 (4) (1985), 189-244.

[12] M. Wilkinson, Critical properties of electron eigenstates in incommensurate systems, Proc. R. Soc. Lond., A 391, (1984), 305-350.

[13] M. Wilkinson, An example of phase holonomy in WKB theory, J. Phys. A, 17 (1984), 3459-3476.

[14] M. Wilkinson, Von Neumann lattices of Wannier functions for Bloch electrons in a magnetic field, Proc. R. Soc. Lond., A 403, (1986), 135-166.

[15] M. Wilkinson, An exact renormalisation group for Bloch electrons in a magnetic field, J. Phys. A., to appear.

J. Sjöstrand
Department of Mathematics,
Université de Paris Sud,
F-91405 Orsay,
France

B. Helffer
Department of Mathematics,
Université de Nantes,
2, Chemin de la Houssinière
F-44072 Nantes,
France.

S. TARAMA

On the initial value problem for the weakly hyperbolic operators in the Gevrey classes

§1. In this note we consider the initial value problem in the Gevrey classes for weakly hyperbolic equations with coefficients not smooth with respect to the time variable.

Let P be a weakly hyperbolic operator:

$$P = D_t^m + \sum_{\substack{j+|\alpha|\leqq m \\ j\leqq m-1}} a_{j,\alpha}(t,x)D_t^j D_x^\alpha$$

with

(H.0) $a_{j,\alpha}(t,x) \in C^0([0,T[,\Gamma^{(s_0)}(R^n))$ $(s_0 > 1)$

where we denote by $\Gamma^{(s)}(R^n)$ the Gevrey class with the index s on R^n, i.e. $f \in \Gamma^{(s)}(R^n)$ if and only if for any compact set $K \subset R^n$ and any multi-index α, we have

$$\sup_{x\in K} |\partial_x^\alpha f(x)| \leqq C_1 C_2^{|\alpha|} |\alpha|!^s$$

with constants C_1 and C_2 independent of α.

We assume that the principal part $P_m = D_t^m + \sum_{j+|\alpha|=m} a_{j,\alpha}(t,x)D_t^j D_x^\alpha$ of P satisfies the following:

(H.1) the characteristic polynomial $\tau^m + \sum_{j+|\alpha|=m} a_{j,\alpha}(t,x)\tau^j\xi^\alpha$ has only real zeros $\lambda_i(t,x,\xi)$ $(\lambda_i(t,x,\xi) \leqq \lambda_{i+1}(t,x,\xi))$ for $(t,x,\xi) \in [0,T[\times R^n \times R^n$.

Let M be the maximal multiplicity of the zeros.

(H.2) a) $a_{j,\alpha}(t,x) \in C^{k_0,k_1}([0,T[\,, C^\infty(R^n))$ $(|\alpha|+j=m)$

with k_0 nonnegative integer and $0 < k_1 \leqq 1$.

b) $\sup_{(t,x)\in[0,T[\times R^n} |a_{j,\alpha}(t,x)| < +\infty$ for $|\alpha| + j = m$.

In the following we assume also that

(H.3) all the coefficients of P are independent of x for $|x| \geqq R_0 > 0$.

For the operator P satisfying (H.1, 2 and 3), we consider the initial value problem $(0 \leqq t_0 < t_1 < T)$

$$(C)_{t_0,t_1} \quad \begin{cases} Pu = f \quad [t_0,r_1] \times R^n \\ D_t^j u(t_0,x) = \phi_j \quad (j = 0,\ldots,m-1) \end{cases}$$

with data $\phi_j \in \Gamma^{(s)}(R^n)$ and a second member $f \in C^0([t_0,t_1], \Gamma^{(s)}(R^n))$.

This problem is already treated in many papers (see for example [2], [3], [6], [7], [9] and [10] in the reference). In this note, in order to obtain the a priori estimates for solutions of $(C)_{t_0,t_1}$, we approximate the given operator by the operator $\tilde{P}$ whose symbol is $\prod_{i=1}^{m} (\tau-\tilde{\lambda}_i(t,x,\xi))$, where $\tilde{\lambda}_i(t,x,\xi)$ is a regularization of the characteristic root $\lambda_i(t,x,\xi)$ and show that the difference $P-\tilde{P}$ is in some sense dominated by $\tilde{P}$. (see Lemma 3). We note that the similar idea can be found in other papers (for example W. Ichinose [4]).

In order to estimate the difference $P-\tilde{P}$, we use the regularity of characteristic roots which is based on the following lemma essentially due to M.D. Bronshtein [1] (see S. Tarama [11], S. Wakabayashi [12]).

<u>LEMMA 0</u>: Let $p(\tau,t) = \tau^m + \sum_{j=1}^{m} a_j(t)\tau^{m-j}$ be polynomial having only real roots with parameter t. Assume that $a_j(t) \in C^{k_0,k_1}(]0,1[)$ with $k_0 \geqq 0$ integer and $0 < k_1 \leqq 1$ and that $\lambda_{i+M}(t)-\lambda_i(t) > 0$ $i = 1,\ldots,m-M$ where we denote by $\lambda_i(t)$ the roots of $p(\tau,t)$ such that $\lambda_{i+1}(t) \geqq \lambda_i(t)$, that is to say, the multiplicity of roots is at most M. Then for any $0 < \varepsilon < 1/2$, we have

$$|\lambda_j(t)-\lambda_j(s)| \leq C_\varepsilon|t-s|^\kappa \text{ for } t,s \in [\varepsilon,1-\varepsilon] \text{ and } j = 1,\ldots,m$$

with $\kappa = \min\{1, (k_0 + k_1)/M\}$, where the constant C_ε depends only on ε, $\sum_{i=1}^{m} \|a_j\|$ with the norm in $C^{k_0,k_1}([\varepsilon/2,1-\varepsilon/2])$ and

$$\min_{\substack{i=1,\ \ldots\ , m-M \\ t \in [\varepsilon/2,\ 1-\varepsilon/2]}} |\lambda_{i+M}(t)-\lambda_i(t)|.$$

Applying this lemma 0 for the polynomial $P_m(t,x,\tau,\xi)$, we obtain

<u>LEMMA 1</u>: For any $\varepsilon > 0$, $t,s \in [0,T-\varepsilon]$, $x \in R^n$ and $\xi \in R^n$,

$$|\lambda_i(t,x,\xi)-\lambda_i(s,x,\xi)| \leq C_\varepsilon|t-s|^\kappa|\xi|(|t|^{-\kappa/2}+|s|^{-\kappa/2}),$$

$$|\lambda_i(t,x,\xi)-\lambda_i(t,y,\xi)| \leq C_\varepsilon|x-y||\xi| \quad ,$$

$$|\lambda_i(t,x,\xi)-\lambda_i(t,x,\eta)| \leq C_\varepsilon|\xi-\eta||\xi| \quad .$$

<u>PROOF</u>: Put $p(t,x,\tau,\xi) = \tau^m + \sum_{j+|\alpha|=m} a_{j,\alpha}(t^2,x)\tau^j\xi^\alpha$.

As $p(t,x,\tau,\xi)$ is hyperbolic polynomial satisfying (H.1, 2 and 3) in $]-T,T[\times R^n \times R^n$, this lemma follows from the lemma 0.

<u>REMARK</u>: If $k_0 + k_1 \leq 1$, then $\tilde{p}(t,x,\tau,\xi) = \tau^m + \sum_{j+|\alpha|=m} a_{j,\alpha}(|t|,x)\tau^j\xi^\alpha$ satisfies (H.1, 2 and 3) in $]-T,T[\times R^n \times R^n$, then we have

$$|\lambda_i(t,x,\xi)-\lambda_i(s,x,\xi)| \leq C_\varepsilon|t-s|^\kappa|\xi| \quad .$$

Now we regularize the characteristic roots by the smooth function $\chi(t) \in C_0^\infty(]-1,0[)$ such that $\int\chi(t)dt = 1$ and $\int\chi(t)t^\ell dt = 0$ for $\ell = 1,\ldots, \max\{m,k_0\}$.

Put with $p_0 > 0$ and $0 < p_1 \leq 1/2$

$$\tilde{\lambda}_i(t,x,\xi) = \int\Phi(t-s,x-y,\xi-\eta,\xi)\lambda_i(s,y,\eta)dsdyd\eta$$

where

$$\Phi(s,y,\eta,\xi) = \chi(\langle\xi\rangle^{p_0}s)\chi_n(\langle\xi\rangle^{p_1}y)\chi_n(\langle\xi\rangle^{p_1-1}\eta)\langle\xi\rangle^{p_0+2np_1-n}$$

with

$$\chi_n(t_1,\dots,t_n) = \prod_{i=1}^{n} \chi(t_i) \text{ and } \langle\xi\rangle = (1+|\xi|^2)^{1/2}.$$

Then according to the lemma 1, we have

<u>LEMMA 2</u>: For $0 < t \leqq T-\varepsilon$ with $\varepsilon > 0$,

$$|\lambda_i(t,x,\xi)-\tilde{\lambda}_i(t,x,\xi)| \leqq C_\varepsilon \langle\xi\rangle^{1-\min\{\kappa p_0,p_1\}} t^{-\kappa/2}$$

$$\tilde{\lambda}_i(t,x,\xi) \in S^1_{1-p_1,p_1}(t)$$

$$t^{\kappa/2}\partial_t \tilde{\lambda}_i(t,x,\xi) \in S^{1+(1-\kappa)p_0}_{1-p_1,p_1}(t)$$

$$\partial_{x_j} \tilde{\lambda}_i(t,x,\xi) \in S^1_{1-p_1,p_1}(t)$$

$$\partial_{\xi_j} \tilde{\lambda}_i(t,x,\xi) \in S^1_{1-p_1,p_1}(t)$$

where we say that a symbol $a(t,x,\xi) \in S^m_{\rho,\delta}(t)$ if, for any multi-indices α and β and any $\varepsilon > 0$,

$$\sup_{\substack{0<t<T-\varepsilon \\ (x,\xi)\in R^{2n}}} |\partial_x^\alpha\partial_\xi^\beta a(t,x,\xi)|\langle\xi\rangle^{-m-\delta|\alpha|+\rho|\beta|} < +\infty.$$

(see Kumano-go [5] for the properties of symbols and the calculus of the pseudo-differential operators associated to these symbols.)

<u>REMARK</u>: We denote by $\tilde{\lambda}^d_j(t,x,\xi)$ the regularization of the root $\lambda_j(t,x,\xi)$ with the function $d^{-2n}\Phi(s,y/d,\eta/d,\xi)$. Then we have the estimates

$$|\tilde{\lambda}^d_j(t,x,\xi)-\lambda_j(t,x,\xi)| \leqq C_\varepsilon(t^{-\kappa/2}\langle\xi\rangle^{1-\kappa p_0} + d\langle\xi\rangle^{1-p_1}),$$

$$|\partial_x^\alpha\partial_\xi^\beta\tilde{\lambda}^d_j(t,x,\xi)| \leqq C_{\varepsilon,\alpha,\beta}d^{-|\alpha|-|\beta|+1}\langle\xi\rangle^{1-p_1+|\alpha|p_1-|\beta|(1-p_1)} \text{ for } |\alpha|+|\beta|\geqq 1,$$

and

$$|\partial_x^\alpha \partial_\xi^\beta \partial_t \tilde{\lambda}_j^d(t,x,\xi)| \leq C_{\varepsilon,\alpha,\beta}\, d^{-|\alpha|-|\beta|} t^{-\kappa/2} \langle\xi\rangle^{1-\kappa p_0+|\alpha|p_1-|\beta|(1-p_1)}$$

with the constants C_ε, $C_{\varepsilon,\alpha,\beta}$ independent of d.

This kind of the modification is necessary when we treat the problem in the Gevrey class with the index $s < \min\{1+\kappa, M/(M-1)\}$. (See §3).

Now we rewrite the principal part P_m of the operator P, using the symbol $\tau-\tilde{\lambda}_i(t,x,\xi)$. For any subset $I \subset \{1,\dots,m\}$, we define the symbol $\Lambda_I(t,x,\tau,\xi)$ by

$$\Lambda_I(t,x,\tau,\xi) = \prod_{i\in I} (\tau-\tilde{\lambda}_i(t,x,\xi)),\ \Lambda_\phi = I$$

and the operator $\Lambda_I(t,x,D_t,D_x)$ by

$$\Lambda_I(t,x,D_t,D_s) = D_t^{|I|} + \sum_{j=0}^{|I|-1} a_j^I(t,x,D_x) D_t^j$$

where $a_j^I(t,x,D_x)$ is the pseudo-differential operator whose symbol is

$$\sum_{\substack{J\subset I \\ J=|I|-j}} \prod_{i\in J} (-\tilde{\lambda}_i(t,x,\xi))$$

Then we obtain

<u>LEMMA 3</u>:

a) $$P_m(t,x,\tau,\xi) = \Lambda_{\{1,\dots,m\}} + \sum_{|I|\leqq m-1} C_I(t,x,\xi)\Lambda_I(t,x,\tau,\xi) + \sum_{i=1}^{m} c_i(t,x,\xi)\tau^{m-i}$$

with

$$t^{\kappa(m-|I|)/2}\, C_I(t,x,\xi) \in S_{1-p_1,p_1}^{(1-\min\{\kappa p_0,p_1\})(m-|I|)}(t)$$

and

$$c_i(t,x,\xi) = \sum_{|\alpha|=i} (a_{m-i,\alpha}(t,x)-\tilde{a}_{m-i,\alpha}(t,x,\xi))\xi^\alpha \in S_{1,0}^{i-(k_0+k_1)p_0}(t).$$

b) For k and I such that $k\in \{1,\dots,m\}\setminus I$,

$$(D_t-\tilde{\lambda}_k(t,x,D_x))\Lambda_I(t,x,D_t,D_x) = \Lambda_{I\cup\{k\}}(t,x,D_t,D_x) + C_{k,I}(t,x,D_t,D_x)$$

with

$$C_{k,I}(t,x,\tau,\xi) = \sum_{J \subsetneqq I} C_J^{k,I}(t,x,\xi)\Lambda_J(t,x,\tau,\xi)$$

where

$$t^{\kappa/2}C_J^{k,I}(t,x,\xi) \in S_{1-p_1,p_1}^{1+(1-\kappa)p_0+p_1(|I|-|J|-1)}(t).$$

The proof is given at §2.

By the way, as the multiplicity of characteristic roots is at most M, the set $\{\Lambda_I(t,x,\tau,\xi): m-M \leqq |I| \leqq m-1\}$ generates the space of polynomials in τ with degree less or equal to m-1. Then we have

<u>LEMMA 4</u>: For any integer k ($0 \leqq k \leqq m-1$), there exist the symbols

$$a_J^k(t,x,\xi) \in S_{1-p_1,p_1}^{m-1-|J|}(t) \quad (m-M \leqq |J| \leqq m-1)$$

such that for $|\xi| \geqq R_0$,

$$\langle\xi\rangle^{m-1-k}\tau^k = \sum_{m=M\leqq|J|\leqq m-1} a_J^k(t,x,\xi)\Lambda_J(t,x,\tau,\xi).$$

Now assume

$$p_1 \leqq \kappa p_0 \text{ and } p_0 = 1/(1+\kappa) \text{ with } \kappa = \min\{1, (k_0+k_1)/M\}.$$

Then we have

<u>PROPOSITION 5</u>: Let be $\delta = 1-p_1$ and $\langle D\rangle = (1+|\Delta_x|)^{1/2}$.

Then we have

1) $$\frac{d}{dt}\Big(\sum_{I \subsetneqq \{1,\ldots,m\}} \|t^{-\kappa(m-1-|I|)/2}\langle D\rangle^{\delta(m-1-|I|)}\Lambda_I(t,x,D_t,D_x)u\|^2\Big)^{1/2}$$

$$\leqq C_\varepsilon\{\Big(\sum_{I \subsetneqq \{1,\ldots,m\}} \|t^{-\kappa(m-|I|)/2}\langle D\rangle^{\delta(m-|I|)}\Lambda_I(t,x,D_t,D_x)u\|^2\Big)^{1/2} + \|P_m u\|\}$$

for $0 < t \leqq T-\varepsilon$ with $\varepsilon > 0$.

2) For any multi-index $\nu > 0$ and $0 < t \leqq T-\varepsilon$,

$$\|P_{m(\gamma)}u\| \leqq$$

$$C_{\varepsilon,\gamma}\Big(\sum_{I\subsetneqq\{1,\ldots,m\}} \|t^{-\kappa(m-|I|)/2}\langle D\rangle^{\delta(m-|I|)+(1-\delta)|\gamma|}\Lambda_I(t,x,D_t,D_x)u\|^2\Big)^{1/2}$$

where $P_{m(\gamma)}(t,x,\tau,\xi) = (\frac{\partial}{\partial x})^{\gamma}\, P_m(t,x,\tau,\xi)$.

3) For any (α,j) $(|\alpha|+j \leqq m-1)$,

$$\|D_t^j D_x^{\alpha}u\| \leqq C_{\varepsilon}\Big(\sum_{\substack{I\subset\{1,\ldots,m\}\\ m-M\leqq|I|\leqq m-1}} \|\langle D\rangle^{m-1-|I|}\Lambda_I(t,x,D_t,D_x)u\| +$$

$$+ \sum_{|I|\leqq m-1-M} \|\langle D\rangle^{-1}\Lambda_I(t,x,D_t,D_x)u\|\Big).$$

<u>PROOF</u>: Taking into account the Lemma 3, 4 and the following lemma whose proof is given at §2:

<u>LEMMA 6</u>: For $a(t,x,\xi) \in S^m_{1-p,p_1}$

$$a(t,x,D_x)\Lambda_I(t,x,D_t,D_x) = C_I(t,x,D_t,D_x)$$

with

$$C_I(t,x,\tau,\xi) = a(t,x,\xi)\Lambda_I(t,x,\tau,\xi) + \sum_{J\subsetneqq I} a_J(t,x,\xi)\Lambda_J(t,x,\tau,\xi)$$

where

$$a_J(t,x,\xi) \in S^{m+p_1(|I|-|J|)}_{1-p_1,p_1},$$

we have for I and $k \in \{1,\ldots,m\}/I$

$$(D_t-\tilde{\lambda}_k(t,x,D_x))\Lambda_I(t,x,D_t,D_x) = \Lambda_{I\cup\{k\}}(t,x,D_t,D_x)$$

$$+ \sum_{J\subsetneqq I} t^{-\kappa/2}C_J^{I,k}(t,x,D_x)\Lambda_J(t,x,D_t,D_x)$$

with

$$C_J^{I,k}(t,x,\xi) \in S_{1-p_1,p_1}^{\delta(|I|-|J|+1)}(t)$$

and

$$P_m(t,x,D_t,D_x) = \Lambda_{\{1,\dots,m\}} + \sum_{|I|\leqq m-1} t^{-(m-|I|)\kappa/2} C_I(t,x,D_x)\Lambda_I(t,x,D_t,D_x)$$

with

$$C_I(t,x,\xi) \in S_{1-p_1,p_1}^{\delta(m-|I|)}(t).$$

From this follows the assertion 1) of Proposition 5. The others 2) and 3) are also proved by the similar way. □

Furthermore assume

$$\delta = 1-p_1 \geqq 1-1/M.$$

Then we see, from 3) of Propsotion 5, that for $|\alpha| + j \leqq m-1$

$$*)\qquad D_t^j D_x^\alpha = \sum_{|J|\leqq m-1} a_J^{j,\alpha}(t,x,D_x)\Lambda_J(t,x,D_t,D_x)$$

with $a_J^{j,\alpha}(t,x,\xi) \in S_{1-p_1,p_1}^{(m-|J|)\delta}(t)$.

Taking into account Proposition 5 and *), we obtain the

<u>THEOREM</u>: If $s_0 \leqq s < \min\{2, 1 + (k_0 + k_1)/M, M/(M-1)\}$, there exists one and only one solution $u \in C^m([t_0,t_1], \Gamma^{(s)}(R^n))$ of the problem $(C)_{t_0,t_1}$ with any data $\varphi_j(x) \in \Gamma^{(s)}(R^n)$ and any second member $f \in C^0([t_0,t_1],\Gamma^{(s)}(R^n))$ and there exists a finite propagation speed.

If $s_0 \leqq s = \min\{2, 1 + (k_0 + k_1)/M, M/(M-1)\}$, then for any data $\varphi_j \in \Gamma^{(s)}(R^n)$ and any second member $f \in C^0([t_0,t_1], \Gamma^{(s)}(R^n))$ there exists one solution u of the problem $(C)_{t_0,t_1}$ on the domain which may depend on the φ_j and f.

Indeed we can estimate $\partial_x^\alpha u$ for any α by using the Proposition 5 and *) to $P\partial_x^\alpha u + \sum_{\gamma<\alpha} \binom{\alpha}{\gamma} P_{(\alpha-\gamma)}\partial_x^\gamma u = \partial_x^\alpha f$. And we obrain the a priori estimate of solutions. (See §3). According to Nuij [8], we can find a sequence of the strictly hyperbolic operators $P^{(n)}$ with smooth coefficients satisfying uniformly (H.0,1,2 and 3) which converges to the given operator.

Because there are uniform estimates for the solution of the problem $(C)_{t_0,t_1}$ for each operator $P^{(n)}$, we can find the subsequence of solutions which converges to the desired solution. Note that any solution of each strictly hyperbolic operator $P^{(n)}$ has a uniform finite propagation speed. Hence the theorem is proved.

§2. PROOFS OF LEMMAS 3 AND 6

PROOF OF LEMMA 3:

$$P_m(s,y,\tau,\eta) = \prod_{i=1}^{m} (\tau-\lambda_i(s,y,\eta))$$

$$= \prod_{i=1}^{m} (\tau-\tilde{\lambda}_i(t,x,\xi) + \tilde{\lambda}_i(t,x,\xi)-\lambda_i(s,y,\eta))$$

$$= \sum_{I\subset\{1,\dots,m\}} C_I(t,x,\xi,s,y,\eta)\Lambda_I(t,x,\tau,\xi)$$

where

$$C_I(t,x,\xi,s,y,\eta) = \prod_{i\in\{1,\dots,m\}\setminus I} (\tilde{\lambda}_i(t,x,\xi)-\lambda_i(s,y,\eta)).$$

Applying $\int\Phi(t-s,x-y,\xi-\eta,\xi)dsdyd\eta$, we obtain

$$\tau^m + \sum_{j+|\alpha|=m} \tilde{a}_{j,\alpha}(t,x,\xi)\tau^j\xi^\alpha = \sum_{I\subset\{1,\dots,m\}} \tilde{C}_I(t,x,\xi)\Lambda_I(t,x,\tau,\xi)$$

$$\tilde{C}_I(t,x,\xi) = \int\Phi(t-s,x-y,\xi-\eta,\xi)C_I(t,x,\xi,s,y,\eta)dsdyd\eta.$$

From Lemma 2 we see that a) of Lemma 3 is valid, where we used

$$\int\chi(\langle\xi\rangle^{p_0}(t-s))a(s)\langle\xi\rangle^{p_0}ds = a(t) +$$

$$\int\chi(\langle\xi\rangle^{p_0}(t-s))(t-s)^{k_0} \times$$

$$\{1/(k_0-1)!\int_0^1 (a^{(k_0)}(\theta(t-s)+s)-a^{(k_0)}(t))(1-\theta)^{k_0}d\theta\}\langle\xi\rangle^{p_0}ds.$$

The assertion b) follows from Lemma 6.

PROOF OF LEMMA 6: By the calculus of the pseudo-differential operators (see Kumano-go [5]), we know

$$a(t,x,D_x)\Lambda_I(t,x,D_t,D_x) = C_I(t,x,D_t,D_x)$$

where $C_I(t,x,\tau,\xi)$ is obtained by the oscilatory integral

$$Os\text{-}(2\pi)^{-n}\int e^{-i\eta y}a(t,x,\xi+\eta)\Lambda_I(t,x+y,\tau,\xi)dyd\eta$$

$$= Os\text{-}(2\pi)^{-n}\int e^{-i\eta y}a(t,x,\xi+\eta)\prod_{i\in I}(\tau-\tilde{\lambda}_i(t,x,\xi)+\tilde{\lambda}_i(t,x,\xi)-\tilde{\lambda}_i(t,x+y,\xi))$$

$$= \sum_{J\subset I} Os\text{-}(2\pi)^{-n}\int e^{-i\eta y}a(t,x,\xi+\eta)\prod_{i\in I\setminus J}(\tilde{\lambda}_i(t,x,\xi)-\tilde{\lambda}_i(t,x+y,\xi))dyd\eta \times$$

$\Lambda_J(t,x,\tau,\xi)$. From which follows Lemma 6.

§3. In this section we show the a priori estimate of solution in the case of $s_0 \leqq s \leqq \min\{2,1+(k_0+k_1)/M, M/(M-1)\}$. In the following we consider the estimate for $t \in [0,T-\varepsilon]$ with some fixed $\varepsilon > 0$.

By the assumptions (H.0 and 3), there exist constants C_0 and C_1 such that, for any multi-index α,

$$\sum_{j+|\beta|\leqq m-1} |\partial_x^\alpha a_{j.\beta}(t,x)| \leqq C_1 C_0^{|\alpha|}|\alpha|!^{s_0}$$

$$\sum_{j+|\beta|=m} |\partial_x^\alpha a_{j.\beta}(t,x)| \leqq C_1 C_0^{(|\alpha|-M)_+}(|\alpha|-M)_+!^{s_0}$$

where $(x)_+ = \max\{0,x\}$.

In the case $s = \min\{2,1+(k_0+k_1)/M,M/(M-1)\}$, from the estimate *), we have with $p_1 = \kappa p_0 = 1-1/s$ and $\delta = 1/s$

$$\|(P_{(\gamma)}-P_{m(\gamma)})v\| \tag{3.1}$$

$$\leqq |\gamma|!^{s_0}C\, C_0^{|\gamma|}\Big(\sum_{|I|\leqq m-1}\|\langle D\rangle^{\delta(m-|I|)}\Lambda_I v\|\Big).$$

In the following we use the following inequality:

$$\binom{\alpha}{\gamma} \langle\xi\rangle^{r(1-\delta)} \leq \langle\xi\rangle^{r} + \binom{\alpha}{\gamma}^{1/\delta}. \tag{3.2}$$

which follows from

$$\chi \leq 1 + \chi^{1/\delta} (\chi \geq 0).$$

For $1 \leq |\gamma| \leq M$, we apply Proposition 5.2) to the estimate $P_{m(\gamma)}\nu$, then by (3.2) with $r = |\gamma|$ we have

$$\begin{aligned} \|P_{m(\gamma)}\nu\| &\leq C\binom{\alpha}{\gamma}^{-1} \times \\ &\Big\{ \sum_{|\beta|\leq|\gamma|} \sum_{|I|\leq m-1} \|t^{-\kappa/2(m-|I|)} \langle D\rangle^{\delta(m-|I|)} \Lambda_I \partial_x^{\beta}\nu\| \\ &+ \binom{\alpha}{\gamma}^{1/\delta} \sum_{|I|\leq m-1} \|t^{-\kappa/2(m-|I|)} \langle D\rangle^{\delta(m-|I|)} \Lambda_I \nu\| \Big\}. \end{aligned} \tag{3.3}$$

From inequality (3.2) with $r = m - |I|$,

we have

$$\binom{\alpha}{\gamma} \| \langle D\rangle^{m-|I|} \Lambda_I \nu\| \leq$$

$$\left(\|\langle D\rangle^{(m-|I|)\delta+m-|I|} \Lambda_I \nu\| + \binom{\alpha}{\gamma}^{1/\delta} \| \langle D\rangle^{(m-|I|)\delta} \Lambda_I \nu\| \right).$$

Then, for $|\gamma| \geq M + 1$, by the Proposition 5.3)

$$\begin{aligned} \binom{\alpha}{\gamma} \|P_{m(\gamma)}\nu\| &\leq C\, C_0^{|\gamma|-M} (|\gamma|-M)!^{s_0} \times \\ &\Big(\sum_{m-M\leq|I|m-1} \|\langle D\rangle^{\delta(m-|I|)+m-|I|} \Lambda_I \nu\| \\ &+ \sum_{m-M\leq|I|\leq m-1} \binom{\alpha}{\gamma}^{1/\delta} \|\langle D\rangle^{\delta(m-|I|)} \Lambda_I \nu\| \\ &+ \sum_{|I|\leq m-M-1} \binom{\alpha}{\beta} \|\Lambda_I \nu\| \Big). \end{aligned}$$

Then, from (3.1), (3.3) and (3.4) follows the estimate:

$$\binom{\alpha}{\gamma} \| P_{(\gamma)} \partial_x^{\alpha-\gamma} u \| \tag{3.5}$$

$$\leq \binom{\alpha}{\gamma} \left(\left|\binom{\alpha}{\gamma}\right|\right)^{-1} C\{C_0^{(|\gamma|-M)} + (|\gamma|-M)_+!^{s_0} \times$$

$$\Big(\sum_{|\beta| \leq \min\{M,\ |\gamma|\}} \sum_{|I| \leq m-1} \| t^{-\kappa(m-|I|)/2} \langle D\rangle^{\delta(m-|I|)} \Lambda_I \partial_x^{\beta+\alpha-\gamma} u \| \Big)$$

$$+ C_0^{|\gamma|} |\gamma|!^{s_0} \left(\left|\binom{\alpha}{\gamma}\right|\right)^{1/\delta} \Big(\sum_{|I| \leq m-1} \| t^{-\kappa(m-|I|)/2} \langle D\rangle^{\delta(m-|I|} \Lambda_I \partial_x^{\alpha-\gamma} u \| \Big)\} .$$

By the same reasoning of (3.2), we have

$$\langle\xi\rangle^{\delta} \leq k + 2(k+1)^{1-1/\delta} \langle\xi\rangle \tag{3.6}$$

In summary, by using the notations

$$\|u\|^2 = \sum_{|I| \leq m-1} \| t^{-\kappa(m-1-|I|)/2} \langle D\rangle^{\delta(m-1-|I|)} \Lambda_I u \|^2$$

and

$\|u\|_{\ell} = \max_{|\alpha|=\ell} \|\partial_x^{\alpha} u\|$, from Proposition 5 and (3.5) and (3.6), follows the estimate, with $s = 1/\delta$;

$$\frac{d}{dt} \|\partial_x^{\alpha} u\| \leq |\alpha|!^{s} \{C_2(|\alpha|+1) t^{-\kappa/2} \|u\|_{|\alpha|+1} / (|\alpha|+1)!^{s}$$

$$+ \sum_{|\alpha| \geq k \geq 0} C_2 C_0^{k} (|\alpha|-k) t^{-\kappa/2} \|u\|_{(|\alpha|-k)} / (|\alpha|-k)!^{s} + C\|\partial_x^{\alpha} P u\|$$

Then we obtain

PROPOSITION 7: If u satisfies

$$\partial_t^j u(0,x) = 0 \qquad j = 0,1,\ldots,m-1$$

and with $h \in]0,1[$

$$\sum_{\ell=0}^{\infty} C_0^{-\ell}\ell!^{-s}\ell h^{\ell} \|u\|_{\ell}(t) < +\infty$$

for $t \in [0,T-\varepsilon]$, then there exists a constant ν_0 such that

$$\sum_{\ell=0}^{\infty} C_0^{-\ell}\ell!^{-s}h^{\ell}(1-\nu_0 t^{1-\kappa/2})^{\ell+1} \|u\|(t)$$

$$\leq C(\sum_{\ell=0}^{\infty} C_0^{-\ell}\ell!^{-s}h^{\ell} \int_0^t (1-\nu_0\tau^{1-\kappa/2})^{\ell+1} \times \max_{|\alpha|=\ell} \|\partial_x^{\alpha}(Pu)\|(\tau)d\tau$$

for $0 \leqq t \leqq \min\{T-\varepsilon,(2\nu_0)^{-(1-\kappa/2)^{-1}}\}$.

PROOF: Note that, for almost all t,

$$\frac{d}{dt}\|u\|_{\ell} \leqq \max_{|\alpha|=\ell} \{\frac{d}{dt}\|\partial_x^{\alpha}u\|\}.$$

Put

$$F_{\ell} = (1-\nu_0 t^{1-\kappa/2})^{\ell+1}\ell!^{-s}h^{\ell}(C_0)^{-\ell}\|u\|_{\ell}$$

$$G_{\ell} = (1-\nu_0 t^{1-\kappa/2})^{\ell+1}\ell!^{-s}h^{\ell}(C_0)^{-\ell}\max_{|\alpha|=\ell}\|\partial_x^{\alpha}(Pu)\| .$$

Then taking into account (3.8) and (3.9), we have

$$\frac{d}{dt}F_{\ell} \leqq \{(1+\ell)(1-\nu_0 t^{1-\kappa/2})^{-1}(-\nu_0(1-\kappa/2)t^{-\kappa/2})\}F_{\ell}$$

$$+ C_2(\ell+1)t^{-\kappa/2}h^{-1}C_0(1-\nu_0 t^{1-\kappa/2})^{-1}F_{\ell+1}$$

$$+ \sum_{j\leqq\ell} t^{-\kappa/2}C_2 h^{\ell-j}(1-\nu_0 t^{1-\kappa/2})^{\ell-j}\, jF_j + CG_{\ell}$$

$$\frac{d}{dt}(\sum_{\ell} F_{\ell}) \leqq \sum_{\ell}\{[(1-\nu_0 t^{1-\kappa/2})^{-1}(-\nu_0)(1-\kappa/2)t^{-\kappa/2}$$

$$+ C_2 t^{-\kappa/2}h^{-1}C_0(1-\nu_0 t^{1-\kappa/2})^{-1}$$

$$+ C_2(1-h)^{-1}t^{-\kappa/2}](\ell+1)F_{\ell}+CG_{\ell}\}.$$

If we choose $\nu_0 \geqq (1-\kappa/2)^{-1}\{C_2h^{-1}C_0+C_2(1-h)^{-1}\}$, we obtain the estimate of Proposition 7.

In the case of $s_0 \leqq s < \min\{1+\kappa, M/(M-1)\}$, we must use $\tilde{\lambda}_i^d(t,x,\xi)$ with $d = h^{1/s}$, (see the Remark after the Lemma 2). Then we see that the constant ν_0 in Proposition 7 is independent of h, from which we can obtain the a priori esitmate in the interval [0,T'] with T' independent of u.

Indeed, in this case we choose

$$p_0 = 1/(1+\kappa) \text{ and } p_1 = 1-1/s \text{ and then}$$

$$p_1 < \kappa p_0, \ 1-p_1 > p_1 \quad \text{and} \quad (1-p_1)\ell > \ell - 1 \quad \text{for } \ell \leqq M. \tag{3.7}$$

In the following we use only $\tilde{\lambda}_j^{h^{1/s}}$ as the regularization of λ_j. We write $\tilde{\lambda}_j$ in the place of $\tilde{\lambda}_j^{h^{1/s}}$.

According to the remark after Lemma 2 and the proof of Lemmas 3 and 6 and (3.7) we see that, with some constant C independent of h and $\delta = 1/s$,

$$\|P_m(t,x,D_t,D_x)u\| \tag{3.8}$$

$$\leqq \|\Lambda_{\{1,\dots,m\}}u\| + \sum_{|I|\leqq m-1} C\,\|t^{-\kappa(m-|I|)/2}\langle D\rangle^{\delta(m-|I|)} \times h^{\delta(m-|I|)}\Lambda_I u\| + \sum_{|I|\leqq m-1} C_h\,\|t^{-\kappa(m-|I|)/2}\Lambda_I u\|,$$

for $|\gamma| \leqq M$ (3.9)

$$\|P_{m(\gamma)}(t,x,D_t,D_x)u\|$$

$$\leqq \sum_{|I|\leqq m-1} C\,\|t^{-\kappa(m-|I|)/2}\langle D\rangle^{\delta(m-|I|)+(1-\delta)|\gamma|}h^{\delta(m-|I|-|\gamma|)}\Lambda_I u\| + \sum_{|I|\leqq m-1} C_h\,\|t^{-\kappa(m-|I|)/2}\Lambda_I u\|,$$

for I and $k \in \{1,\dots,m\}\setminus I$, (3.10)

$$\|(D_t-\tilde{\lambda}_k(t,x,D_x)\Lambda_I u-\Lambda_{I\cup\{k\}}u\|$$

$$\leq \sum_{J \subsetneqq I} C \, \| t^{-\kappa/2} h^{\delta(|I|-|J|+1)} \langle D\rangle^{\delta(|I|-|J|+1)} \Lambda_J u \| \; +$$

$$\sum_{J \subsetneqq I} C_h \, \| t^{-\kappa/2} \Lambda_J u \| ,$$

and

for $k = 0,\dots,m-1$. (3.11)

$$\| D_t^k \langle D\rangle^{m-k-1} u \| \leq \sum_{|I|\leq m-1} C \, \| h^{\delta(m-|I|)} \langle D\rangle^{\delta(m-|I|)} \Lambda_I u$$

$$+ \sum_{|I|\leq m-1} C_h \, \| \langle D\rangle^{-1} \Lambda_I u \| .$$

From (3.8) and (3.10) follows

$$\frac{d}{dt} \Big(\sum_{|I|\leq m-1} \| t^{-\kappa(m-1-|I|)/2} \, (h\langle D\rangle)^{\delta(m-1-|I|)} \Lambda_I u \|^2 \Big)^{1/2} \tag{3.12}$$

$$\leq C \sum_{|I|\leq m-1} \| t^{-\kappa(m-|I|)/2} (h\langle D\rangle)^{\delta(m-|I|)} \Lambda_I u \|$$

$$+ C_h \Big(\sum_{|I|\leq m-1} \| t^{-\kappa(m-|I|)/2} (h\langle D\rangle)^{\delta(m-|I|-1)} \Lambda_I u \| + \| P_m u \| ,$$

and, using (3.9), (3.11) and

$$(|\tbinom{\alpha}{\gamma}|) \langle\xi\rangle^{(1-\delta)r} h^{-\delta r} \leq \langle\xi\rangle^{r} + (|\tbinom{\alpha}{\gamma}|)^{1/\delta} h^{-r}, \tag{3.2$'$}$$

we obtain

$$\tbinom{\alpha}{\gamma} \| P_{(\gamma)} \, \partial_x^{\alpha-\gamma} u \|$$

$$\leq \tbinom{\alpha}{\gamma} (|\tbinom{\alpha}{\gamma}|)^{-1} C \{ C_0^{(|\gamma|-M)} + (|\gamma|-M)_+ !^{s_0}$$

$$\Big(\sum_{|\beta|\leq \min\{M,|\gamma|\}} \sum_{|I|\leq m-1} t^{-\kappa/2(m-|I|)} (h\langle D\rangle)^{\delta(m-|I|)} \Lambda_I \partial_x^{\beta+\alpha-\gamma} u \| \Big)$$

$$+ C_0^{|\gamma|} |\gamma| !^{s_0} (|\tbinom{\alpha}{\gamma}|)^{1/\delta} \, h^{-|\gamma|} \Big(\sum_{|I|\leq m-1} \| t^{-\kappa/2(m-|I|)} \quad \times$$

$$(h\langle D\rangle)^{\delta(m-|I|)}\Lambda_I\partial_x^{\alpha-\gamma}u\|\,)\}$$

$$+\binom{\alpha}{\gamma}C_0^{|\gamma|}|\gamma|!^{s_0}\,C_h(\sum_{|I|\leqq m-1}\|t^{-\kappa/2(m-|I|)}(h\langle D\rangle)^{\delta(m-|I|-1)}\times$$

$$\Lambda_I\,\partial_x^{\alpha-\gamma}u\,\|)\}.$$

Then using

$$(h\langle\xi\rangle)^{\delta}\leqq k+2(k+1)^{1-1/\delta}h\langle\xi\rangle,$$

we have

$$\frac{d}{dt}\,\|\partial_x^{\alpha}\,u\,\|h^{|\alpha|})\leqq|\alpha|\;!^{s}\{C\sum_{|k|\leqq|\alpha|+1}C_0^{(|\alpha|-k)}$$

$$\times\,kt^{-\kappa/2}\,\|u\|_k h^k/k!^s$$

$$+\,C_h\sum_{|k|\leqq|\alpha|}C_0^{|\alpha|\;-k}t^{-\kappa/2}\,\|u\|_k h^k/k!^s\},$$

$$+\,C\,\|\partial_x^{\alpha}\,Pu\|h^{|\alpha|}$$

where we use the notations:

$$\|v\|^2=\sum_{|I|\leqq m-1}\|t^{-\kappa(m-1-|I|)/2}(h\langle D\rangle)^{\delta(m-1-|I|)}\Lambda_I v\|^2$$

and

$$\|v\|_{\ell}=\max_{|\alpha|=\ell}\|\partial_x^{\alpha}v\|\,.$$

From this we obtain the a priori estimate

$$\sum_{\ell=0}^{\infty}C_0^{-\ell}\ell!^{-s}h^{\ell}e^{-\nu_1 t^{1-\kappa/2}}(1-\nu_0 t^{1-\kappa/2})^{\ell}\,\|u\|_{\ell}(t)$$

$$\leqq C(\sum_{\ell=0}^{\infty}C_0^{-\ell}\ell!^{-s}h^{\ell}\int_0^t e^{-\nu_1\tau^{1-\kappa/2}}(1-\nu_0\tau^{1-\kappa/2})^{\ell}\times$$

$$\max_{|\alpha|=\ell}\|\partial_x^{\alpha}\,(Pu)\,\|(\tau)d\tau)$$

for $0 \leqq t \leqq \min\{(2\nu_0)^{-2/(2-\kappa)}, T-\varepsilon\}$, where ν_0 is constant independent of h.

References

[1] M.D. Bronshtein, Smoothness of roots of polynomials depending on parameters, Sibirsk. Mat. Z., 20 (1979) 493-501.

[2] F. Colombini, E. De Giorgi and S. Spagnolo, Sur les équations hyperboliques avec des coefficients qui ne dépendent que du temps Ann. Scuola Norm. Sup. Pisa, 6 (1979), 511-559.

[3] F. Colombini, E. Jannelli and S. Spagnolo, Well-posedness in the Gevrey classes of the Cauchy problem for a nonstrictly hyperbolic equation with coefficients on time, Ann. Norm. Sup. Pisa, 10 (1983), 291-312.

[4] W. Ichinose, Propagation of singularities for a hyperbolic equation with non-regular characteristic roots, Osaka J. Math., 17 (1980), 703-749.

[5] H. Kumano-go, Pseudo-differnetial Operators, M.I.T. Press, Cambridge, 1981 (Ch. 2 §1).

[6] T. Nishitani, Sur les équations hyperboliques à coefficients holderiens en t et de la classe de Gevrey en x, Bull. Soc. Math. France, 104 (1983), 113-138.

[7] T. Nishitani, Energy inequality for hyperbolic operators in the Gevrey class, J. Math. Kyoto Univ., 23 (1983), 739-773.

[8] W. Nuij, A note on hyperbolic polynomials, Math. Scand., 23 (1968), 69-72.

[9] Y. Ohya and S. Tarama, Le problème de Cauchy à Caractéristiques Multiples dans la classe de Gevrey - coefficients höideriens en t-, in Hyperbolic Equations and Related Topics ed. by S. Mizohata (Proc. Taniguchi Sympo. Katata 1984), Kinokuniya, Tokyo, 1986.

[10] Y. Ohya and S. Tarama, Le problème de Cauchy à Charactéristiques Multiples dans la classe de Gevrey II, in Proc. of the conference on hyperbolic equations and related topics Padova 1985, to appear.

[11] S. Tarama, in preparation.

[12] S. Wakabayashi, Remarks on hyperbolic polynomials, Tukuba J. Math., 10 (1986), 17-28.

Shigeo Tarama
Department of Applied
Mathematics and Physics,
Kyoto University,
606 Kyoto,
Japan.

J. VAILLANT

Systèmes hyperboliques à multiplicité constante et dont le rang peut varier

ABSTRACT: Hyperbolic systems of constant multiplicity and of variable rank

We consider the operator

$$h(x,D) \equiv a(x,D) + b(x) \equiv ID_0 + \sum_{1 \leq i \leq n} a^i(x)D_i + b(x)$$

is a neighbourhood Ω of 0 in R^{n+1}, $x = (x_0,x')$, where a^i and b are $m \times m$ matrices with C^∞ coefficients or if $n = 1$ with real analytic coefficients which will be made precise later in each different case. We assume the characteristic roots of

$$\det (I\xi_0 + \sum_{1 \leq i \leq n} a^i(x)\xi_i) = 0$$

are real and are of constant multiplicity.

We consider the Cauchy problem

$$h(x,D)u(x) = f(x) \in C^\infty(\Omega), \; u(t,x') = g_i(x') \in C^\infty(\Omega_t).$$

We first reduce the operator to the form

$$h(x,D) = ID_0 + a_1(x,D') + b(x,D'),$$

where $a_1(x,\xi')$ is a symbol of order 1 and nilpotent, b is a symbol of order 0, and $\det a(x,\xi) = \det (I\xi_0 + a_1(x,\xi')) = \xi_0^m$.

The basis of the proof is a transformation by an elliptic operator $\Delta(x,D')$ and the theorem of Egorov. If A is the cofactor matrix, we assume

$$A = \xi_0^q A(x,\xi), \text{ with } A(x,0,\xi') \not\equiv 0,$$

$$a(x,\xi) = P(x,\xi) \text{ diag. } [\xi_0^p, \xi_0^{q_1},\ldots,\xi_0^{q_k},1,\ldots,1] \; Q(x,\xi),$$

with det $Q(x,0,\xi') \not\equiv 0$, $\xi_0{}^p$, $\xi_0{}^{q_1} \ldots \xi_0{}^{q_k}$ are invariant factors, and $q = q_1 + q_2 + \ldots q_k$, $p \geqq q_1 \geqq \ldots \geqq q_k$.

We define the "generalized rank" of the system to be constant if and only if det $Q(x,0,\xi') \neq 0$.

We essentially assume C^∞ coefficients when the generalized rank is constant, and when the generalized rank is variable we assume that $n = 1$ and the coefficients are analytic.

We define new conditions of Levi; we distinguish three cases, namely, $q = 0$, $0 < q < m/2$ and $q \geqq m/2$. These Levi conditions are invariant under transformations by ellitpic operators $\Delta(x,D')$.

If the generalized rank is constant, these Levi conditions are necessary and sufficient in each of the three following cases: $q = 0$; $q < m/2$ and $m \leqq 4$; $q \geqq m/2$ and $m = 4$.

In the case of variable generalized rank we discuss several cases if $m \leqq 4$. In these cases also the Levi conditions are still necessary and sufficient. We use here a method of differnetial designularization.

0. INTRODUCTION

L'étude des conditions nécessaires et suffisantes, pour que le problème de Cauchy relatif à un système d'opérateurs différentiels soit bien posé, lorsque les caractéristiques sont de multiplicité constante et le rang est constant, a été entreprise par plusieurs auteurs; nous citerons Kajitani, Petkov, Vaillant; dans les bibliographies de leurs travaux, on trouvera des références plus complètes. Lorsque le rang peut varier, l'étude a été entamée par Matsumoto, dans le cas de la multiplicité 2.

L'objet du travail exposé ici est d'obtenir des "conditions de Levi", explicites sur les coefficients de la matrice, dans des coordonnées convenables, valables, lorsque le rang varie, et pour des multiplicités quelconques. Il est nécessaire pour cela de préciser d'abord la définition du rang constant ou variable, en microlocalisant la matrice et en considérant ses facteurs invariants. Lorsque les coefficients sont C^∞ et le rang généralisé constant, on obtient des conditions, que nous pensons être générales, et dont nous démontrons la nécessité et la suffisance dans les cas connus et dans des cas de multiplicité plus grande non encore considérés.

Lorsque le rang varie, on considère des coefficients analytiques; le cas de coefficients C^∞ nécessite des hypothèses supplémentaires, comme dans [9]. On suppose aussi la dimension de l'espace égale à 2. Dans ces conditions, on démontre dans des cas significatifs, jusqu'à la multiplicité 4, que les conditions introduites sont nécessaires et suffisantes.

Les démonstrations détaillees paraîtront dans d'autres publications.

I. HYPOTHESES; REDUCTION ET NOTATIONS

a) $x = (x_0, x') = (x_0, x_1, \ldots, x_n) \in \Omega$;

Ω est un voisinage ouvert de 0 dans $\mathbf{R}^{n+1}$.

$a^i(x)$, $1 \leq i \leq n$ et $b(x)$ sont des matrices $m \times m$ fonctions analytiques de x; on considère l'opérateur différentiel matriciel:

$$h(x,D) = a(x,D) + b(x) = I\, D_0 + \sum_{1 \leq i \leq n} a^i(x)\, D_i + b(x),$$

où $D_0 = \dfrac{\partial}{\partial x_0}$, $D_i = \dfrac{\partial}{\partial x_i}$.

REMARQUE I.1: L'hypothèse d'analyticité des coefficients est indispensable pour la démonstration des conditions suffisantes, lorsque le rang est variable; on peut obtenir les conditions nécessaires en supposant seulement les coefficients C^∞ et, dans le cas du rang constant, avec quelques précautions, on peut obtenir les conditions suffisantes avec l'hypothèse C^∞.

On considère le problème de Cauchy

$$h(x,D)\, u = f$$

$$u_{|x_0=t} = g_t(x')$$

f est le second membre donné, g_t la donnée de Cauchy sur $x_0 = t$ et u l'inconnue.

DEFINITIONS I.1: Le problème de Cauchy est bien posé en $(\underline{t},\underline{x}') \in \Omega$ si et seulement si il existe un voisinage ouvert ω de $(\underline{t},\underline{x}')$, $\omega \subset \Omega$, tel que: $\forall f\ C^\infty(\omega)$, $\forall g_t \in C^\infty(\omega_t)$, où : $\omega_t = \omega \cap \{x_0 = t\}$, il existe une solution C^∞, unique du problème de Cauchy dans ω; le problème de Cauchy est bien posé dans Ω, si il est bien posé en tout point de Ω.

On dira que le problème de Cauchy est bien posé au voisinage de 0 ou brièvement localement bien posé, si il existe un voisinage ouvert de 0, Ω, où le problème de Cauchy est bien posé.

REMARQUE I.2: Le dernière définition est justifiée par la considération du cas du rang variable où les coefficients sont analytiques; en fait on les considérera comme des germes en 0.

On note $\xi = (\xi_0,\xi') = (\xi_0,\xi_1,\ldots,\xi_n) \in \mathbb{R}^{n+1}$ la variable duale de x et on considère le déterminant caractéristique:

$$\text{dét}(\xi_0\, I + \sum_{1\leq i\leq n} a^i(x)\xi_i).$$

On fait *l'hypothèse* qu'il est *hyperbolique* et que les racines caractéristiques sont de *multiplicité constante*, c'est-à-dire que les racines λ_j de

$$\text{dét}(I\xi_0 + \sum_{1\leq i\leq n} a^i(x)\xi_i) = 0,$$

considérée comme équation en ξ_0 sont réelles et telles que:

$$\lambda_j(x,\xi') \neq \lambda_k(x,\xi'),$$

$\forall x$, $\xi' \neq 0$, $j \neq k$; elles ont alors pour tout $(x,\xi'), \xi' \neq 0$ la régularité des coefficients; la multiplicité de λ_j est notée m_j; on a : $m = \sum_{1 \leq j \leq r} m_j$.

b) On fait une première réduction de l'opérateur h.

<u>PROPSOITION I.1</u>: $\forall(\underline{x}, \underline{\xi}')$, $\underline{\xi}' \neq 0$, il existe un voisinage conique de ce point et un symbole matriciel $\Delta_0(x,\xi')$ défini dans ce voisinage, homogène de degré 0 en ξ', de déterminant non nul, régulier et tel que:

$$\Delta_0^{-1}(x,\xi') \, a(x,\xi) \, \Delta_0(x,\xi') = \tilde{a}(x,\xi),$$

$\tilde{a}(x,\xi) =$

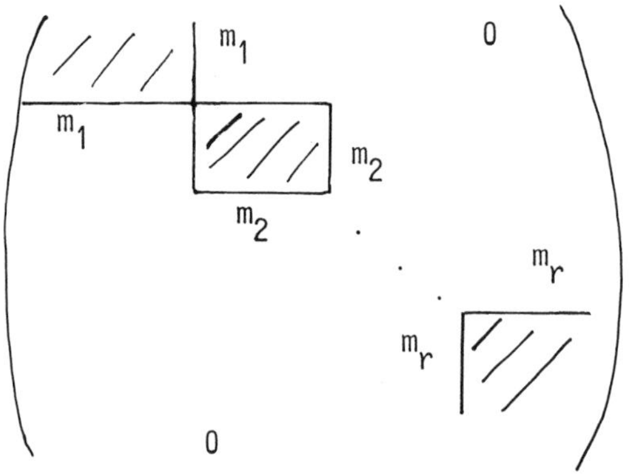

;

en dehors des blocs, la matrice est nulle; le bloc d'indice j est de la forme:

$$\tilde{a}_{(j)}(x,\xi) = \begin{pmatrix} \xi_0 - \lambda_j & & 0 \\ & \ddots & \\ 0 & & \xi_0 - \lambda_j \end{pmatrix} + \tilde{a}_{(j)}(x,\xi'),$$

ou $\tilde{a}_{(j)}$ est homogène de degré 1 en ξ', régulière et nilpotente.

On désigne par $\Delta(x,D')$ un opérateur pseudo-différential d'ordre zéro, du symbole $\Delta(x,\xi')$ developpable en symboles homogènes, tel que:

$$\Delta(x,\xi') = \Delta_0(x,\xi') + \Delta_1(x,\xi') + \ldots + \Delta_\ell(x,\xi') + \ldots ,$$

où Δ_ℓ est d'ordre $- \ell$.

<u>PROPOSITION I.2</u>: $\forall(\underline{x},\underline{\xi}')$, $\underline{\xi}' \neq 0$, il existe un opérateur pseudodifférentiel $\Delta(x,D')$, tel que:

$$h(x,D)\ \Delta(x,D') = \Delta(x,\Delta')\tilde{h}(x,D),$$

(modulo un opérateur de $L^{-\infty}$), où:

$\tilde{h}(x,D) = \tilde{a}(x,D) + \tilde{b}(x,D')$; $\tilde{a}(x,\xi)$ est obtenu dans la proposition 1; $\tilde{b}(x,\xi')$ est défini dans un voisinage conique, d'ordre 0 et de la forme:

$\tilde{b}(x,\xi') =$

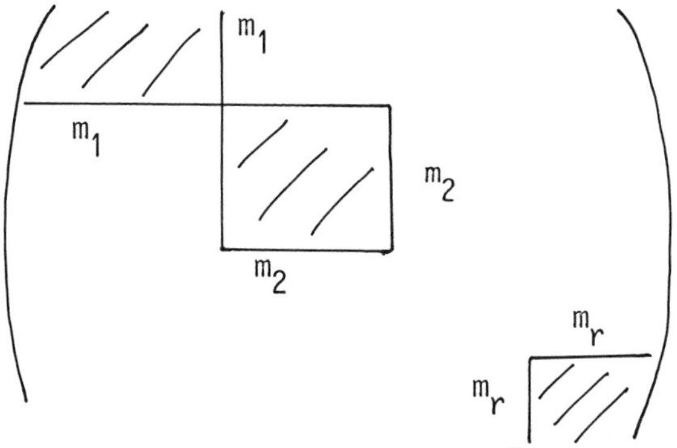

;

en dehors des blocs, la matrice est nulle; les $\tilde{b}$ sont développables en symboles homogènes et réguliers.

<u>PREUVE</u>: On a essentiellement à résoudre des équations matricielles du type

$$\tilde{a}_{(j)}\delta - \delta\,\tilde{a}_{(k)} = \varepsilon, \quad j \neq k,$$

où ε est une matrice connue et δ est inconnue; on a encore:

$$(\lambda_k - \lambda_j)\delta + \tilde{a}_j\delta - \delta\tilde{a}_k = \varepsilon;$$

que l'on résout, en tenant compte du fait que: $\lambda_j \neq \lambda_k$ et du degré de nilpotence de $\tilde{a}_j$ et $\tilde{a}_k$.

REMARQUE I.3: On peut ramener l'étude des conditions nécessaires et suffisantes pour que le problème de Cauchy soit bien posé à celles concernant l'opérateur $\tilde{h}$.

Aussi, nous changerons de notations et supposerons désormais l'opérateur initial de la forme:

$$h(x,D) = a(x,D) + b(x,D') = (D_0 - \lambda(x,D') I + a(x,D') + b(x,D'),$$

où $a(x,\xi')$ est matriciel $m \times m$ homogène, d'ordre 1 et nilpotent, $b(x,\xi')$ matriciel $m \times m$ d'ordre 0 développable en symboles homogènes,

$$\text{dét } [(\xi_0 - \lambda(x,\xi')) I + a(x,\xi')] = (\xi_0 - \lambda(x,\xi'))^m.$$

c) On réduira à nouveau microlocalement l'opérateur, de façon à se ramener à une racine caractéristique λ nulle. On utilise pour cela le théorème d'Egorov, cf. Hörmander [5], Petkov [7]. Nous résumerons cette application.

PROPOSITION I.3: On définit un opérateur intégral de Fourier par:

$$Gu(x) = (2\pi)^{-n-1} \int e^{i[\varphi(x,\eta')+x_0\eta_0]} g(x,\eta')\hat{u}(\eta)d\eta,$$

où $g = 1$ dans le voisinage conique considéré, φ verifie:

$$D_0\varphi - \lambda(x,D'\varphi) = 0$$

$$\varphi(\underline{x}_0, x',\eta') = x' \, . \, \eta'.$$

On note H un inverse de G et on considere H h G; son symbole principal est de la forme: $\tilde{\tilde{a}}(x,\xi) = I\xi_0 + \tilde{\tilde{a}}(x,\xi')$, où:

$$\det \tilde{\tilde{a}}(x,\xi) = \xi_0^m, \quad \tilde{\tilde{a}} \text{ est nilpotent. On a :}$$

$$H\ h\ G = \tilde{\tilde{h}}(x,D) = \tilde{\tilde{a}}(x,D) + \tilde{\tilde{b}}(x,D), \text{ modulo } L^{-\infty},$$

ou $\tilde{\tilde{b}}(x,D)$ est d'ordre 0.

On peut remplacer $\tilde{\tilde{b}}(x,D)$ par un opérateur $\tilde{\tilde{b}}'(x,D')$ où D_0 n'intervient pas en généralisant un lemme obtenu dans un cas plus simple par Chazarain [2] (théorème 5.1).

PROPOSITION I.4: On note ici $\Delta(x,D)$ un opérateur elliptique développé en symboles homogènes sous la forme:

$$\Delta(x,\xi) = I + \Delta_1(x,\xi) + \dots + \Delta_j(x,\xi) + \dots$$

il existe un opérateur Δ tel que:

$$\tilde{\tilde{h}}(x,D) = [\tilde{\tilde{a}}(x,D) + \tilde{\tilde{b}}'(x,D')] \text{ o } \Delta(x,D),$$

modulo un opérateur de $L^{-\infty}$ où $\tilde{\tilde{b}}'(x,D')$ est un opérateur d'ordre 0.

PREUVE: On développe $\tilde{\tilde{b}}'$ en symboles homogènes: $\tilde{\tilde{b}}'_0 + \dots$ On détermine d'abord $\tilde{\tilde{b}}'_0$ et Δ_1 en fonction de $\tilde{\tilde{b}}_0$.

La première équation s'écrit:

$$\tilde{\tilde{b}}_0 = (I\xi_0 + \tilde{\tilde{a}})\,\Delta_1 + \tilde{\tilde{b}}'_0$$

et a pour solution, en posant: $\partial^0 = \dfrac{\partial}{\partial\xi_0}$:

$$\tilde{\tilde{b}}'_0(x,\xi') = \tilde{\tilde{b}}_0(x,o,\xi') - \tilde{\tilde{a}}(x,\xi')\partial^0\tilde{\tilde{b}}_0(x,o,\xi') + \dots + \frac{(-1)^{m-1}}{(m-1)!}\,\tilde{\tilde{a}}^{m-1}(x,\xi')\,(\partial^o)^{m-1}\tilde{\tilde{b}}_0(x,o,\xi')$$

$$\begin{aligned}\Delta_1(x,\xi') = {} & \partial^o\tilde{\tilde{b}}_o(x,o,\xi') + (\xi_o I - \tilde{\tilde{a}}(x,\xi'))\,\frac{(\partial_o)^2\tilde{\tilde{b}}_o(x,o,\xi')}{2!} + \\ & \dots + [\xi_o^{m-2}\, I - \tilde{\tilde{a}}(x,\xi')\xi_0^{m-3} + \dots + (-1)^{m-2}\tilde{\tilde{a}}^{m-2}(x,\xi')]\,\frac{(\partial_0)^{m-1}\tilde{\tilde{b}}_0(x,o,\xi')}{(m-1)!} \\ & + [\xi_0^{m-1}\, I - \tilde{\tilde{a}}(x,\xi')\xi_0^{m-2} + \dots + (-1)^{m-1}\tilde{\tilde{a}}^{m-1}(x,\xi')]\,\frac{\int_0^1 (\partial_0)^m\,\tilde{\tilde{b}}_0(x,t\xi_0,\xi')dt}{m!}\,.\end{aligned}$$

REMARQUE I.4: Dans l'étude des conditions nécessaires et suffisantes, qui est essentiellement micro locale et, comme on n'a pas perdu de régularité, on peut remplacer l'opérateur h(x,D) par $\tilde{\tilde{a}}(x,D) + \tilde{\tilde{b}}(x,D')$.

Nous changerons à nouveau de notation en tenant compte de cette remarque.

NOTATIONS

1) $h(x,D) = a(x,D) + b(x,D')$

$$= ID_0 + \mathcal{a}(x,D') + b(x,D'),$$

où $\mathcal{a}(x,\xi')$ est matriciel $m \times m$ homogène d'ordre 1 et nilpotent, analytique pour (x,ξ'), $\xi' \neq 0$; $b(x,\xi')$ est d'ordre 0 développable en symboles homogènes et analytiques pour (x,ξ'), $\xi' \neq 0$,

$$\text{dét}(I\xi_0 + \mathcal{a}(x,\xi')) = \xi_0^m.$$

2) Dans le cas de R^2, on obtient:

$$h(x,D) = ID_0 + \mathcal{a}^1(x)D_1 + b_0(x) + b_1(x)\, D_1^{-1} + \dots + b_j(x)D_1^{-j} + \dots ,$$

où les coefficients sont analytiques.

REMARQUE I.5: Dans le cas de R^2 les demonstrations précédentes et les opérateurs Δ se simplifient.

II. RANG GÉNÉRALISE, OPÉRATEURS L ET CONDITIONS DE LÉVI

a) On note A la matrice des cofacteurs de a de sorte que:

$$aA = Aa = \xi_0^m\, I$$

On note q la plus haute puissance de ξ_0 qui divise A, de sorte que:

$$A(x,\xi) = \xi_0^q\, A(x,\xi), \qquad 0 \leq q \leq m-1,$$

$$A(x,o,\xi') \not\equiv 0.$$

On pose $p = m-q$.

Une étude plus précise des mineurs de a sera nécessaire; elle nécessitera l'introduction d'entiers $q_1,q_2,\dots,q_k$ que nous allons définir.

On considère l'anneau localisé de l'anneau des polynômes en ξ_0 ayant pour coefficients des germes de fonctions analytiques en (x,ξ'), par rapport a l'idéal engendre par ξ_0 [10]; c'est l'anneau des fractions construites à partir de ces polynômes et dont le dénominateur n'est pas divisible par ξ_0; il est principal et ses idéaux sont engendrés par les puissances de ξ_0.

Dans cet anneau, $a(x,\xi)$ est équivalente a une matrice diagonale, dont les élements sont les facteurs invariants de a, c'est-à-dire qu'il existe P et Q telles que:

$$a(x,\xi) = P(x,\xi)\begin{pmatrix} \xi_0^{q_0} & & & & & & \\ & \xi_0^{q_1} & & & & & \\ & & \ddots & & & & \\ & & & \xi_0^{q_k} & & & \\ & & & & 1 & & \\ & & & & & \ddots & \\ & & & & & & 1 \end{pmatrix} Q(x,\xi)$$

$$\text{dét}\, P(x,o,\xi') \not\equiv 0, \quad \text{dét}\, Q(x,o,\xi') \not\equiv 0;$$

$$q_0 \geqq q_1 \geqq \dots \geqq q_k;\ q_0 + q_1 + \dots + q_k = m;$$

on sait que $\xi_0^{q_1+\dots+q_k}$ est la plus haute puissance de ξ_0 qui divise A; on en déduit que: $q = q_1 + \dots + q_k$ et que: $q_0 = p$.

On remarque que:

$$\text{dét}\, P(x,\xi)\ \text{dét}\, Q(x,\xi) \equiv 1.$$

Les calculs de [10] p. 240, 243 permettent des précisions utiles.

$Q(x,\xi)$ peut être choisi comme une matrice de polynômes que l'on peut calculer explicitement et son déterminant s'exprime à l'aide des mineurs (p. 242); on peut aussi calculer P explicitement et les dénominateurs qui apparaîssent sont les facteurs de dét $Q(x,\xi)$.

Remarquons efin que $\xi_0^{q_j+\ldots+q_k}$, $j \leq k$ est la plus haute puissance de ξ_0 qui divise tous les mineurs d'ordre m-j de a.

DEFINITION II.1: La suite $(q_1,\ldots,q_k)$ est le rang génerali$s$é de a.

b) Nouse allons définir en coordonnées naturelles sur $T^*(R^{n+1})$ des opérateurs différentiels en x, qui seront importants dans la définition des conditions de Levi. ($\xi' \neq 0$).

On notera: $\partial^\alpha = \dfrac{\partial^{|\alpha|}}{\partial\xi_0^{\alpha_0}\ldots\partial\xi_n^{\alpha_n}}$, $\alpha = (\alpha_0,\alpha')$; $D^\alpha = \dfrac{\partial^{|\alpha|}}{\partial x_0^{\alpha_0}\ldots\partial x_n^{\alpha_n}}$

DEFINITION II.2:

$$L_0(x,\xi,D_x) = A(x,\xi)[\sum_{|\alpha|=1} \partial^\alpha a(x,\xi)D^\alpha + b_0(x,\xi')]$$

$$L_1(x,\xi,D_x) = A(x,\xi)[\frac{1}{2!}\sum_{|\alpha'|=2} \partial^{\alpha'} a(x,\xi)D^{\alpha'} + \sum_{|\alpha'|=1} \partial^{\alpha'} b_0(x,\xi')D^{\alpha'} + b_1(x,\xi')]$$

....

$$L_{m-2}(x,\xi,D_x) = A(x,\xi)[\frac{1}{(m-1)!}\sum_{|\alpha'|=m-1} \partial^{\alpha'} a(x,\xi)D^{\alpha'} +$$

$$+ \frac{1}{(m-2)!}\sum_{|\alpha'|=m-2} \partial^{\alpha'} b_0(x,\xi')D^{\alpha'}$$

$$+ \frac{1}{(m-3)!}\sum_{|\alpha'|=m-3} \partial^{\alpha'} b_1(x,\xi')\, D^{\alpha'} + \ldots + b_{m-2}(x,\xi')].$$

c) Conditions de Lèvi. On distinguera trois cas

1er cas: $q = 0$, $A = A$.

DEFINITION II.3: La première condition de Lévi L_1 s'énonce:

$L_0(A)$ est divisible par ξ_0^{m-1} : $L_0(A) = \xi_0^{m-1}\,\Lambda_1$.

Elle équivaut à l'annulation de (m-1) symboles matriciels $m \times m$ en (x,ξ'), qui sont les (m-1) premiers coefficients de $L_0(A)$ ordonné en ξ_0.

$\Lambda_1(x,\xi)$ est une matrice $m \times m$, polynomiale en ξ_0, homogène en ξ', et symbole d'ordre m-1.

La deuxième condition L_2 s'énonce: $L_0(\Lambda_1) - \xi_0\, L_1(A)$ est divisible par ξ_0^{m-1}:

$$L_0(\Lambda_1) - \xi_0 L_1(A) = \xi_0^{m-1}\, \Lambda_2$$

On définit les suivantes par récurrence.

La $(m-1)^{\text{ème}}$ et dernière condition L_{m-1} s'énonce:

$$L_0(\Lambda_{m-2}) - \xi_0 L_1(\Lambda_{m-3}) + \ldots + (-1)^{m-2}\xi_0^{m-2}\, L_{m-2}(A) = \xi_0^{m-1}\, \Lambda_{m-1}$$

On peut faire sur ces conditions les mêmes remarques que pour la première.
On notera $(L) = (L_1,\ldots,L_{m-1})$ l'ensemble de ces conditions de Levi.
Il sera utile d'en donner une expression sous la forme de calculs d'opérateurs.

NOTATION: A un symbole homogène $\Lambda(x,\xi)$ polynomial en ξ_0, on associera un opérateur pseudodifférentiel $\Lambda'(x,D)$, différentiel en D_0 de symbole développable en symboles homogènes dont Λ est le symbole principal. On note σ l'opération prendre le symbole principal.

EXEMPLE: à $A(x,\xi)$, on associe $A'(x,D)$; $\sigma(A') = A$.

DEFINITION II.4:

$$L_0^{\#} \equiv \sigma\,\{A'(x,D)\,[h(x,D)\,A'(x,D) - I\,D_0^m]\}.$$

REMARQUE: $L_0^{\#}$ dépend du choix de A'.

DEFINITION: La condition $L_1^{\#}$ s'énonce.

Il existe un opérateur A' tel que $L_0^{\#}$ soit divisible par ξ_0^{m-1}. On obtient le

LEMME: Les conditions L_1 et $L_1^{\#}$ sont équivalentes.

REMARQUE: $L_1^{\#}$ ne dépend donc pas du choix de A'.

Si L_1 est réalisée, on note $\Lambda_1^{\#}$ le quotient d'ordre m-1, de $L_0^{\#}$ par ξ_0^{m-1}:

$$L_0^{\#} = \xi_0^{m-1}\ \Lambda_1^{\#}.$$

REMARQUE: $\Lambda_1^{\#} = \Lambda_1 + R_1$, où R_1 est un symbole d'ordre m-1 polynomial en ξ_0.

On se propose de définir par récurrence, des conditions $L^{\#}$ équivalentes aux conditions L.

Soit $1 \leq s_0 \leq m-3$. On admet que, si pour tout s, $1 \leq s \leq s_0$, les conditions $L_1,\dots,L_{s-1}$ sont vérifiées, alors pour tout $1 \leq s \leq s_0$, il existe des symboles homogènes d'ordre m-1, polynomiaux en ξ_0, $\Lambda_1^{\#},\dots,\Lambda_{s-1}^{\#}$, $R_1,\dots,R_s$ et des operateurs A', $\Lambda_1^{\#'},\dots,\Lambda_{s-1}^{\#'}$ tels que:

$$L_{s-1}^{\#} = \sigma\{A'[h(\Lambda_{s-1}^{\#'} - \Lambda_{s-2}^{\#'} D_0 + \dots + (-1)^{s-2} \Lambda_1^{\#'} D_0^{s-2}$$

$$+ (-1)^{s-1} A' D_0^{s-1}) - (-1)^{s-1} D_0^{m+s-1}]\}$$

$$= L_0(\Lambda_{s-1}) - \xi_0\ L_1(\Lambda_{s-2}) + \dots + (-1)^{s-1}\ \xi_0^{s-1}\ L_{s-1}(A) + \xi_0^{m-1}\ R_s.$$

Supposons alors que L_{s_0} est vérifiée; on pose:

$$\Lambda_{s_0}^{\#} = \Lambda_{s_0} + R_{s_0}.$$

On définit:

$$L_{s_0}^{\#} = \sigma\{A'[h(\Lambda_{s_0}^{\#'} - \Lambda_{s_0-1}^{\#'} D_0 + \dots + (-1)^{s_0-1} \Lambda_1^{\#'} D_0^{s_0-1}$$

$$+ (-1)^{s_0} A' D_0^{s_0}) - (-1)^{s_0} D_0^{m+s_0}]\}.$$

Il reste alors à prouver que:

$$L_{s_0}^{\#} = L_0\ (\Lambda_{s_0}) - \xi_0 L_1(\Lambda_{s_0-1}) + \dots + (-1)^{s_0}\ \xi^{s_0}\ L_{s_0}(A) + \xi_0^{m-1}\ R_{s_0+1},$$

où R_{s_0+1} est homogène d'ordre m-1, polynomiale en ξ_0. On obtient ce résultat. On en déduit la

PROPOSITION II.1:

$$\forall s,\ 1 \leqq s \leqq m-2$$

$$(L_1,\ldots,L_s) \text{ équivaut à } (L_1^{\#},\ldots,L_s^{\#}).$$

REMARQUE II.1: Pour m = 2, on a une seule condition de Lévi, qui coïncide avec celles données précedemment par plusieurs auteurs. cf. [1], [9].

Pour m = 3, les deux conditions obtenues coïncident avec celles de [1].

Invariance par un opérateur elliptique

On notera:

$$\Delta(x,D') = \Delta_0(x,D') + \Delta_1(x,D') + \ldots + \Delta_j(x,D') + \ldots$$

un opérateur elliptique d'ordre 0, développable en symboles homogènes, à coefficients réguliers et par $\Delta^{(i)}(x,D')$ un inverse de $\Delta(x,D')$ modulo $L^{-\infty}$;

$$\Delta_0^{(i)}(x,\xi') = \Delta_0^{-1}(x,\xi').$$

A tout opérateur pseudo-différentiel P(x,D) on fera correspondre son transformé par $\Delta(x,D')$:

$$\tilde{P}(x,D) = \Delta^{(i)}(x,D')\, P(x,D)\, \Delta(x,D').$$

EXEMPLE: $\tilde{h} = \Delta^{(i)} h\Delta = \tilde{a}(x,D) + \tilde{b}(x,D') = ID_0 + \tilde{a}(x,D') + \tilde{b}(x,D')$, modulo $L^{-\infty}$; $\tilde{a}(x,\xi) = \Delta_0^{-1}\, a\, \Delta_0 = I\xi_0 + \Delta_0^{-1}\, a\, \Delta_0$ a les mêmes facteurs invariants que a.

PROPOSITION II.2: Si $h(x,D) = ID_0 + a(x,D') + b(x,D')$, satisfait les conditions de Levi $(L_1,\ldots,L_s)$, son transformé $\tilde{h}(x,D)$ les satisfait aussi.

PREUVE: On utilise la proposition II.1.

REMARQUE II.2: Les conditions de Levi sont invariantes dans un changement de coordonnées spatial: $x' \mapsto y'$; une autre publication prouve leur invariance: $x \mapsto y$.

Case de $\mathbb{R}^2$

Les conditions de Levi se simplifient. Comme nous étudierons ce cas lorsque le rang est variable, où des difficultés nouvelles apparaissent, *nous supposerons de plus*:

$$b_1 = \dots = b_j = \dots = 0$$

c'est-à-dire que, en simplifiant les notations:

$$h(x,D) = ID_0 + a(x)D_1 + b(x),$$

où a et b sont des germes analytiques en 0.

Dans ces conditions:

$$L_1 = \dots = L_{m-2} = 0,$$

et on posera:

$$L_0 = L ,$$

$$L^r = \underbrace{L \circ L \circ \dots \circ L}_{r \text{ fois}}$$

Les conditions de Levi, s'écrivent

$L(A)$ est divisible par ξ_0^{m-1}

$L^2(A)$ est divisible par $\xi_0^{2(m-1)}$

....

$L^{m-1}(A)$ est divisible par $\xi_0^{(m-1)^2}$

EXEMPLE: 1) $m = 2$ On a une condition de Levi; en l'explicitant, on obtient:

$$a(\dot{a} + ba)(x) = 0,$$

où l'on pose $\dot{a} = D_0\, a$.

2) $m = 3$ On posera $\beta = b - D_1 a$, $M = a^2$, $\dot{a} + ba = \delta$,

$$N = \dot{M} + \beta M + a\delta.$$

Les conditions s'écrivent:

$$(aN)(x) = 0$$

et

$$[a(\dot{N} + \beta N) + M(\dot{\delta} + b\delta)](x) = 0.$$

<u>2eme cas</u>: $p > q > 0 \Longleftrightarrow 0 < q < \frac{m}{2}$.

<u>DEFINITIONS</u>:

<u>La première condition de Levi</u> L_1 <u>s'énonce</u>:

$L_0(A)$ est divisible par ξ_0^{p-q}

$$L_0(A) = \xi_0^{p-q}\, \Lambda_1$$

<u>La deuxième condition de Levi</u> L_2 <u>s'énonce</u>:

$L_0(\Lambda_1) - \xi_0^q\, L_1(A)$ est divisible par ξ_0^{p-1}

$$L_0(\Lambda_1) - \xi_0^q\, L_1(A) = \xi_0^{p-1}\, \Lambda_2$$

<u>La troisième condition de Levi s'énonce</u>:

$L_0(\Lambda_2) - \xi_0 L_1(\Lambda_1) + \xi_0^{q+1}\, L_2(A)$ est divisible par ξ_0^{p-1}.

On définit les suivantes de même.

La dernière condition de Levi s'énonce:

$$L_0(\Lambda_{m-2}) - \xi_0 L_1(\Lambda_{m-3}) + \dots + (-1)^{m-3}\, \xi_0^{m-3}\, L_{m-3}\,(\Lambda_1)$$
$$+ (-1)^{m-2}\, \xi_0^{m+q-3}\, L_{m-2}(A)$$

est divisible par ξ_0^{p-1}.

On remarque, que dans la dernière, le terme $(-1)^{m-2}\, \xi_0^{m+q-3}\, L_{m-2}(A)$ est inutile, car: $m+q-3 \geqq p-1$.

PROPOSITION II.3: Ces conditions sont invariantes par transformation par un opérateur elliptique, (comme dans la proposition II.2).

REMARQUE II.3: Pour m = 3, on a : q = 1, p = 2.

Les conditions de Levi s'écrivent:

$$L_0(A) = \xi_0\, \Lambda_1$$

$$L_0(\Lambda_1) = \xi_0\, \Lambda_2.$$

Pour m = 4, on a : q = 1, p = 3. Les conditions s'écrivent:

$$L_0(A) = \xi_0^2\, \Lambda_1$$

$$L_0(\Lambda_1) - \xi_0\, L_1(A) = \xi_0^2\, \Lambda_2$$

$$L_0(\Lambda_2) - \xi_0\, L_1(\Lambda_1) = \xi_0^2\, \Lambda_3$$

Cas de R^2

Ces conditions s'énoncent:

$L(A)$ est divisible par ξ_0^{p-q}

$L^2(A)$ est divisible par $\xi_0^{p-q+p-1}$

....

$L^{m-1}(A)$ est divisible par $\xi_0^{p-q+(m-2)(p-1)}$

EXEMPLES

1) $m = 3$

$L(A)$ est divisible par ξ_0

$L^2(A)$ est divisible par ξ_0^2

2) $m = 4$

$L(A)$ est divisible par ξ_0^2

$L^2(A)$ est divisible par ξ_0^4

$L^3(A)$ est divisible par ξ_0^6

3eme cas: $q \geqq p \Longleftrightarrow q \geqq \frac{m}{2}$.

On sait que : $q \leqq m-1$.

Examinons d'abord le cas $q = m-1$. Dans ce cas, la considération des facteurs invariants montre que: $a = \xi_0\, I$ et l'opérateur est fortement hyperbolique. Nous laisserons ce cas dans la suite et supposerons $q \leqq m-2$. Ce n'est donc possible que si $m \geqq 4$.

Nous étudierons ici seulement le cas $m = 4$.

Il y a deux possibilités pour les facteurs invariants:

$$\text{i)}\quad a \sim \begin{pmatrix} \xi_0^2 & & & 0 \\ & \xi_o & & \\ 0 & & \xi_o & \\ & & & 1 \end{pmatrix} \quad \text{et} \quad \text{ii)}\quad a \sim \begin{pmatrix} \xi_0^2 & & & 0 \\ & \xi_o^2 & & \\ 0 & & 1 & \\ & & & 1 \end{pmatrix}$$

Les conditions de Levi s'écrivent:

Cas i) $L_0(A)$ est divisible par ξ_0; $L_0(A) = \Lambda_1\, \xi_0$

$L_0(\Lambda_1)$ est divisible par ξ_0; $L_0(A) = \Lambda_2 \xi_0$

$L_0(\Lambda_2) - L_0\, L_1(A)$ est divisible par ξ_0

Cas ii) $L_0L_0(A)$ est divisible par ξ_0 ; $L_0^2(A) = \Lambda_1\xi_0$

$L_0(\Lambda_1)$ est divisible par ξ_0; $L_0(\Lambda_1) = \Lambda_2\xi_0$

$L_0(\Lambda_2)$ - $(L_0L_1L_0(A)$ est divisible par ξ_0.

Proposition II.4: Ces conditions sont invariantes par transformation par un opérateur elliptique.

Cas de $\mathbb{R}^2$ - On obtient:

cas i) $L(A)$ est divisible par ξ_0
$L^2(A)$ est divisible par ξ_0^2
$L^3(A)$ est divisible par ξ_0^3

cas ii) $L^2(A)$ est divisible par ξ_0
$L^3(A)$ est divisible par ξ_0^2
$L^4(A)$ est divisible par ξ_0^3

Rang généralisé constant

DEFINITION: On dira que le rang généralisé est constant (dans un voisinage conique), si il existe une matrice $Q(x,0,\xi')$, (fin du IIa) et références), telle que:

$$\det Q(x,0,\xi') \neq 0, \quad \forall(x,\xi'), \ \xi' \neq 0.$$

CONSEQUENCE: La suite $(q_1,\ldots,q_k)$ caractérisait aussi les facteurs invariants, pour tous les (x,ξ'), $\xi' \neq 0$, tels que dét $Q(x,0,\xi') \neq 0$, dans l'anneau localisé de l'anneau des polynômes et ξ_0, à coefficients constants. Si le rang généralisé est constant, cette caractérisation est valable pour tous les (x,ξ'), $\xi' \neq 0$ (du voisinage conique considéré).

PROPOSITION II.5: Si le rang généralisé est constant, il existe un symbole matriciel $\Delta_0(x,\xi')$, homogène de degré 0 en ξ', de déterminant non nul, régulier et tel que:

$$\Delta_0^{-1}(x,\xi')\ a(x,\xi)\ \Delta_0(x,\xi') = \tilde{\tilde{a}}(x,\xi),$$

où:

$$\tilde{\tilde{a}}(x,\xi) = I\ \xi_0 + \tilde{\tilde{\tilde{a}}}(x,\xi')$$

et:

$$\tilde{\tilde{\tilde{a}}}(x,\xi') = \begin{pmatrix} J_p & & & 0 \\ & J_{q_1} & & \\ 0 & & \ddots & \\ & & & J_k \end{pmatrix} |\xi'| = J|\xi'|$$

est un tableau de matrices de Jordan J correspondant aux facteurs invariants définis par p, $q_1,\ldots,q_k$.

La proposition suivante montre que, par le choix d'un opérateur elliptique convenable elliptique Δ de symbole principal Δ_0, on peut aussi simplifier les coefficients b grâce à un résultat d'Arnold, appliqué par Petkov [8] au problème considére; (pour q = 0, on avait déjà le résultat par Kajitani [4]). Nous désignerons par $\tilde{\tilde{\tilde{b}}}$ la forme simplifiee en renovoyant à [8] pour les détails du cas général et nous expliciterons ensuite le cas q = 0.

PROPOSITION II.6: Si le rang généralisé est constant, il existe un opérateur pseudodifférentiel $\Delta(x,D')$ tel que:

$h(x,D)\ \Delta(x,D') = \Delta(x,D')\ \tilde{\tilde{\tilde{h}}}(x,d)$, modulo un opérateur régularisant, où:

$\tilde{\tilde{\tilde{a}}}(x,\xi')$ a la forme de la proposition II.5 et où:

$\tilde{\tilde{\tilde{b}}}(x,\xi')$ est de la forme normale d'Arnold.

REMARQUE II.4: On peut aussi supposer les $\tilde{\tilde{\tilde{b}}}$ diagonaux nuls, en généralisant une réduction valable dans le cas scalaire [2], [5], [11].

PROPOSITION II.7: q = 0. On supprimera les tildas pour économiser les notations

$$h(x,D) = ID_0 + J|D'| + b(x,\xi');$$

la forme réduite de b est:

$$b = \begin{pmatrix} 0 & \dots & & 0 \\ \vdots & & 0 & \vdots \\ 0 & \dots & & 0 \\ b_1^m & \dots & b_{m-1}^m & 0 \end{pmatrix} \qquad b = b_0 + b_1 + \dots + b_j + \dots$$

Les conditions de Levi sont equivalentes aux conditions suivantes:

$$\text{ordre de } b_A^m \leqq A-m, \quad 1 \leqq A \leqq m-1.$$

REMARQUE II.5: Sous cette forme, ces conditions sont celles de Kajitani [4]. Compte tenu de l'invariance des conditions de Levi par transformation par un opérateur elliptique, elles sont équivalentes aux conditions de Lévi sur l'opérateur h initial (avant la transformation précédente).

REMARQUE II.6: On a transformé de même les conditions de Levi pour $p > q > 0$ et $q \geqq p$, dans tous les cas où $m \leqq 4$ et certains cas où $m = 5$. Les détails seront donnés ailleurs.

PROPOSITION II.8: On ne suppose pas le rang généralisé constant. Si les conditions de Levi sont satisfaites pour l'opérateur h, elles sont satisfaites pour l'opérateur adjoint de h : h^t.

PREUVE: L'ensemble des (x,ξ') où le rang généralisé est constant est partout dense et les conditions de Levi expriment l'annulation de fonctions continues. Il suffit donc de démontrer la proposition dans le cas du rang généralisé constant. D'autre part, dire que h^t verifie les conditions de Levi équivaut à dire que l'adjoint du transformé de h par une transformation du type de la proposition II.2 les vérifie.
On choisit une transformation comme dans la proposition II.6 et on ramène l'opérateur à une forme simplifiée, (comme dans la proposition II.7 pour $q = 0$).

Sous cette forme, on voit que, si l'opérateur vérifie les conditions de Levi, son adjoint les vérifie aussi.

III: NÉCESSITÉ DES CONDITIONS DE LEVI

THÉORÈME III: Si $q = 0$, m quelconque, ou bien si $m \leq 4$, pour que le problème de Cauchy soit bien posé, il faut que les conditions de Levi soient satisfaites.

RÉSUMÉ DE LA PREUVE: Il suffit de considérer le cas du rang généralisé constant.

Dans ce cas pour $q = 0$, la nécessité a été démontré par Kajitani [4]. Pour les autres cas, si $m \leq 3$, la preuve est celle de Petkov [6], cf. aussi Berzin, Vaillant [1]; pour $m = 4$ on a des calculs analogues de développements asymptotiques; la base de la méthode étant l'utilisation du théorème du graphe fermé.

REMARQUE III: Pour la nécessité, on peut considérer des opérateurs h à coefficients C^∞, (en précisant convenablement les hypothèses.)

IV. SUFFISANCE DES CONDITIONS DE LEVI

On distinguera deux cas.

1) Cas du rang généralisé constant

THEOREME IV.1: Si le rang généralisé est constant, si $q = 0$, ou si $m \leq 4$, pour que le problème de Cauchy soit bien posé, il faut et il suffit que les conditions de Levi soient satisfaites.

RÉSUMÉ DE LA PREUVE: L'existence résulte de la construction de parametrix, en utilisant les réductions précédentes. L'unicité résulte de la proposition II.8 (cf. Berzin, Vaillant [1], Petkov [7]).

REMARQUE: En précisant un peu les hypothèses, ce résultat vaut si les coefficients de h sont C^∞.

2) Cas de R^2

On utilisera de façon essentielle l'analyticité des coefficients. Le rang peut varier avec x.
On étudie les cas où m varie de 2 à 4

a) $q = 0$

a_1) $m = 2$.

THEOREME IV.2: Pour que le problème de Cauchy soit localement bien posé, il faut et il suffit que la condition de Levi soit satisfaite.

PREUVE: Le système s'écrit:

$$D_o u_1 + a_1^1 D_1 u_1 + a_2^1 D_1 u_2 + b_1^1 u_1 + b_2^1 u_2 = f_1$$

$$D_o u_2 + a_1^2 D_1 u_1 + a_2^2 D_1 u_2 + b_1^2 u_1 + b_2^2 u_2 = f_2$$

avec les données de Cauchy:

$$u_1(t,x_1) = g_1(x_1), \quad u_2(t,x_1) = g_2(x_1).$$

On utilise d'abord le fait que $a^2 = 0$, $a.a = 0$; on en déduit qu'on peut écrire les coefficients de a sous la forme:

$$a_1^1 = a_2^2 = k\sigma\rho, \qquad a_2^1 = k\sigma^2, \qquad a_1^2 = -k\rho^2,$$

où les germes σ et ρ sont premiers entre eux.

La condition de Levi s'écrit alors:

$$\dot{\sigma}\rho - \sigma\dot{\rho} + b_1^2\sigma^2 - b_2^1\rho^2 + (b_1^1 - b_2^2)\rho\sigma = 0.$$

Le système équivaut au suivant, où pour raccourcir l'écriture, nous avons négligé les f et où ℓ est analytique, dépendant des coefficients:

$$D_o u_1 + k\sigma w + b_1^1 u_1 + b_2^1 u_2 = 0$$

$$D_o u_2 - k\rho w + b_1^2 u_1 + b_2^2 u_2 = 0$$

$$D_o w - (\ell - k\rho D_1\sigma + k\sigma D_1\rho - b_1^1)w + (\rho D_1 b_1^1 + \sigma D_1 b_1^2)u_1 + (\rho D_1 b_2^1 + \sigma D_1 b_2^2)u_2 = 0$$

$$u_1(t,x_1) = g_1(x_1),\ u_2(t,x_1) = g_2(x_1),\ w(t,x_1) = \rho D_1 g_1(x_1) + \sigma D_1 g_2(x_1).$$

C'est un système différentiel ordinaire qui permet de déterminer régulièrement u_1 et u_2.

<u>REMARQUE</u>: Ce résultat a d'abord été obtenu par Matsumoto [9] par une autre méthode; la méthode précédente permet les généralisations.

a_2) $m = 3$

On traite d'abord un cas particulier

$$D_o u_1 + a_1^1 D_1 u_1 + a_2^1 D_1 u_2 + a_3^1 D_1 u_3 + b_1^1 u_1 + b_2^1 u_2 + b_3^1 u_3 = f_1$$

$$D_o u_2 + a_2^2 D_1 u_2 + a_3^2 D_1 u_3 + b_1^2 u_1 + b_2^2 u_2 + b_3^2 u_3 = f_2$$

$$D_o u_3 + a_2^3 D_1 u_2 + a_3^2 D_1 u_3 + b_1^3 u_1 + b_2^3 u_2 + b_3^3 u_3 = f_3$$

$$u_1(t,x_1) = g_1(x_1),\quad u_2(t,x_1) = g_2(x_1)\quad u_3(t,x_1) = g_3(x_1).$$

On néglige d'écrire les f, ce qui, ici, ne présente pas d'intérêt.
Du fait que $a^3 \equiv 0$, $a^2 \not\equiv 0$, on déduit:

$$a_1^1 = 0,\quad a_2^2 = - a_3^3,\quad (a_2^2)^2 + a_3^2 a_2^3 = 0.$$

Les conditions de Levi donnent:

$$a_2^2 b_1^2 + a_3^2\, b_1^3 = 0$$

$$a_3^2\, b_1^2 - a_3^3\, b_1^2 + a_3^2\, b_1^2\, (b_1^1 - b_3^3) - a_2^2\, b_1^2\, b_3^2 - a_3^1 (b_1^2)^2 = 0$$

$$a_2^3\, b_1^3 - a_2^3\, b_1^3 + a_2^3\, b_1^3\, (b_1^1 - b_2^2) - a_3^3\, b_2^3\, b_1^3 - a_2^1 (b_1^3)^2 = 0$$

Par un calcul analogue à celui du cas m = 2, on résout le système, en le transformant.

Dans le cas général, nous ferons une hypothèse pour simplifier les calculs.

<u>THEOREME IV.3</u>: Supposons $N_1^1(x)$ et $M_1^1(x)$ premiers entre eux. Pour que le problème de Cauchy soit localement bien posé, il faut et il suffit que les conditions de Levi soient vérifiées.

<u>PREUVE</u>: Comme $M \neq 0$, on a pu supposer, par exemple, $M_1^1 \neq 0$, on en déduit que

$$M_1^2\, N_1^1 = N_1^2\, M_1^1$$

$$M_1^3\, N_1^1 = N_1^3\, M_1^1 ,$$

d'où: M_1^1 divise M_1^2 et M_1^3; le vecteur analytique d(x) défini par $d_1 = 1$, $d_2 = \frac{M_1^2}{M_1^1}$, $d_3 = \frac{M_1^3}{M_1^1}$ forme une base du noyau de $a(x)$ pour x voisin de 0.

On fait un changement de base régulier au voisinage de 0 en prenant d pour un vecteur de base et a prend la forme du cas particulier. Compte tenu de l'invariance des conditions de Levi, le résultat est démontré.

a_3) m = 4.

<u>THEOREME IV.4</u>: Supposons que

$$a(x) = \begin{pmatrix} 0 & \mu_1 & 0 & 0 \\ 0 & 0 & \mu_2 & 0 \\ 0 & 0 & 0 & \mu_3 \\ 0 & 0 & 0 & 0 \end{pmatrix}$$

ou les μ sont des germes analytiques en 0, tels que, puisque q = 0, $\mu_1, \mu_2, \mu_3 \neq 0$; supposons les μ irréductibles. Pour que problème de Cauchy soit localement bien posé, il faut et il suffit que les conditions de Levi soient vérifiées.

PREUVE: Elle est indiquée dans Vaillant [11]

b) $p > q > 0$

THEOREME IV.5: Supposons que:

$$\text{pour } m = 3 \qquad a(x) = \begin{pmatrix} a_1^1 & a_2^1 & a_3^1 \\ 0 & a_2^2 & a_3^2 \\ 0 & a_2^3 & a_3^3 \end{pmatrix}, \; a^2 = 0, \; a \not\equiv 0;$$

$$\text{pour } m = 4 \qquad a(x) = \begin{pmatrix} 0 & \mu_1 & 0 & 0 \\ 0 & 0 & \mu_2 & 0 \\ 0 & 0 & 0 & 0 \\ 0 & 0 & 0 & 0 \end{pmatrix}, \; \mu_1 \mu_2 \not\equiv 0$$

et les μ sont irréductibles.

Pour que le problème de Cauchy soit localement bien posé, il faut et il suffit que les conditions de Levi soient vérifiées.

c) $q \geqq p$

On a donc $m = 4$, $p = 2$, $q = 2$.

THEOREME IV.6: Supposons que

1)

$$a(x) = \begin{pmatrix} 0 & \mu & 0 & 0 \\ 0 & 0 & 0 & 0 \\ 0 & 0 & 0 & 0 \\ 0 & 0 & 0 & 0 \end{pmatrix}, \; \mu \not\equiv 0,$$

ou

2)

$$a(x) = \begin{pmatrix} 0 & \mu_1 & 0 & 0 \\ 0 & 0 & 0 & 0 \\ 0 & 0 & 0 & \mu_2 \\ 0 & 0 & 0 & 0 \end{pmatrix}, \; \mu_1 \mu_2 \not\equiv 0,$$

et b_3^2 est premier avec μ_1 μ_2; les μ sont irreductibles.

Dans ces deux cas pour que le problème de Cauchy soit localement bien posé, il faut et il suffit que les conditions de Levi soient vérifiées.

REMARQUE: (i) On obtient, par les considérations précédentes, la propagation des singularités et la perte de régularité due aux multiplicités.

ii) On pourrait étudier les problèmes précédents dans les classes de Gevrey en reliant les conditions de Levi aux indices de Gevrey.

References

[1] R. Berzin, J. Vaillant, J. Math Pures et Appl., 58, 1979, p. 165-216.

[2] J. Chazarain, Ann. Inst. Fourier, 24, 1974, p. 209-223.

[3] V. Ivrii, V.M. Petkov, Upsehi Math. Nauk., 25, n° 5, 1974, p.3.70.

[4] K. Kajitani, Public. RIMS Kyoto, 15, 1979, p. 519-550.

[5] L. Hörmander, The Analysis of linear Partial Differential equations IV Springer Verlag, 1985.

[6] V.M. Petkov, Trudy Sem. Petrovsk, vyr 1 (1975), p. 211-236.

[7] V.M. Petkov, Trans. Moscow Math. Soc. 1980, Issue I.

[8] V.M. Petkov, Math. Nachr. 93, 1979, p. 117-131.

[9] W. Matsumoto, J. Math. Kyoto Univ. 21, n° 1-2, 1981, p. 47-84 et p. 251-271.

[10] J. Vaillant, Ann. Institut Fourier 151, 1965, p. 225-311.

[11] J. Vaillant, Hyperbolic equations, 1985, par Colombini et Murthy, Longman Scientifical and Technical, UK.

[12] J. Vaillant, C.R. Acad. Sc. Paris, 304, Série 1, 1987, p. 379-394.

[13] J. Vaillant, C.R. Acad. Sc. Paris, t. 305, Série 1, 1987, p. 377-380.

Jean Vaillant
Unite Associée au CNRS 761
Mathématiques, tour 45-46,
5 ème étage
Université de Paris VI
4, place Jussieu
75252 Paris Cedex 05
France.

C. WAGSCHAL

Problème de Cauchy ramifié: opérateurs a caractéristiques tangentes ayant un contact d'ordre constant

ABSTRACT: Ramified Cauchy problem: Operators with tangential characteristics having a contact of constant order

In this paper we study the ramified Cauchy problem for second order operators with characteristics of variable multiplicity. In an earlier note [C.R. Acad. Sc. Paris, t. 303, 1986] we have proved a theorem for operators whose principal symbol is $\xi_0(\xi_0 + qx_0^{q-1}\xi_1)$: in this case, the solution is shown to be ramified, in a certain sense, around the two characteristic hypersurfaces. Here, we prove that this theorem is still true for operators of the following type.

In C^{n+1}, coordinates of which are denoted by $x = (x_0,x') = (x_0,x_1,\ldots,x_0)$, we consider a second order operator a(x,D) with holomorphic coefficients and the principal symbol is assumed to be

$$g(x,\xi) = (\xi_0-\lambda_0(x,\xi'))(\xi_0-\lambda_1(x,\xi'))$$

where the functions λ_i are holomorphic in a neighbourhood of the point $\bar{x} = 0$, $\bar{\xi}' = (1,0,\ldots,0)$ and satisfy

$$\lambda_0(\bar{x},\bar{\xi}') = \lambda_1(\bar{x},\bar{\xi}').$$

We denote by K_0 and K_1 the characteristic hypersurfaces issued from $T:x_0 = x_1 = 0$ and we suppose that K_0 and K_1 are tangent, along T with constant order of contact. Then we consider the Cauchy problem [$S:x_0 = 0$]

$$\begin{cases} a(x,D)u(x) = 0, \\ D_0^k u(x)|_S = w_k(x'), \ k = 0,1, \end{cases}$$

where the functions w_k are ramified around T. Under a technical assumption, we prove that the solution is ramified around $K_0 \cup K_1$ as in the note already mentioned.

Soit $a(x,D)$ un opérateur différentiel linéaire du second ordre à coefficients holomorphes au voisinage de l'origine de $\mathbb{C}^{n+1}$ $[x = (x_0,x_1,\ldots,x_n) = (x_0,x')]$. On suppose que le symbole principal $g(x,\xi)$ s'écrit $[\xi = (\xi_0,\xi')]$

$$g(x,\xi) = (\xi_0-\lambda_0(x,\xi'))(\xi_0-\lambda_1(x,\xi'))$$

ou les fonctions λ_i sont holomorphes au voisinage du point $\bar{x} = 0$, $\bar{\xi}' = (1,0,\ldots,0)$ et

$$\lambda_0(\bar{x},\bar{\xi}') = \lambda_1(\bar{x},\bar{\xi}'). \tag{0.1}$$

On considère le probleme de Cauchy ramifié

$$\begin{cases} a(x,D)u(x) = 0, \\ D_0^h u(x)|_S = w_h(x'),\ h = 0,1, \end{cases} \tag{0.2}$$

où $S : x_0 = 0$, les fonctions w_h sont supposées ramifiées autour de l'hyperplan de S

$$T : x_0 = x_1 = 0.$$

Autrement dit, les fonctions w_h sont holomorphes au voisinage (relativement à S) d'un point $y \in S-T$ et se prolongent analytiquement le long de tout chemin $\gamma : I \to \Omega - T$ d'origine y, $I = [0,1]$, où Ω désigne un voisinage ouvert connexe de l'origine dans S.

Le problème de Cauchy (0.2) admet une unique solution holomorphe au voisinage de y; cette solution u définit un germe de fonction holomorphe en ce point y dont on veut effectuer le prolongement analytique.

Notons k_i $(i = 0,1)$ la solution du problème de Cauchy

$$D_0 k_i(x) = \lambda_i(x,D'k_i(x)),\ k_i|_S = x_1,$$

et K_0, K_1 les hypersurfaces caractéristiques issues de T

$$K_i : k_i(x) = 0.$$

On notera que K_0 et K_1 sont tangentes en 0 d'après (0.1).

1. THEOREME DE REPRESENTATION

Pour effectuer le prolongement analytique du germe u, on utilise un théorème de représentation intégrale de ce germe; rappelons ce théorème ([4, théorème 1.2]).

Pour x voisin de y, la solution u de (0.2) peut s'écire

$$u(x) = \sum_{m=0}^{\infty} I_m(x), \tag{1.1}$$

avec

$$I_m(x) = \int_{S_m(x_0)} u_m(\phi_m(\tau,x'),\tau,x')d\tau_1 \wedge \ldots \wedge d\tau_m \tag{1.2}$$

ou $S_m(x_0)$ désigne le m-simplexe singulier de $\mathbb{C}^{m+1}$

$$S_m(x_0) : s \in \Delta_m^0 \to x_0(s, 1-(s_1+\ldots+s_m)) \in \mathbb{C}^{m+1},$$

$$\Delta_m^0 = \{s \in \mathbb{R}^m;\ 0 \leqq s_j,\ s_1+\ldots+s_m \leqq 1\},\ s = (s_j)_{1\leqq j\leqq m},$$

où les fonctions $u_m(t,\tau,x') \equiv u_m(t,\tau_1,\ldots,\tau_{m+1},x')$ sont holomorphes au voisinage du point $t = y_1$, $\tau = 0$, $x' = y'$ et se prolongent analytiquement à $\mathcal{R}_\omega \times \Omega_r^m$, $\mathcal{R}_\omega$ désignant le revêtement universel d'un disque pointé $\dot{D}_\omega = \{t \in \mathbb{C};\ 0 < |t| < \omega\}$ et

$$\Omega_r^m = \{(\tau,x') \in \mathbb{C}^{m+1} \times \mathbb{C}^n;\quad \|\tau\| + \|x'\| < r\},$$

$$\|\tau\| = \sum_{\ell=1}^{m+1} |\tau_\ell| \quad , \quad \|x'\| = \sum_{j=1}^{n} |x_j|;$$

en outre, il existe $c > 0$ et, pour tout compact $K \subset \mathcal{R}_\omega$, il existe $c_K > 0$ tel que

$$|u_m(t,\tau,x')| \leqq c_K c^m m!,\ \text{pour } (t,\tau,x') \in K \times \Omega_r^m; \tag{1.3}$$

enfin, les fonctions holomorphes $\phi_m : \Omega_r^m \to \mathbb{C}$ sont définies par

$$\phi_0(x) = \phi_0(\tau_1,x')|_{\tau_1=x_0} = k_0(x), \qquad (1.4)_0$$

$$m \geq 1 \begin{cases} D_{\tau_{m+1}}\, \phi_m(\tau,x') = \lambda_m\, (\tau_1+\ldots+\tau_{m+1},x',D'\phi_m(\tau,x')),\tau = (\tau_1,\ldots,\tau_{m+1}), \\ \qquad\qquad (1.4)_{m\geq 1} \\ \phi_m(\tau_1,\ldots,\tau_m,0,x') = \phi_{m-1}(\tau_1,\ldots,\tau_m,x'), \end{cases}$$

où $\lambda_m = \lambda_0$ si m est pair et $\lambda_m = \lambda_1$ si m est impair.

NOTE. Pour $m = 0$, $I_0(x) = u_0(k_0(x),x)$.

REMARQUE 1.1: La série (1.1) converge normalement au voisinage de y d'après (1.3).

REMARQUE 1.2: Le simplexe $S_m(x_0)$ est tracé dans l'hyperplan de $\mathbb{C}^{m+1}$

$$H_m^0(x_0) = \{\tau \in \mathbb{C}^{m+1};\ \tau_1 + \ldots + \tau_{m+1} = x_0\}.$$

Considérons d'autre part les faces de Δ_m^0

$$\Delta_m^j = \{s \in \mathbb{R}^m;\ s_j = 0\},\ 1 \leq j \leq m,$$

$$\Delta_m^{m+1} = \{s \in \mathbb{R}^m;\ s_1 + \ldots + s_m = 1\},$$

et les hyperplans de $\mathbb{C}^{m+1}$

$$H_m^j(x_0) = \{\tau \in \mathbb{C}^{m+1};\ \tau_j = 0\},\ 1 \leq j \leq m+1.$$

On a alors $S_m(x_0)(\Delta_m^j) \subset H_m^j(x_0)$, $0 \leq j \leq m+1$: le simplexe $S_m(x_0)$ est donc un cycle relatif de $\left(H_m^0(x_0),\ \bigcup_{j=1}^{m+1} H_m^j(x_0)\right)$ et l'intégrale (1.2) ne dépend que de la classe d'homologie relative de ce cycle.

Lorsque l'opérateur est à caractéristiques multiples de multiplicité constante ($\lambda_0 \equiv \lambda_1$), on constate (cf. [4]) que (1.1) peut s'écrire

$$u(x) = I_0'(x) = u_0'(k_0(x),x)$$

et plus généralement lorsque les racines caractéristiques sont en involution, c'est-à-dire lorsque $\{\xi_0-\lambda_0(x,\xi'), \xi_0-\lambda_1(x,\xi')\} = 0$, que (1.1) peut s'écrire

$$u(x) = I_0'(x) + I_1'(x)) = u_0'(k_0(x),x)) + \int_0^{x_0} u_1'(\phi_1(\sigma,x),\sigma,x)d\sigma$$

où $\phi_1(\sigma,x) = \phi_1(\sigma,x_0-\sigma,x')$.

Dans la suite, nous allons nous intéresser à la situation géométrique la plus simple, celle où les hypersurfaces K_0 et K_1 sont tangentes le long de T avec un contact d'ordre constant.

2. CARACTERISTIQUES CONFONDUES : $K_0 = K_1$

Nous supposons dans ce paragraphe que $K_0 = K_1$. Ceci n'implique pas que les racines caractéristiques soient en involution comme le montre un exemple aussi simple que $g(x,\xi) = \xi_0(\xi_0-x_0x_1\xi_1)$ pour lequel $K_0 = K_1 = \{x \in \mathbb{C}^{n+1}; x_1 = 0\}$ et $\{\xi_0,\xi_0 - x_0x_1\xi_1\} = x_1\xi_1$.

Grâce à un changement de coordonnées locales, on peut supposer que $K = (K_0 = K_1)$ est l'hyperplan $x_1 = 0$; autrement dit,

$$\lambda_1(x;1,0,\ldots,0) = 0, \text{ pour } x_1 = 0. \tag{2.1}$$

PROPOSITION 2.1: Il existe $s > 0$ et des fonctions holomorphes $\psi_m:\Omega_s^m \to \mathbb{C}$ telles que

$$\phi_m(\tau,x') = x_1(1 + \psi_m(\tau,x')), \text{ pour } (\tau,x') \in \Omega_s^m.$$

En outre, il existe une fonction majorante F d'une seule variable telle que

$$\psi_m(\tau,x') << F([\tau] + [x']), \text{ pour tout } m \in \mathbb{N}.$$

où $[\tau] = \sum_{j=1}^{m+1} \tau_j$, $[x'] = \sum_{k=1}^{n} x_k$.

PREUVE. On pose a priori $\phi_m = x_1(1 + \psi_m)$. Pour $m \geqq 0$, on a

$$x_1 D_{\tau_{m+1}} \psi_m = \lambda_m(\tau_1 + \ldots + \tau_{m+1}, x'; x_1D_1\psi_m + \psi_m + 1, x_1D''\psi_m) \tag{2.2}$$

$$\phi_m(\tau_1,\ldots,\tau_m,0,x') = \psi_{m-1}(\tau_1,\ldots,\tau_m,x'),\ m \geqq 1, \tag{2.3}$$

$$\psi_0(0,x') = 0, \tag{2.4}$$

où $D'' = (D_j)_{2\leqq j\leqq n}$. Posons, avec $\xi'' = (\xi_j)_{2\leqq j\leqq n}$,

$$\mu_i(x,y,\xi') = \mu_i(x;x_1\xi_1+y+1,x_1\xi''),\ i = 0,1;$$

ces fonctions μ_i sont holomorphes au voisinage du point $x = 0$, $y = 0$, $\xi' = 0$ et vu (2.1)

$$\mu_i(x,y,\xi')|_{x_1=0} = \lambda_i(x;y + 1,0,\ldots,0)|_{x_1=0} = 0;$$

il existe donc des fonctions $\nu_i(x,y,\xi')$ holomorphes au voisinage du point $x = 0$, $y = 0$, $\xi' = 0$ telles que

$$\mu_i(x,y,\xi') = x_1\nu_i(x,y,\xi').$$

On peut donc écrire (2.1) sous la forme [$\nu_m = \nu_0$ si m est pair, $\nu_m = \nu_1$ si m est impair]

$$D_{\tau_{m+1}}\psi_m = \nu_m(\tau_1 + \ldots + \tau_{m+1},x';\ \psi_m,D'\psi_m). \tag{2.5}$$

Les équations (2.3), (2.4) et (2.5) déterminent les fonctions ψ_m, la fonction ψ_m étant holomorphe au voisinage de l'origine de $\mathbb{C}^{m+1}\times\mathbb{C}^n$.

Soit M une fonction majorante des fonctions ν_i de la forme

$$\nu_i \ll M(x_0 + \ldots + x_n,y,\xi').$$

Cherchons une fonction majorante Θ d'une seule variable telle que

$$\psi_m \ll \Theta(t),\quad \text{où } t = \rho[\tau] + [x'], \tag{2.6}$$

où $\rho \geqq 1$ est un paramètre. Etant donné que $\psi_m(0) = 0$, on peut supposer $\Theta(0) = 0$. D'après (2.3) et (2.4), on a d'autre part $D'\psi_m(0) = 0$; si (2.6) est vérifié, on a donc

$D'\psi_m << [D\Theta](t)$ où $[D\Theta] = D\Theta - D\Theta(0)$.

Pour que (2.6) soit vérifié, il suffit donc que

$$M([\tau] + [x'], \Theta(t), [D\Theta](t)) << \rho\, D\Theta(t)$$

et vu que $\rho \geqq 1$, il suffit a fortiori que

$$M(t,\Theta(t), [D\theta](t)) << \rho\, D\Theta(t). \qquad (2.7)$$

Vu le lemme 2.4 de [3], on peut choisir M de la forme

$$M(t,y,\xi') = c_1\phi_R(t)\, \frac{R}{R-y} \prod_{j=1}^{n} \frac{R}{R-\xi_j}\ ,\ c_1 > 0,\ R > 0;$$

cherchons alors Θ de la forme $\Theta(t) = c_2 D_t^{-1}\phi_R(t)$ où $c_2 > 0$ est une constante à déterminer. On a alors $[D\Theta] = c_2[\phi_R] << c_2\phi_R$ et d'après le lemme 2.5 de [3] $\Theta << c_2c_3\phi_R$ $(c_3 > 0)$. Si on choisit c_2 tel que

$$0 < c_2c_3 < R \text{ et } 0 < c_2 < R, \qquad (2.8)$$

on peut utiliser la proposition 2.3 de [3] qui prouve que

$$M(t,\Theta(t),[D\Theta](t)) << c_1c_4\phi_R(t)$$

où

$$c_1 = \left(K + \frac{c_2c_3}{R-c_2c_3}\right)\left(K + \frac{c_2}{R-c_2}\right)^n$$

et on en déduit (2.7) dès que $\rho \geqq \mathrm{Max}(1,c_1c_4)$. Ceci achéve la démonstration de (2.6) et par conséquent de la proposition.

C.Q.F.D.

Vu que $\psi_m(0) = 0$, on peut supposer $F(0) = 0$. En choisissant $s > 0$ assez petit, on peut donc supposer

$$|\psi_m(\tau,x')| \leqq \frac{1}{2},\ \text{pour } (\tau,x') \in \Omega_s^m. \qquad (2.9)$$

Considérons alors les fonctions

$$v_m(t,\tau,x') = u_m(t(1 + \psi_m(\tau,x')),\tau,x');$$

v_m est holomorphe au voisinage du point $t = y_1$, $\tau = 0$, $x' = y'$ et se prolonge analytiquement à $\mathcal{R}_{2\omega/3} \times \Omega_a^m$ où $a = \mathrm{Min}(r,s)$. Il résulte également de (2.9) que, pour tout compact $K \subset \mathcal{R}_{2\omega/3}$, il existe un compact $K' \subset \mathcal{R}_\omega$ telque pour tout $m \geqq 0$

$$(t,\tau,x') \in K \times \Omega_a^m \Rightarrow (t(1 + \psi_m(\tau,x')),\tau,x')) \in K' \times \Omega_a^m,$$

d'où, d'après (1.3),

$$|v_m(t,\tau,x')| \leqq c_k' c^m m!, \text{ pour } (t,\tau,x') \in K \times \Omega_a^m.$$

Il en résulte que la fonction

$$\mathcal{U}(t,x) = \sum_{m=0}^{\infty} \int_{S_m(x_0)} v_m(t,\tau,x')d\tau_1 \wedge \dots \wedge d\tau_m$$

est holomorphe sur $\mathcal{R}_{2\omega/3} \times \Omega_a$, où $\Omega_a = \{x \in \mathbb{C}^{n+1};\ \sum_{j=0}^{n} |x_j| < a\}$, et que $u(x) = \mathcal{U}(x_1,x)$, soit.

<u>COROLLAIRE 2.2</u>: Lorsque $K_0 = K_1 = K$, la solution du problème de Cauchy (0.2) peut s'écrire $u(x) = \mathcal{U}(k(x),x)$ où $k(x) = 0$ est une équation locale de K et $\mathcal{U}$ une fonction holomorphe sur $\mathcal{R}_{\tilde{\omega}}$, $\times \Omega_a$. La solution u est donc ramifiée autour de K.

3. CARACTERISTIQUES TANGENTES AVEC UN CONTACT D'ORDRE CONSTANT

On suppose désormais que

K_0 et K_1 sont tangentes le long de T avec un contact d'ordre constant. (3.1)

On peut alors choisir des coordonnees locales telles que

$$K_0 : x_1 = 0,\ K_1 : x_1 - x_0^q = 0, \text{ où } q \geqq 2. \tag{3.2}$$

Il en résulte que $k_0(x) = x_1 a(x)$, $k_1(x) = (x_1 - x_0^q)b(x)$ où a et b sont holomorphes au voisinage de l'origine et $a(0) = b(0) = 1$; modulo un changement de coordonnées locales, on peut supposer par exemple que la fonction b est identiquement égale à 1. On a ainsi

$$k_0(x) = x_1 a(x),\ a(0) = 1 \text{ et } k_1(x) = x_1 - x_0^q. \tag{3.3}$$

Il en résulte que

$$\begin{cases} \lambda_0(x;1,0,\dots,0) = 0, \text{ pour } x_1 = 0, \\ \lambda_1(x;1,0,\dots,0) = -\,q x_0^{q-1}. \end{cases} \tag{3.4}$$

Les opérateurs les plus simples vérifiant (3.1) sont les opérateurs dont le symbole principal est

$$g(x,\xi) = \xi_0(\xi_0 + q x_0^{q-1}\xi_1); \tag{3.5}$$

nous avons fait une étude approfondie de ces opérateurs dans [5]: on a construit un ouvert $\mathcal{U}_a$ du revêtement universel de $\mathbb{C}^{n+1} - K_0 \cup K_1$ tel que la solution du problème de Cauchy ramifié se prolonge analytiquement à $\mathcal{U}_a$ ([5, théorème 6.1]).

On se propose de vérifier que le même théorème subsiste sous l'hypothèse (3.4) lorsque

$$\text{les fonctions } \xi' \to \lambda_1(x,\xi') \text{ sont linéaires.} \tag{3.6}$$

Autrement dit, avec les notations de [5], on a le

<u>THEOREME 3.1</u>: Avec les hypothèses (3.4) et (3.6), il existe $a > 0$ tel que la solution du problème de Cauchy (0.2) se prolonge analytiquement à $\mathcal{U}_a$.

<u>REMARQUE 3.1</u>: On peut affaiblir (3.6)) de la façon suivante et supposer seulement (cf. remarque 4.1) qu'il existe k, $1 \leq k \leq n$, tel que les fonctions

$\lambda_i(x;\xi_1,\dots,\xi_k,0,\dots,0)$ soient linéaire en $(\xi_1,\dots,\xi_k)$ et ne dépendent que des variables $(x_0,x_1,\dots,x_k)$. Pour $k = n$, cette hypothèse coincide avec (3.6). Par exemple, le symbole

$$g(x,\xi) = \xi_0^2 + qx_0^{q-1}\xi_0\xi_1 - \left(\frac{q}{2}\,x_0^{q-1}\right)^2 \xi_2^2$$

$$= (\xi_0 + qx_0^{q-1}(\xi_1 + (\xi_1^2+\xi_2^2)^{1/2}))(\xi_0 + qx_0^{q-1}(\xi_1 - (\xi_1^2+\xi_2^2)^{1/2})),$$

où $(\xi_1^2 + \xi_2^2)^{1/2}$ désigne la détermination telle que $1^{1/2} = 1$, vérifie cette hypothèse avec $k = 1$.

La preuve du théorème 6.1 de [5] ne repose que sur la représentation (1.1) de la solution; nous allons montrer qu'on a la même représentation avec les hypothèses (3.4) et (3.6). Plus précisément, les fonctions k_1, ϕ_m associées au symbole (3.5) étant notées k_1^0, ϕ_m^0 nous allons vérifier que

$$I_m(x) = \int_{S_m(x_0)} u_m(\phi_m(\tau,x'),\tau,x')d\tau_1 \wedge \dots \wedge d\tau_m \tag{3.7}$$

$$= \int_{S_m(x_0)} u'_m(\phi_m^0(\tau',x'),\tau',x')d\tau'_1 \wedge \dots \wedge d\tau'_m$$

où les fonctions u'_m possèdent les mêmes propriétès que les fonctions u_m; ceci prouvera le théorème 3.1.

Rappelons ([5]) que les fonctions k_1^0, ϕ_m^0 sont données par les formules suivantes

$$k_0^0(x) = x_1, \quad k_1^0(x) = x_1 - x_0^q,$$

$$\phi_{2m-1}^0 = \phi_{2m}^0 = x_1 + \sum_{j=1}^{2m} (-1)^{j+1}(\tau_1 + \dots + \tau_j)^q, \quad m \geq 1.$$

4. ETUDE DES FONCTIONS ϕ_m

Avec les hypotheses (3.4) et (3.6), on peut préciser la structure des fonctions ϕ_m comme suit. D'après (3.4), il existe une fonction μ_0 holomorphe au voisinage de l'origine $\mathbb{C}^{n+1}$ telle que

$$\lambda_0(x;1,0,\ldots,0) = x_1\mu_0(x),$$

posons

$$\nu_0(x;\xi_0,\xi') = \mu_0(x)\xi_0 + \lambda_0(x,\xi).$$

On a alors la

PROPOSITION 4.1: Avec les hypothèses (3.4) et (3.6), on a

$$\phi_{2m}(\tau,x') = \tag{4.1}$$

$$x_1\phi_{2m,m}(\tau,x') + \sum_{j=0}^{m-1}\left[(\tau_1+\ldots+\tau_{2j+1})^q-(\tau_1+\ldots+\tau_{2j+2})^q\right]\phi_{2m,j}(\tau,x'),$$

$$\phi_{2m+1}(\tau,x') = \tag{4.2}$$

$$x_1\phi_{2m+1,m}(\tau,x') + \sum_{j=0}^{m}[(\tau_1+\ldots+\tau_{2j+1})^q-(\tau_1+\ldots+\tau_{2j+2})^q]\phi_{2m+1,j}(\tau,x'),$$

où les fonctions $\phi_{m,j}$ holomorphes au voisinage de l'origine de $\mathbb{C}^{m+1} \times \mathbb{C}^n$ sont solutions des problèmes de Cauchy

$$D_{\tau_{2m+1}}\phi_{2m,m} = \nu_0(\tau_1+\ldots+\tau_{2m+1},x';\phi_{2m,m},D'\phi_{2m,m}),\ m \geq 0, \tag{4.3}$$

$$D_{\tau_{m+1}}\phi_{m,j} = \lambda_m(\tau_1+\ldots+\tau_{m+1},x',D'\phi_{m,j}),\ \text{pour } 0 \leq j \leq [\tfrac{m}{2}]-1, \tag{4.4}$$

$$\phi_{0,0} = 1,\ \text{pour } \tau_1 = 0, \tag{4.5}$$

$$\phi_{2m,m} = \phi_{2m-1,m-1},\ \text{pour } \tau_{2m+1} = 0,\ m \geq 1, \tag{4.6}$$

$$\phi_{2m,j} = \phi_{2m-1,j},\ \text{pour } \tau_{2m+1} = 0,\ 0 \leq j \leq m-1, \tag{4.7}$$

$$\phi_{2m+1,j} = \phi_{2m,j},\ \text{pour } \tau_{2m+2} = 0,\ 0 \leq j \leq m. \tag{4.8}$$

PREUVE: On cherche a priori ϕ_m de la forme (4.1) ou (4.2). On note d'abord que les conditions (4.5) à (4.8) impliquent que $\phi_0 = x_1$ pour $\tau_1 = 0$ et

$\phi_m = \phi_{m-1}$ pour $\tau_{m+1} = 0$ lorsque $m \geqq 1$. Pour que ϕ_{2m} vérifie l'équation du premier ordre (1.4), il suffit que les fonctions $x_1\phi_{2m,m}$ et $\phi_{2m,j}$ $(0 \leqq j \leqq m-1)$ vérifient cette équation. Pour $\phi_{2m,j}$ $(0 \leqq j \leqq m-1)$, ceci conduit à (4.4) et pour $\phi_{2m,m}$, on a

$$x_1 D_{\tau_{2m+1}} \phi_{2m,m} = \lambda_0(\tau_1+\ldots+\tau_{2m+1}, x', x_1 D_{\tau_{2m+1}} + \phi_{2m,m}, x_1 D'' \phi_{2m,m})$$

$$= x_1 \mu_0(\tau_1+\ldots+\tau_{2m+1}, x')\phi_{2m,m} + x_1 \lambda_0(\tau_1+\ldots+\tau_{2m+1}, x', D'\phi_{2m,m}),$$

c'est à dire (4.3).

Pour que ϕ_{2m+1} vérifie l'équation du premier ordre (1.4), il suffit que les fonctions $[x_1 + (\tau_1 + \ldots + \tau_{2m+1})^q - (\tau_1 + \ldots + \tau_{2m+2})^q]\phi_{2m+1,m}$ et $[(\tau_1+\ldots+\tau_{2j+1})^q - (\tau_1+\ldots+\tau_{2j+2})^q]\phi_{2m+1,j}$ $(0 \leq j \leq m-1)$ vérifient cette équation. Les fonctions entre crochets étant solutions de cette equation (d'après (3.4) pour le facteur figurant devant $\phi_{2m+1,m}$), il suffit que les fonctions $\phi_{2m+1,j}$ $(0 \leqq j \leqq m)$ soient solutions, c'est à dire vérifient (4.4).

C.Q.F.D.

REMARQUE 4.1: La proposition précédente vaut encore avec les hypothèses plus faibles de la remarque 3.1, à condition de substituer aux fonctions λ_i les fonctions $\lambda_i(x;\xi_1,\ldots,\xi_k,0,\ldots,0)$, les fonctions ϕ_m ne dépendent alors que des variables $(x_1,\ldots,x_k)$ (et τ).

Précisons les propriétés des fonctions $\phi_{m,j}$.

PROPOSITION 4.2: Il existe une fonction majorante F d'une seule variable telle que

$$\phi_{m,j}(\tau,x') \ll F([\tau] + [x']), \text{ pour tout } m,j.$$

PREUVE: Posons $\lambda_i(x,\xi') = \sum_{j=1}^{m} \lambda_{i,j}(x)\xi_j$. Il existe $c_0 > 0$ et $R' > 0$ tel que

$$\mu_0, \lambda_{i,j} \ll M([x]), \text{ où } M(t) = \frac{c_0}{R'-t}, \ [x] = \sum_{k=0}^{n} x_k.$$

Soit $0 < R < R'$ et soit ρ un parametre $\geqq 1$; vérifions que

$$\phi_{m,j} \ll \Theta(\rho[\tau] + [x']), \text{ pour tout } m,j, \tag{4.9}$$

où $\Theta(t) = \frac{c}{R-t}$, c étant une constante ≥ 0 à déterminer. Pour que (4.9) soit vérifié, il suffit d'après (4.3) à (4.8) que

$$\begin{cases} M(t)[\Theta(t) + nD\Theta(t)] \ll \rho\, D\Theta(t), \\ \qquad 1 \ll \Theta(t). \end{cases}$$

D'après la proposition 6.1 de [1], on a

$$\Theta(t) \ll RD\Theta(t) \quad \text{et} \quad M(t)D\Theta(t) \ll c_0(R'-R)^{-1}D\Theta(t),$$

d'où

$$M(t)[\Theta(t) + nD\Theta(t)] \ll c_1 D\Theta(t),\ c_1 = c_0(R'-R)^{-1}(R+n);$$

les conditions précédentes sont donc vérifiées dès que

$$\rho \geqq \text{Max}\,(1,c_1),\ c \geqq R;$$

ceci prouve (4.9) et la proposition.

C.Q.F.D.

Il résulte de cette proposition que, pour $s > 0$ suffisamment petit, les fonctions $\phi_{m,j}$ sont holomorphes sur Ω_s^m; en outre, vu que $\phi_{m,j}(0) = 1$, étant donné $\varepsilon, 0 < \varepsilon < 1$, on peut supposer

$$|\phi_{m,j}-1| \leqq \varepsilon \quad , \text{ sur } \Omega_s^m. \tag{4.10}$$

Nous utiliserons enfin la

<u>PROPOSITION 4.3</u>: On a

$$\phi_{2m,j} = \phi_{2m,j-1}, \quad \text{pour} \quad \tau_{2j+1} = 0,\ 1 \leq j \leqq m, \tag{4.11}$$

$$\phi_{2m-1,j-1} = \phi_{2m-1,j}, \text{ pour } \tau_{2j+1} = 0,\ 1 \leq j \leqq m-1. \tag{4.12}$$

PREUVE: Notons d'abord que (4.11) pour $j = m$ résulte de (4.6) et (4.7). On vérifie ensuite (4.11) et (4.12) par récurrence sur m. Pour $1 \leqq j \leqq m-1$, les fonctions $\phi_{2m,j}$ et $\phi_{2m,j-1}$ (resp. $\phi_{2m-1,j-1}$ et $\phi_{2m-1,j}$) vérifient la même équation du premier ordre et les données de Cauchy coincident pour $\tau_{2j+1} = 0$ d'après l'hypothèse de récurrence.

5. LES DIFFEOMORPHISMES Θ_m

Notons $\theta_{m,j}$, $0 \leqq j \leqq m'-1$ où $m' = [\frac{m}{2}]$, les fonctions holomorphes au voisinage de l'origine de $\mathbb{C}^{m+1} \times \mathbb{C}^n$ définies par

$$\theta^q_{m,j} = \phi_{m,j}\,\phi^{-1}_{m,m'},\ \theta_{m,j}(0) = 1. \tag{5.1}$$

D'après (4.10), ces fonctions sont bien définies sur Ω^m_s et il existe une constante $\delta(= (\frac{1+\varepsilon}{1-\varepsilon})^{\frac{1}{q}})$ telle que

$$|\theta_{m,j}| \leqq \delta \ \text{ sur } \Omega^m_s, \text{ pour tout } m,j. \tag{5.2}$$

A l'aide de ces fonctions $\theta_{m,j}$, définissons des difféomorphismes locaux au voisinage de l'origine de $\mathbb{C}^m$ de la façon suivante. On pose

$$\begin{aligned} \tau'_1+\ldots+\tau'_{2j+1} &= (\tau_1+\ldots+\tau_{2j+1})\theta_{m,j}(\tau,x'),\ 0 \leqq j \leqq m'-1,\\ \tau'_1+\ldots+\tau'_{2j+2} &= (\tau_1+\ldots+\tau_{2j+2})\theta_{m,j}(\tau,x'),\ 0 \leqq j \leqq m'-1,\\ \tau'_1+\ldots+\tau'_k &= \tau_1+\ldots+\tau_k,\ \text{pour } 2m'+1 \leqq k \leqq m+1, \end{aligned} \tag{5.3}$$

relations que nous notons $\tau' = \Theta_m(\tau,x')$. Etant donné que $\phi_{m,j} = 1$ pour $\tau = 0$ d'après (4.5) à (4.8), on a $\theta_{m,j} = 1$ pour $\tau = 0$; il en résulte que l'application $\tau \to \Theta_m(\tau,x')$ a pour dérivée en $\tau = 0$ l'application identique de $\mathbb{C}^{m+1}$. On en déduit l'existence d'un voisinage (relativement à S) ouvert V de l'origine de $\mathbb{C}^m$ et des voisinages ouverts $\mathcal{O}$ et $\mathcal{O}'$ de 0 dans $\mathbb{C}^{m+1}$ tel que $\Theta_m(.,x')$ soit, pour tout $x' \in V$, un difféomorphisme de $\mathcal{O}$ sur $\mathcal{O}'$. Notons $\Xi_m(.,x')$ le difféomorphisme réciproque.

Notons dès maintenant que

$$\phi_m(\tau,x') = \phi^o_m(\tau',x')\phi_{m,m'}(\tau,x'), \text{ pour } \tau = \Xi_m(\tau',x'). \tag{5.4}$$

PROPOSITION 5.1: Les relations (5.3) sont équivalentes à des relations de la forme

$$\tau'_j = \tau_j\Theta_{m,j}(\tau,x'),\ 1 \leqq j \leqq m+1, \tag{5.5}$$

où les fonctions $\Theta_{m,j}$ sont holomorphes au voisinage de l'origine de $C^{m+1} \times \mathbb{C}^n$. On a $\Theta_{m,j} = 1$ pour $\tau = 0$ et il existe une fonction majorante F d'une seule variable telle que

$$\Theta_{m,j} \ll F([\tau] + [x']), \text{ pour tout } m,j. \tag{5.6}$$

PREUVE: Les équations (5.3) peuvent s'écrire

$$\tau'_1 = \tau_1\theta_{m,0},$$

$$\tau'_{2j+2} = \tau_{2j+2}\theta_{m,j},\quad 0 \leqq j \leqq m'-1,$$

$$\tau'_{2j+1} = \tau_{2j+1}\theta_{m,j} + (\tau_1+\ldots+\tau_{2j})(\theta_{m,j}-\theta_{m,j-1}),\ 1 \leqq j \leqq m'-1,$$

et

$$\tau'_{m+1} = \tau_{m+1} + (\tau_1+\ldots+\tau_m)(1-\theta_{m,m'-1}), \text{ lorsque m est pair,}$$

$$\begin{cases} \tau'_m = \tau_m+(\tau_1+\ldots+\tau_{m-1})(1-\theta_{m,m'-1}), \\ \tau'_{m+1} = \tau_{m+1}, \text{ lorsque m est impair.} \end{cases}$$

Pour démontrer l'existence des fonctions $\Theta_{m,j}$, il faut donc vérifier que

$$\theta_{m,j} = \theta_{m,j-1}, \text{ pour } \tau_{2j+1} = 0,\ 1 \leqq j \leqq m'-1,$$

et

$$\theta_{m,m'-1} = 1, \begin{cases} \text{pour } \tau_{m+1} = 0, \text{ lorsque m est pair,} \\ \text{pour } \tau_m = 0, \text{ lorsque m est impair,} \end{cases}$$

et ceci est exactement (4.11) et (4.12).

Vu que $\theta_{m,j} = 1$ pour $\tau = 0$, on a bien $\Theta_{m,j} = 1$ pour $\tau = 0$.

Vérifions enfin (5.6). D'après (5.2), il existe une fonction majorante G d'une seule variable telle que $G(0) = 1$ et

$$\theta_{m,j} << G([\tau] + [x']), \text{ pour tout } m,j.$$

On a alors

$$(\theta_{m,j} - \theta_{m,j-1})\tau_{2j+1}^{-1} << 2DG([\tau] + [x']),$$

$$(\theta_{m,m'-1} - 1)\tau_{m+1}^{-1} \ll DG([\tau] + [x']), \text{ si m est pair,}$$

$$(\theta_{m,m'-1} - 1)\tau_{m}^{-1} \ll DG([\tau] + [x']), \text{ si m est impair,}$$

et ceci permet de conclure.

C.Q.F.D.

<u>COROLLAIRE 5.2</u>: On a, pour tout $m \geqq 1$,

$$\Theta_m(H_m^j(x_o),x') \subset H_m^j(x_o), \quad 0 \leqq j \leqq m+1.$$

<u>PREUVE</u>: Pour $j = 0$, cette inclusion résulte de la dernière équation (5.3) ($k = m+1$); pour $1 \leqq j \leqq m+1$, elle résulte de (5.5).

C.Q.F.D.

Autrement dit, $\Theta_m(.,x')\circ S_m(x_o)$ et $S_m(x_o)$ appartiennent à la même classe d'homologie relative de $(H_m^0(x_o), \bigcup_{j=1}^{m+1} H_m^j(x_o))$ dans l'espace $\mathbb{C}^{m+1}$.

Pour ce qui concerne les diffeomorphismes réciproques $\Xi_m = (\Xi_{m,j})_{1\leqq j\leqq m+1}$, on a la

PROPOSITION 5.3: Il existe une fonction majorante G d'une seule variable et des fonctions majorantes $G_{m,j}(\tau',x')$, $1 \leqq j \leqq m+1$, telles que $G(0) = 0$ et

$$\sum_{j=1}^{m+1} G_{m,j}(\tau',x') \ll G([\tau'] + [x']), \quad \text{pour tout } m, \tag{5.7}$$

$$\Xi_{m,j} \ll G_{m,j}, \text{ pour tout } m,j. \tag{5.8}$$

PREUVE: Ecrivons (5.5) sous la forme

$$\tau'_j = \tau_j - \tau_j\Theta'_{m,j}(\tau,x'),\ \Theta'_{m,j} = 0 \text{ pour } \tau = 0,\ 1 \leqq j \leqq m+1;$$

d'après (5.6), il existe une fonction majorante que nous notons encore F telle que $F(0) = 0$ et

$$\Theta'_{m,j} \ll F([\tau] + [x']), \text{ pour tout } m,j.$$

Notons $\tau_j = G_{m,j}(\tau',x')$ la solution des équations

$$\begin{aligned} &\tau'_j = \tau_j - \tau_j F([\tau] + [x']),\ 1 \leqq j \leqq m+1, \\ &\qquad G_{m,j}(0) = 0. \end{aligned} \tag{5.9}$$

On vérifie aisément que

$$0 \ll G_{m,j} \text{ et } \Xi_{m,j} \ll G_{m,j}.$$

En outre, $[\tau] = \sum_{j=1}^{m+1} G_{m,j}(\tau',x')$ est d'après (5.9) la solution de l'équation

$$[\tau'] = [\tau] - [\tau]F([\tau] + [x']);$$

cette fonction est donc de la forme $G([\tau],[x'])$ où G est holomorphe au voisinage de l'origine de $\mathbb{C}^2$ et ceci prouve (5.7).

C.Q.F.D.

COROLLAIRE 5.4: Il existe r' > 0 tel que les fonctions $\Xi_{m,j}$ soient holomorphes sur $\Omega_{r'}^{m}$, et

$$\sum_{j=1}^{m+1} |\Xi_{m,j}(\tau',x')| < r, \text{ pour } (\tau',x') \in \Omega_{r'}^{m} \ ; \tag{5.10}$$

en outre, si $J_m(\tau',x')$ désigne le jacobien de l'application $\tau' \to \Xi_m(\tau',x')$, il existe c > 0 tel que

$$|J_m(\tau',x')| \leq c^{m+1}, \text{ pour } (\tau',x') \in \Omega_{r'}^{m} \ . \tag{5.11}$$

PREUVE: D'après (5.7) et (5.8), $\Xi_{m,j}$ est holomorphe sur $\Omega_{r'}^{m}$, si r' est inférieur au rayon de convergence de G et

$$\sum_{j=1}^{m+1} |\Xi_{m,j}(\tau',x')| \leq G(\|\tau'\| + \|x'\|),$$

ceci prouve (5.10). On a par ailleurs

$$J_m(\tau',x') \ll \prod_{k=1}^{m+1} \left(\sum_{j=1}^{m+1} \frac{\partial G_{m,j}}{\partial \tau_k} (\tau',x') \right) \ll (DG([\tau'] + [x']))^{m+1},$$

d'où $|J_m(\tau',x')| \leq DG(r')^{m+1}$.

C.Q.F.D.

6. PREUVE DU THEOREME 3.1:

Il s'agit de vérifier (3.7). Effectuons dans l'intégrale définissant I_m le changement de variable $\tau' = \Theta_m(\tau,x')$. Vy (5.4) et le corollaire 5.2, on obtient (3.7) avec

$$u_m'(t,\tau',x') = u_m(t\phi_{m,m},(\tau,x'),\tau,x')J_m(\tau',x') \text{ où } \tau = \Xi_m(\tau',x').$$

D'après (5.10) et (4.10) (où nous prenons $\varepsilon = \frac{1}{2}$), la fonction u_m' est holomorphe au voisinage de $t = y_1$, $\tau' = 0$, $x' = y'$ et se prolonge analytiquement à $R_{\frac{2\omega}{3}} \times \Omega_{r'}^{m}$, en utilisant (5.11) on constate, en raisonnant comme au paragraphe 2, qu'il existe c > 0 et, pour tout compact $K \subset R_{\frac{2\omega}{3}}$, une constante $c_k > 0$ telle que

$$|u'_m(t,\tau',x')| \leq c_k c^m m!, \text{ pour } (t,\tau',x') \in K \times \Omega^m_{r'}.$$

Ceci prouve (3.7) et par conséquent le théorème 3.1.

Bibliographie

[1] Y. Hamada, J. Leray et C. Wagschal. Systèmes d'équations aux dérivées partielles à caractéristiques multiples: problème de Cauchy ramifié; hyperbolicité partielle, J. Math. pures et appl. 55, 1976, p. 297-352.

[2] J. Leray. Le calcul différentiel et intégral sur une variété analytique complexe, Bull. Soc. math., 87, 1959, p. 81-180.

[3] C. Wagschal. Le problème de Goursat non linéaire, J. Math. pures et appl. 58, 1979, p. 309-337.

[4] C. Wagschal. Problème de Cauchy ramifié a caractéristiques multiples holomorphes de multiplicité variable, J. Math. pures et appl. 62, 1983, p. 99-127.

[5] C. Wagschal. Problème de Cauchy ramifié pour une classe d'opérateurs à caractéristiques tangentes, J. Math. pures et appl., à paraître.

Claude Wagschal
Laboratoire Central des Ponts et Chaussées,
58, Boulevard Lefebvre,
75732, Paris Cedex 15,
France.

S. WAKABAYASHI & K. KAJITANI

Hyperbolic operators in Gevrey classes

1. INTRODUCTION

The hyperbolic Cauchy problem is well-posed in some Gevrey classes without assumptions on lower order terms (see Bronshtein [4]). In such Gevrey classes we can deal with the hyperbolic Cauchy problem as if the operators had constant coefficients. To make it possible Bronshtein [3] proved Lipschitz continuity of the characteristic roots. Using pseudo-differential operators of infinite order, we can reduce the Cauchy problem in Gevrey classes to the problem in the Sobolev spaces and simplify Bronshtein's proof (see [9], [10]).

In this article we shall discuss two problems. In §2 we shall give results on propagation of signularities for microhyperbolic operators in Gevrey classes. In §3 we shall deal with the hyperbolic mixed (initial-boundary value) problem in Gevrey classes. These two problems can be reduced to the problems in the Sobolev spaces. Investigating properties of hyperbolic polynomials (functions), we can solve the problems in the Sobolev spaces. In §4 we shall give an outline of the proof of the results in §3. We should remark that Beals [2] proved well-posedness of the hyperbolic mixed problem in some function spaces by the theory of semigroups.

2. MICROHYPERBOLIC OPERATORS IN GEVREY CLASSES

Let Ω be an open set in $\mathbf{R}^n$, and let $\kappa > 1$. We say that f belongs to the Gevrey class $\mathcal{E}^{(\kappa)}(\Omega)$ if $f \in C^\infty(\Omega)$ and if for any $K \subset\subset \Omega$ and any $h > 0$ there is $C > 0$ such that

$$|D^\alpha f(x)| \leq Ch^{|\alpha|}\ |\alpha|!^\kappa \tag{2.1}$$

for any $x \equiv (x_1,\ldots,x_n) \in K$ and any multi-index $\alpha = (\alpha_1,\ldots,\alpha_n)$, where $D = (D_1,\ldots,D_n) = i^{-1}(\partial_1,\ldots,\partial_n)$, $\partial_j = \partial/\partial x_j$ and $|\alpha| = \Sigma_{j=1}^n \alpha_j$. Here $A \subset\subset B$ implies that the closure $\bar{A}$ of A is compact and included in the interior $\overset{\circ}{B}$ of B. We denote $\mathcal{D}^{(\kappa)}(\Omega) = \mathcal{E}^{(\kappa)}(\Omega) \cap C_0^\infty(\Omega)$. We can introduce locally convex topologies into $\mathcal{E}^{(\kappa)}(\Omega)$ and $\mathcal{D}^{(\kappa)}(\Omega)$ in a usual manner, respectively, (see, e.g., [14]). We denote by $\mathcal{D}^{(\kappa)\prime}(\Omega)$ and $\mathcal{E}^{(\kappa)\prime}(\Omega)$ the strong dual spaces of $\mathcal{D}^{(\kappa)}(\Omega)$ and $\mathcal{E}^{(\kappa)}(\Omega)$, respectively. The elements of $\mathcal{D}^{(\kappa)\prime}(\Omega)$ are called ultradistributions on Ω. $\mathcal{E}^{(\kappa)\prime}(\Omega)$ is identified with the subspace of $\mathcal{D}^{(\kappa)\prime}(\Omega)$ whose elements have compact supports in Ω. We also write $\mathcal{E}^{(\kappa)}, \ldots,$ instead of $\mathcal{E}^{(\kappa)}(\mathbf{R}^n)$, ... (see , e.g., [14]). Let us define symbol classes $S^m_{(\kappa)}$, where $m \in \mathbf{R}$. We say that a symbol $p(x,\xi)$ belongs to $S^m_{(\kappa)}$ if $p(x,\xi) \in C^\infty(T^*\mathbf{R}^n)$ and for any compact subset K of $\mathbf{R}^n$ and any $A > 0$ there is $C \equiv C_{K,A} > 0$ such that

$$|p^{(\alpha)}_{(\beta)}(x,\xi)| \leq CA^{|\alpha|+|\beta|}(|\alpha|+|\beta|)!^{\kappa}\langle\xi\rangle^{m-|\alpha|}$$

for $x \in K$, $\xi = (\xi_1,\ldots,\xi_n) \in \mathbf{R}^n$ and any multi-indeces α and β, where $\langle\xi\rangle = (1+|\xi|^2)^{1/2}$ and $p^{(\alpha)}_{(\beta)}(x,\xi) = \partial^\alpha_\xi D^\beta_x p(x,\xi)$. We impose the following conditions:

(A-1) $p(x,\xi) \in S^m_{(\kappa_1)}$, where $\kappa_1 > 1$ and $m \in \mathbf{R}$. And $p(x,D)$ is properly supported.

(A-2) There is a symbol $p_m(x,\xi)$, which is positively homogeneous of degree m in ξ, such that $p(x,\xi)-\sigma(\xi)p_m(x,\xi) \in S^{m-1}_{(\kappa_1)}$, where $\sigma(\xi) \in \mathcal{E}^{(\kappa_1)}$ and $\sigma(\xi) = 1$ for $|\xi| \geq 1$ and $\sigma(\xi) = 0$ for $|\xi| \leq 1/2$.

DEFINITION 2.1: Let $z^0 = (x^0,\xi^0) \in T^*\mathbf{R}^n \setminus 0$ and $\theta \in T_{z^0}(T^*\mathbf{R}^n) \simeq \mathbf{R}^{2n}$. We say that $p(x,\xi)$ (or $p_m(x,\xi)$) is microhyperbolic with respect to θ at z^0 if there are a neighbourhood $\mathcal{U}$ of z^0 in $T^*\mathbf{R}^n \setminus 0$, $\ell \in \mathbf{N} \cup \{0\}$ and positive constants c and t_0 such that

$$|\Sigma^\ell_{j=0}\ (-it\theta)^j p_m(x,\xi)/j!| \geq ct^\ell$$

for $(x,\xi) \in \mathcal{U}$ and $0 \leq t \leq t_0$, where $\theta = (\theta_x, \theta_\xi)$ is regarded as a vector field $\theta = \theta_x \cdot (\partial/\partial x) + \theta_\xi \cdot (\partial/\partial\xi)$.

REMARK: (i) The above definition coincides with the definition given in [22]. (ii) When $p_m(x,\xi)$ is real analytic, the above definition coincides with the definition of partially microhyperbolicity given by Kashiwara and Kawai [13].

Let Ω be an open conic set in $T^*\mathbf{R}^n \setminus 0$. We assume that

(A-3) $p_m(x,\xi)$ is microhyperbolic at each point in Ω.

For $z^0 \in T^*\mathbf{R}^n \setminus 0$ we can write

$$p_m(z^0 + s\delta z) = s^{\mu}(p_{mz^0}(\delta z) + o(1)) \text{ as } s \to 0,$$

where $p_{mz^0}(\delta z) \not\equiv 0$ in $\delta z \in T_{z^0}(T^*\mathbf{R}^n)$, if there are multi-indices α and β such that $p_{m(\beta)}^{(\alpha)}(z^0) \neq 0$. $p_{mz^0}(\delta z)$ is called the localization polynomial of $p_m(z)$ at z^0 and $\mu \equiv \mu(z^0)$ is called the multiplicity of $p_m(z)$ at z^0. If $p_m(z)$ is microhyperbolic with respect to θ at z^0, then $p_{mz^0}(\delta z)$ is hyperbolic with respect to θ, i.e.,

$$p_{mz^0}(\delta z - is\theta) \neq 0 \quad \text{for} \quad \delta z \in T_{z^0}(T^*\mathbf{R}^n) \text{ and } s > 0$$

(see, e.g., [7]). Therefore, we can define $\Gamma(p_{mz^0},\theta)$ as the connected component of the set $\{\delta z \in T_{z^0}(T^*\mathbf{R}^n);\ p_{mz^0}(\delta z) \neq 0\}$ which contains θ, when $p_m(z)$ is microhyperbolic with respect to θ at z^0. For some properties of hyperbolic polynomials and $\Gamma(p_{mz},\theta)$ we refer to Atiyah, Bott and Gårding [1].

DEFINITION 2.2: Let $\kappa \geqq \kappa_1$ and $f \in \mathcal{D}^{(\kappa_1)\prime}$. $WF_{(\kappa)}(f)$ is defined as the complement in $T^*\mathbf{R}^n \setminus 0$ of the collection of all (x^0,ξ^0) in $T^*\mathbf{R}^n \setminus 0$ such that there are a neighbourhood U of x^0 and a conic neighbourhood Γ of ξ^0 such that for every $\phi \in \mathcal{D}^{(\kappa_1)}(U)$ and every $A > 0$ there is a positive constant C satisfying

$$|F[\phi f](\xi)| \leqq C \exp[-A|\xi|^{1/\kappa}] \text{ for } \xi \in \Gamma ,$$

where $F[f](\xi) \equiv \hat{f}(\xi)$ denotes the Fourier transform of f (see [6]).

Moreover, we assume that

(A-4) $\mu(\Omega) \equiv \sup_{z\in\Omega}\mu(z) < +\infty$, and $\kappa_1 \leq \kappa(\Omega)$, where $\kappa(\Omega) = \mu(\Omega)/(\mu(\Omega)-1)$ if $\mu(\Omega) \geq 2$ and $\kappa(\Omega) = 2$ if $\mu(\Omega) \leq 1$.

THEOREM 2.3: Assume that (A-1)-(A-4) are valid, and let $\theta:\Omega \ni z \mapsto \theta(z) \in T_z(\Omega)$ be a continuous vector field such that $p_m(z)$ is microhyperbolic with respect to $\theta(z)$ at each $z \in \Omega$. We assume that $\kappa_1 \leq \kappa \leq \kappa(\Omega)$. If $u \in \mathcal{D}^{(\kappa_1)'}$, $z^0 \in WF_{(\kappa)}(u) \cap \Omega$ and $WF_{(\kappa)}(pu) \cap \Omega = \emptyset$, then there are $a \in (-\infty,0) \cup \{-\infty\}$ and a Lipschitz continuous function $z(t)$ defined on $(a,0]$ with values in Ω such that $z(t) \in WF_{(\kappa)}(u)$ for $t \in (a,0]$, $(d/dt)z(t) \in \Gamma(p_{mz(t)}, \theta(z(t)))^\sigma \cap \{\delta z\,;\,|\delta z| = 1\}$ for a.e. $t \in (a,0]$, and $z(0) = z^0$, and $\lim_{t\to a+0} z(t) \in \partial\Omega$ if $a > -\infty$, where $\Gamma^\sigma = \{(\delta x,\delta\xi) \in T_z(T^*\mathbf{R}^n);\ \delta x\cdot\delta\eta-\delta y\cdot\delta\xi \geq 0$ for any $(\delta y,\delta\eta) \in \Gamma\}$ for $z \in T^*\mathbf{R}^n\setminus 0$ and $\Gamma \subset T_z(T^*\mathbf{R}^n)$ and $\partial\Omega$ denotes the boundary of Ω in $T^*\mathbf{R}^n$.

For the proof of Theorem 2.3 and further results we refer to [11].

3. THE HYPERBOLIC MIXED PROBLEM IN GEVREY CLASSES

We denote $\mathbf{R}^n_+ = \{x \in \mathbf{R}^n;\ x_n > 0\}$. Let $P(x,D)$ and $B_j(x',D)$ $(1 \leq j \leq \ell)$ be a partial differential operator of order m defined on $\mathbf{R}^n$ and partial differential operators of order m_j defined on $\mathbf{R}^{n-1}$, respectively, i.e., $P(x,D) = \Sigma_{|\alpha|\leq m}\ a_\alpha(x)D^\alpha$, $B_j(x',D) = \Sigma_{|\alpha|\leq m_j}\ b_{j\alpha}(x')D^\alpha$, where $m \in \mathbf{N}$, $m_j \in \mathbf{N} \cup \{0\}$ and $x' = (x_1,\ldots,x_{n-1}) \in \mathbf{R}^{n-1}$. We shall consider the mixed initial boundary value problem

$$(\mathrm{MP})_t \quad \begin{cases} P(x,D)u(x) = f(x) & (x \in \mathbf{R}^n_+), \\ B_j(x',D)u(x)|_{x_n=0} = g_j(x') & (x' \in \mathbf{R}^{n-1},\ 1 \leq j \geq \ell), \\ \operatorname{supp} u \subset \{x \in \mathbf{R}^n_+;\ x_1 \geq t\}, & \end{cases}$$

where $t \in \mathbf{R}$ is fixed, $\operatorname{supp} f \subset \{x \in \mathbf{R}^n_+;\ x_1 \geq t\}$ and $\operatorname{supp} g_j \subset \{x' \in \mathbf{R}^{n-1};\ x_1 \geq t\}$ $(1 \leq j \leq \ell)$. Define

$$p_m(x,\xi) = \Sigma_{|\alpha|=m}\ a_\alpha(x)\xi^\alpha, \quad b_j(x',\xi) = \Sigma_{|\alpha|=m_j}\ b_{j\alpha}(x')\xi^\alpha.$$

We assume that

(M-1) the hyperplane $\{x_1 = \text{constant}\}$ is noncharacteristic for P, i.e., $P_m(x,\theta) \neq 0$ for any $x \in \mathbf{R}^n$, where $\theta = (1,0,\ldots,0) \in \mathbf{R}^n$.

Then Sakamoto [18] gave necessary and sufficient conditions for the mixed problem $(MP)_0$ to be C^∞ well-posed when $P(x,\xi)$ and $B_j(x',\xi)$ $(1 \leq j \leq \ell)$ have constant coefficients (see, also, [5], [19]). Assume that $a_\alpha(x) \in C^\infty(\mathbf{R}^n)$ and $b_{j\alpha}(x') \in C^\infty(\mathbf{R}^{n-1})$. Then, if the mixed problem $(MP)_t$ is C^∞ well-posed for any $t \in \mathbf{R}$, $P_m(x,\xi)$ is hyperbolic with respect to θ for $x \in \overline{\mathbf{R}^n_+}$, i.e., $P_m(x,\xi - i\gamma\theta) \neq 0$ for $x \in \overline{\mathbf{R}^n_+}$, $\xi \in \mathbf{R}^n$ and $\gamma > 0$ (see [15], [17], [20]). So, instead of (M-1) we assume that

(M-1)' $P_m(x,\xi - i\gamma\theta) \neq 0$ for $x \in \mathbf{R}^n$, $\xi \in \mathbf{R}^n$ and $\gamma > 0$.

Moreover, we impose the following conditions:

(M-2) The boundary $\{x \in \mathbf{R}^n;\ x_n = 0\}$ is noncharacteristic for P, i.e., $P_m(x',0;0,\ldots,0,1) \neq 0$ for $x' \in \mathbf{R}^{n-1}$, where $P_m(x;\xi) = P_m(x,\xi)$.

(M-3) The number ℓ of the boundary conditions is equal to the number of the roots with positive imaginary part of $P_m(0;-i,0,\ldots,0,\xi_n) = 0$.

We note that (M-3) must be satisfied if $P(x,\xi)$ and $B_j(x',\xi)$ have constant coefficients and $(MP)_0$ is C^∞ well-posed. Now we can define the Lopatinski determinant $R(x',\zeta')$ for $\{P_m(x,\xi), b_j(x',\xi)\}$. It follows from (M-1)' and (M-2) that $p(x',\xi) \equiv P_m(x',0,\xi)$ can be written as follows:

$$p(x',\zeta) = p(x',0,\ldots,0,1)p_+(x',\zeta)p_-(x',\zeta),$$

$$p_+(x',\zeta) = \Pi_{j=1}^{\ell} (\zeta_n - \lambda_j^+(x',\zeta')),$$

$$p_-(x',\zeta) = \Pi_{j=1}^{m-\ell} (\zeta_n - \lambda_j^-(x',\zeta')),$$

$$\lambda_j^\pm(x',\zeta') \in \mathbf{C}_\pm,$$

for $x' \in \mathbf{R}^{n-1}$ and $\zeta' \in \mathbf{C}_- \times \mathbf{R}^{n-2}$, where $\mathbf{C}_\pm = \{s \in \mathbf{C};\ \pm\mathrm{Im}\ s > 0\}$. Define

$$L(x',\zeta') = \left(\frac{1}{2\pi i} \oint \frac{\zeta_n^{k-1} b_j(x',\zeta)}{p_+(x',\zeta)} d\zeta_n\right)_{\substack{j\downarrow 1,\ldots,\ell,\\ k\rightarrow 1,\ldots,\ell}}$$

$$R(x',\zeta') = \det L(x',\zeta')$$

for $x' \in \mathbf{R}^{n-1}$ and $\zeta' \in \mathbf{C}_- \times \mathbf{R}^{n-2}$. We assume that

(M-4) $R(x',-i\theta') \neq 0$ for $x' \in \mathbf{R}^{n-1}$.

If $a_\alpha(x) \in C^\infty(\mathbf{R}^n)$ and $b_{j\alpha}(x') \in C^\infty(\mathbf{R}^{n-1})$, then for the mixed problem we can obtain a similar result to one obtained by Lax and Mizohata, i.e. under assumptions (M-1)-(M-4) it is necessary for $(MP)_t$ to be C^∞ well-posed for any $t \in \mathbf{R}$ that $R(x',\zeta')$ satisfies the Lopatinski condition for $x' \in \mathbf{R}^{n-1}$, i.e.

(M-4)' $R(x',\xi'-i\gamma\theta') \neq 0$ for $x' \in \mathbf{R}^{n-1}$, $\xi' \in \mathbf{R}^{n-1}$ and $\gamma > 0$

(see [8], [20]). So, instead of (M-4) we assume (M-4)'. We remark that under the assumptions (M-1) and (M-2) the mixed problem $(MP)_0$ for $\{P_m(x^{0\prime},0,D),\ b_j(x^{0\prime},D)\}$ is C^∞ well-posed for each fixed $x^{0\prime} \in \mathbf{R}^{n-1}$ if and only if

(M-1)" $P_m(x^{0\prime},0,\xi-i\gamma\theta) \neq 0$ for $\xi \in \mathbf{R}^n$ and $\gamma > 0$

and (M-3) and (M-4)' are satisfied (see [18], [19]). By Seidenberg's lemma we have the following

<u>LEMMA 3.1</u>: Under the assumptions (M-1)', (M-2), (M-3) and (M-4)' there are $\mu_\pm \in \mathbb{Q}$ and a function $c(x') > 0$ such that $c(x')^{-1}$ is locally bounded and

$$|p_+(x',\xi-i\gamma\theta)R(x',\xi'-i\gamma\theta')| \geq c(x')\gamma^{\mu+},$$

$$|p_-(x',\xi-i\gamma\theta)| \geq c(x')\gamma^{\mu-}$$

for $x' \in \mathbf{R}^{n-1}$, $\xi \in \mathbf{R}^n$ with $|\xi'| = 1$ and $0 < \gamma \leq 1$.

We also assume that

(M-5) $a_\alpha(x) \in E^{(\kappa)}$ and $b_{j\alpha}(x') \in E^{(\kappa)}(\mathbf{R}^{n-1})$ and

$1 < \kappa \leqq \min\{(\mu_\pm + 2)/(\mu_\pm + 1),\ (r_0 + \mu_\pm)/(r_0 + \mu_\pm - 1)\}$,

where r_0 is the maximum of the multiplicities of the roots of $P_m(x,\xi) = 0$ when $(x,\xi_2,\ldots,\xi_n) \in \mathbf{R}^n \times (\mathbf{R}^{n-1}\setminus\{0\})$.

To state the assumption on finite dependence domains, we define

$$\Gamma_x = \begin{cases} \Gamma(P_m(x,\cdot),\theta) & \text{for } x_n > 0, \\ \Gamma(P_m(x,\cdot),\theta) \cap \Gamma_{x'}(R,\theta) & \text{for } x_n = 0, \\ \mathbf{R}^n & \text{for } x_n < 0, \end{cases}$$

where $\Gamma(P_m(x,\cdot),\theta)$ denotes the connected component of the set $\{\xi \in \mathbf{R}^n;\ P_m(x,-i\xi) \neq 0\}$ which contains θ, $\dot{\Gamma}_{x'} = \{\xi' \in \mathbf{R}^{n-1};\ \xi \in \Gamma(p(x',\cdot),\theta)$ for some $\xi_n \in \mathbf{R}\}$ and $\Gamma_{x'}(R,\theta)$ denotes the connected component of the set $\{\xi \in \mathbf{R}^n;\xi' \in \dot{\Gamma}_{x'}$ and $R(x',-i\xi') \neq 0\}$ which contains θ. We note that $R(x',\zeta')$ can be defined for $x' \in \mathbf{R}^{n-1}$ and $\zeta' \in \mathbf{R}^{n-1} - i\,\dot{\Gamma}_{x'}$ (see [18]). Define for $x \in \mathbf{R}^n$

$K_x^\pm = \{x(t) \in \mathbf{R}^n;\ \pm t \geqq 0,$ and $\{x(t)\}$ is a Lipschitz continuous curve in $\mathbf{R}^n$ satisfying $x(0) = x$ and $(d/dt)x(t) \in \Gamma_{x(t)}{}^*$ (a.e. t)$\}$,

where $\Gamma^* = \{x \in \mathbf{R}^n;\ x\cdot\xi \geqq 0$ for any $\xi \in \Gamma\}$. We must note that Leray [16] defined flows similar to $K_x^\pm$, which were called "emissions", and studied the supports of solutions of the Cauchy problem. Assume that

(M-6) $K_x^- \cap \{x_1 \geqq 0\}$ is bounded for each $x \in \mathbf{R}^n$.

We say that $f \in C^\infty(+0;\ \mathcal{D}^{(\kappa)\prime}(\mathbf{R}^{n-1}))$ if for any $\chi \in \mathcal{D}^{(\kappa)}(\mathbf{R}^{n-1})$ there is $\delta > 0$ such that $\chi(x')f(x) \in C^\infty([0,\delta);\ \mathcal{D}^{(\kappa)\prime}(\mathbf{R}^{n-1}_{x'}))$.

<u>THEOREM 3.2</u>: Assume that (M-1)', (M-2), (M-3), (M-4)', (M-5) and (M-6) are satisfied. If $f \in C^\infty(+0;\ \mathcal{D}^{(\kappa)\prime}(\mathbf{R}^{n-1})) \cap \mathcal{D}^{(\kappa)\prime}(\mathbf{R}^n_+)$, $g_j \in \mathcal{D}^{(\kappa)\prime}(\mathbf{R}^{n-1})(1 \leqq j \leqq \ell)$,

$\text{supp}\, f \subset \{x \in \mathbf{R}^n_+;\ x_1 \geq 0\}$ and $\text{supp}\, g_j \subset \{x' \in \mathbf{R}^{n-1};\ x_1 \geq 0\}$, then $(MP)_0$ has a unique solution $u \in C^\infty(+0;\ \mathcal{D}^{(\kappa)\prime}(\mathbf{R}^{n-1})) \cap \mathcal{D}^{(\kappa)\prime}(\mathbf{R}^n_+)$. Moreover,

$$\text{supp}\, u \subset \{x \in \mathbf{R}^n_+;\ x \in K^+_y \text{ for some } y \in \text{supp}\, f \cup (\cup_{j=1}^{\ell} \text{supp}\, g_j \times \{0\})\}.$$

If $f \in \mathcal{E}^{(\kappa)}(\overline{\mathbf{R}^n_+})$ and $g_j \in \mathcal{E}^{(\kappa)}(\mathbf{R}^{n-1})$ $(1 \leq j \leq \ell)$, then $u \in \mathcal{E}^{(\kappa)}(\overline{\mathbf{R}^n_+})$, i.e. $(MP)_0$ is well-posed in $\mathcal{E}^{(\kappa)}$. Here $f \in \mathcal{E}^{(\kappa)}(\overline{\mathbf{R}^n_+})$ means that for any compact subset K of $\overline{\mathbf{R}^n_+}$ and any $h > 0$ there is $C > 0$ satisfying (2.1).

REMARK. Instead of $\mathbf{R}^n_+$, for a general domain $\mathbf{R} \times \Omega$, where Ω is a domain of $\mathbf{R}^{n-1}$, we can obtain the same results as in Theorem 3.2.

4. OUTLINE OF THE PROOF OF THEOREM 3.2

We can assume without loss of generality that $P(x,\xi)$ and $B_j(x',\xi)$ have constant coefficients outside a neighbourhood of the origin, since it can be shown that $(MP)_0$ has finite dependence domains. Put $\Gamma = \cap_{x \in \mathbf{R}^n} \Gamma_x$ and choose $\chi \in \mathcal{D}^{(\kappa)}(\mathbf{R})$ so that $0 \leq \chi(t) \leq 1$, $|\chi'(t)| \leq 3$ and $\chi(t) = 1$ if $|t| \leq 1/2$ and $= 0$ if $|t| \geq 1$. Let $T > 0$ and $\varepsilon > 0$, which will be chosen sufficiently small. We may assume that $\theta + (0,\delta\xi'',\delta\xi_n) \in \Gamma$ for $|(\delta\xi'',\delta\xi_n)| \leq 2(1+1/\varepsilon) < T_5+1 >^{1+\varepsilon}/T_4$, making a change of variables $x_1 = y_1/N$, $x'' = (x_2,\ldots,x_{n-1}) = y''$ and $x_n = y_n$ if necessary, where $\delta\xi'' = (\delta\xi_2,\ldots,\delta\xi_{n-1})$, $T_j = 2^j T$ and $\langle s\rangle = (1+s^2)^{1/2}$. Let M be a compact subset of Γ satisfying $\tilde{\theta}-\theta \in \Gamma$ for $\tilde{\theta} \in M$. Define

$$\zeta(t) \equiv \zeta_\varepsilon(t) = \int_0^t \langle s\rangle^{-(1+\varepsilon)} ds, \quad F(t) = \int_0^t \chi(s) ds.$$

We put for $\tilde{x} \in \overline{\mathbf{R}^n_+}$ with $|\tilde{x}_1| \leq T/2$ and $\tilde{\theta} \in M$

$$\tilde{\zeta}_T(x;\tilde{\theta},\tilde{x}) = \zeta((x_1-\tilde{x}_1)/T_1 + F((x-\tilde{x})\cdot(\tilde{\theta}-\theta)/T_1)\chi((x_1-\tilde{x}_1)/T_3)).$$

Then it is easy to see that there is a cone $\tilde{\Gamma} \subset\subset \Gamma$ satisfying $\nabla_x \tilde{\zeta}_T(x;\tilde{\theta},\tilde{x}) \in \tilde{\Gamma}$ for $x \in \mathbf{R}^n$ and that $\{x \in \mathbf{R}^n;\ \tilde{\zeta}_T(x;\tilde{\theta},\tilde{x}) > 0 \text{ and } |x_1-\tilde{x}_1| \leq T\} = \{x \in \mathbf{R}^n;\ (x-\tilde{x})\cdot\tilde{\theta} > 0 \text{ and } |x_1-\tilde{x}_1| \leq T\}$. Here $\Gamma_1 \subset\subset \Gamma_2$ means that $\Gamma_1 \cap \{\xi \in \mathbf{R}^n;$

$|\xi| = 1\} \subset\subset \Gamma_2 \cap \{\xi \in \mathbf{R}^n;\ |\xi| = 1\}$, where Γ_1 and Γ_2 are cones (with their vertices at the origin) in $\mathbf{R}^n$. Define

$$P^T(x,\xi) = P^T_m(x,\xi) + \chi(|x|/T)(P-P_m)(x,\xi),$$

$$P^T_m(x,\xi) = P_m(\chi(|x|/T)x, \xi_1 - i(1-\chi(x_1/T_5))\langle\xi''\rangle, \chi(x_1/T_5)\xi'', \xi_n),$$

$$B^T_j(x',\xi) = b^T_j(x',\xi) + \chi(|x'|/T)(B_j - b_j)(x',\xi),$$

$$b^T_j(x',\xi)$$

$$= b_j(\chi(|x'|/T)x', \xi_1 - i(1-\chi(x_1/T_5))\langle\xi''\rangle, \chi(x_1/T_5)\xi'', \xi_n).$$

Let us consider

$$(\mathrm{MP})' \quad \begin{cases} P^T(x,D)u(x) = f(x) \quad (x \in \mathbf{R}^n_+), \\ B^T_j(x',D)u(x)|_{x_n=0} = g_j(x') \quad (x' \in \mathbf{R}^{n-1}, \quad 1 \leq j \leq \ell), \\ \operatorname{supp} u \subset \{x_1 \geqq 0\}, \end{cases}$$

where $f \in C^\infty([0,\infty); \mathcal{E}^{(\kappa)\prime}(\mathbf{R}^{n-1}_{x'}))$ has a bounded support in $\{x \in \mathbf{R}^n;\ x_1 \geqq 0\}$ and $g_j \in \mathcal{E}^{(\kappa)\prime}(\mathbf{R}^{n-1})$, $\operatorname{supp} g_j \subset \{x' \in \mathbf{R}^{n-1};\ x_1 \geqq 0\}$. For $b \in \mathbf{R}$, $a \geqq 0$, $h \geqq 1$ and $\gamma \geqq 0$ we put

$$\Lambda(x_1,\xi') \equiv \Lambda_{b,h+\gamma}(x_1,\xi') = \zeta(x_1 - b) < \xi' >^{1/\kappa}_{h+\gamma},$$

$$\Phi(x,\xi') \equiv \Phi^{a,\gamma}_{b,h}(x,\xi') \equiv \Phi^{a,\gamma}_{b,h,T}(x,\xi';\tilde\theta,\tilde x)$$

$$= a\Lambda_{b,h+\gamma}(x_1,\xi') + hx_1 + \gamma\tilde\zeta_T(x;\tilde\theta,\tilde x),$$

where $\langle\xi'\rangle_h = (h^2 + |\xi'|^2)^{1/2}$. From calculus of pseudo-differential operators given in §2 of [11], it follows that

$${}^t(\exp[-a\Lambda_{b,h+\gamma}])(x_1,D')(\exp[a\Lambda_{b,h+\gamma}])(x_1,D')$$

$$= I + \dot q^a_{b,h+\gamma}(x_1,D'),$$

$$q^a_{b,h+\gamma}(x_1,D') : H^s_{\kappa,\delta}(\mathbf{R}^{n-1}) \to H^{s+1-1/\kappa}_{\kappa,\delta}(\mathbf{R}^{n-1}) \text{ continuously,}$$

where I denotes the identity operator, $s \in \mathbf{R}$ and $H^s_{\kappa,\delta}(\mathbf{R}^{n-1}) = \{u;\ \langle\xi'\rangle^s \exp[\delta\langle\xi'\rangle^{1/\kappa}]\hat{u}(\xi') \in L^2(\mathbf{R}^{n-1})\}$. Moreover, there is $h(a,b,s) \geqq 1$ such that $I+q^a_{b,h+\gamma}(x_1,D')$ has the inverse $(I + q^a_{b,h+\gamma}(x_1,D'))^{-1}$ on $H^s(\mathbf{R}^{n-1})$ if $h \geqq h(a,b,s)$, where $H^s(\mathbf{R}^{n-1})$ denotes the Sobolev space on $\mathbf{R}^{n-1}$ of order s. Putting formally

$$v(x) \equiv v^{a,\gamma}_{b,h,T}(x;\tilde{\theta},\tilde{x}) = (I+q^a_{b,h+\gamma}(x_1,D'))^{-1} \times {}^t(\exp[-\Phi^{a,\gamma}_{b,h}])(x,D')u(x),$$

i.e.

$$u(x) = (\exp[\Phi^{a,\gamma}_{b,h}])(x,D')v(x),$$

we can reduce the mixed problem (MP)' to the boundary value problem in the Sobolev spaces

$$(\mathrm{MP})'' \begin{cases} \tilde{P}(x,D)v(x) = \tilde{f}(x) & (x \in \mathbf{R}^n), \\ \tilde{B}_j(x',D)v(x)|_{x_n=0} = \tilde{g}_j(x') & (x' \in \mathbf{R}^{n-1},\ 1 \leq j \leq \ell), \end{cases}$$

where

$$\tilde{P}(x,D) \equiv \tilde{P}^{a,\gamma}_{b,h,T}(x,d;\tilde{\theta},\tilde{x})$$

$$= {}^t(\exp[-\Phi^{a,\gamma}_{b,h}])(x,D')P^T(x,D)(\exp[\Phi^{a,\gamma}_{b,h}])(x,D'),$$

$$\tilde{B}_j(x',D)v(x)|_{x_n=0} \equiv \tilde{B}^{a,\gamma}_{j,b,h,T}(x',D;\tilde{\theta},\tilde{x})v(x)|_{x_n=0}$$

$$= {}^t(\exp[-\Phi^{a,\gamma}_{b,h}])(x,D')B^T_j(x',D)(\exp[\Phi^{a,\gamma}_{b,h}])(x,D')v(x)|_{x_n=0},$$

$$\tilde{f}(x) \equiv \tilde{f}^{a,\gamma}_{b,h,T}(x;\tilde{\theta},\tilde{x}) = {}^t(\exp[-\Phi^{a,\gamma}_{b,h}])(x,D')f(x),$$

$$\tilde{g}_j(x') \equiv \tilde{g}^{a,\gamma}_{j,b,h,T}(x';\tilde{\theta},\tilde{x})$$

$$= {}^t(\exp[-\Phi^{a,\gamma}_{b,h}|_{x_n=0}])(x',D')g_j(x').$$

In fact, from the Paley-Wiener theorem in Gevrey classes and results in §2 of [11] it follows that for any $b < 0$ there is $a(b,f,g_j) > 0$ such that $\tilde{f}(x) \in H^{s+1-m+r_0}(R^n_+)$ and $\tilde{g}_j(x') \in H^{s-m_j}(R^{n-1})$ if $a \geq a(b,f,g_j)$. Moreover, we have $\tilde{f}(x) \in H^{s+1-m+r_0}(R^n_+)$ and $\tilde{g}_j(x') \in H^{s-m_j}(R^{n-1})$ if $f \in \mathcal{E}^{(\kappa)}(\overline{R^n_+})$, $g_j \in \mathcal{E}^{(\kappa)}(R^{n-1})$ and $b \in \mathbf{R}$. We assume that s is an integer $\geq \max(\mu_+, m-r_0, m_1, \ldots, m_\ell)$. Uniqueness of solutions of (MP)" and uniform estimates of $v(x)$ in γ imply that $u(x)$ satisfies the initial condition $\mathrm{supp}\, u \subset \{x_1 \geq 0\}$. Similarly we can obtain an outer estimates of supp u. By results in §2 of [11] and the same arguments as in the proof of Egorov's theorem we can prove that

$$\tilde{P}(x,\xi) = \tilde{p}(x,\xi) + q(x,\xi),$$

$$\tilde{B}_j(x',\xi) = \tilde{b}_j(x',\xi) + c_j(x',\xi),$$

$$\tilde{p}(x,\xi) = J(x_1,\xi';a)P_m(X(x,\xi'), \Xi(x,\xi)),$$

$$\tilde{b}_j(x',\xi) = J(x_1,\xi';a)b_j(X'(x',0,\xi'), \Xi(x',0,\xi))$$

if $h \geq h(a,T^{-1})$, where $J(x_1,\xi';a) \neq 0$, $X(x,\xi') \sim \chi(|x|/T)\{x + ia \times \nabla_\xi \Lambda(x_1,\xi')$, $\Xi(x,\xi) \sim (\xi_1, \chi(x_1/T_5)\xi'', \xi_n) - h\theta - i\gamma\nabla_x\tilde{\zeta}_T(x;\tilde{\theta},\tilde{x}) - i\, a\nabla_x\Lambda(x_1,\xi') - i(1-\chi(x_1/T_5)) < \xi'' > \theta$ and $q(x,\xi)$ and $c_j(x',\xi)$ are lower order terms in some sense.

<u>LEMMA 4.1</u>: For any compact subset $\tilde{M}$ of Γ and any $C_0 > 0$ there are $d > 0$ and $c > 0$ such that

$$|P_m(x-ity,\xi-it\tilde{\theta})| \geq ct^{r_0},$$

$$|P_{m(\beta)}^{(\alpha)}(x-ity,\xi-it\tilde{\theta})| \leq C_{\alpha,\beta}t^{-|\alpha|-|\beta|}|P_m(x-ity,\xi-it\tilde{\theta})|,$$

$$|p_+(x'-ity',\xi-it\tilde{\theta})R(x'-ity',\xi'-it\tilde{\theta}')| \geq ct^{\mu_+},$$

$$|R_{(\beta')}^{(\alpha')}(x'-ity',\xi'-it\tilde{\theta}')|$$

$$\leqq C_{\alpha',\beta'} t^{-|\alpha'|-|\beta'|} |R(x'-ity',\xi'-it\tilde{\theta}')|,$$

$$|p_-(x'-ity',\xi-it\tilde{\theta})| \geqq ct^{\mu_-},$$

$$|p_{\pm(\beta')}^{(\alpha)}(x'-ity',\xi-it\tilde{\theta})|$$

$$\leqq C_{\alpha,\beta'} t^{-|\alpha|-|\beta'|} |p_\pm(x'-ity',\xi-it\tilde{\theta})|,$$

if $x \in \overline{\mathbf{R}^n_+}$ $|x| \leq 1$, $\xi \in \mathbf{R}^n, |\xi'| = 1$, $|\xi_n| \leqq C_0$, $\tilde{\theta} \in \tilde{M}$, $y \in \mathbf{R}^n$, $|y| \leqq d$ and $0 < t \leqq 1$. Here $P_m(x+iy,\xi)$ and $b_j(x'+iy',\xi)$ are almost analytic extensions of $P_m(x,\xi)$ and $b_j(x',\xi)$, respectively, and $R(x' + iy',\xi')$ is the Lopatinski determinant for $\{P_m(x'+iy',0,\xi), b_j(x'+iy',\xi)\}$.

REMARK. For $P_m(x,\xi)$ Lemma 4.1 was proved by Bronshtein [4].

LEMMA 4.2: Let $s \in \mathbf{R}$. Then there is a bounded linear operator $F : H^{s+1-m+r_0}(\mathbf{R}^n) \to H^{s+1}(\mathbf{R}^n)$ such that

$$\| \langle D\rangle_{h+\gamma}^{s+1-m+r_0} \langle D\rangle_h^{m-r_0} Fw(x) \|_{L^2(\mathbf{R}^n)}$$

$$\leqq C_s \| \langle D\rangle_{h+\gamma}^{s+1-m-r_0} w(x) \|_{L^2(\mathbf{R}^n)},$$

$$\tilde{P}(x,D)Fw(x) = w(x) \text{ for } w \in H^{s+1-m+r_0}(\mathbf{R}^n).$$

By Lem ma 4.2 it is sufficient to solve (MP)" with $\tilde{f}(x) = 0$. We put

$$\tilde{L}(x',\xi') = J(x_1,\xi';a)L(X'(x',0,\xi'),\Xi'(x',0,\xi',0)).$$

Note that det $L(X'(0,\xi'),\Xi'(x',0,\xi',0)) = R(X'(x',0,\xi'), \Xi'(x',0,\xi',0))$ and that

$$R(X'(x',0,\xi'),\Xi'(x',0,\xi',0)) = R(0,-i\theta')(i\Xi_1(x',0,\xi',0))^{h_0}$$

if $|x_1| \geqq T_5 + T/4$ and $h \geqq h(a,T^{-1})$, where $h_0 = \Sigma_{j=1}^{\ell} m_j - \ell(\ell-1)/2$.

Define

$$Q_0(x,\xi) = \tilde{p}_+(x,\xi)^{-1}(1,\xi_n,\ldots,\xi_n^{\ell-1})\tilde{L}(x',\xi')^{-1},$$

$$Q_\nu(x,\xi) = -\tilde{p}(x,\xi)^{-1}\ \{\Sigma_{0<|\alpha|\leqq r}\ \alpha!^{-1}\tilde{p}^{(\alpha)}(x,\xi)Q_{\nu-1(\alpha)}(x,\xi)$$

$$+\ \Sigma_{|\alpha|\leqq r}\ \alpha!^{-1}q^{(\alpha)}(x,\xi)Q_{\nu-1(\alpha)}(x,\xi)\},\quad \nu = 1,2,\ldots,$$

where $\tilde{p}(x,\xi) = \tilde{p}(x,0,\ldots,0,1)\tilde{p}_+(x,\xi)\tilde{p}_-(x,\xi)$, the roots of $\tilde{p}_\pm(x,\xi) = 0$ in ξ_n belong to $\mathbf{C}_\pm$ and $r >> 1$. Let $E_s : H^{s+1-m+r_0}(\mathbf{R}^n_+) \to H^{s+1-m+r_0}(\mathbf{R}^n)$ be a bounded linear operator such that

$$E_s w|_{x_n>0} = w \quad \text{for } w \in H^{s+1-m+r_0}(\mathbf{R}^n_+),$$

$$\|\langle D\rangle_{h+\gamma}^{s+1-m+r_0}\ E_s w(x)\|_{L^2(\mathbf{R}^n)}$$

$$\leqq C_s\ \Sigma_{j=0}^{s+1-m+r_0}\ \|\langle D'\rangle_{h+\gamma}^{s+1-m+r_0-j}\ D_n^j w(x)\|_{L^2(\mathbf{R}^n_+)}.$$

Substitute

$$v(x) = Q(x,D')\begin{bmatrix} h_1(x') \\ \vdots \\ h_\ell(x') \end{bmatrix} - FE_s\tilde{P}(x,D)Q(x,D')\begin{bmatrix} h_1(x') \\ \vdots \\ h\ (x') \end{bmatrix} \Big|x_n > 0$$

into (MP)" with $\tilde{f}(x) = 0$, where $N >> 1$,

$$Q(x,D')\begin{bmatrix} h_1(x') \\ \vdots \\ h_\ell(x') \end{bmatrix}$$

$$= \Sigma_{\nu=0}^N \int e^{ix'\cdot\xi'}\ (\tfrac{1}{2\pi i}\oint_{C(x,\xi')} e^{ix_n\cdot\xi_n}\ Q_\nu(x,\xi)\begin{bmatrix} \hat{h}_1(\xi') \\ \vdots \\ \hat{h}_\ell(\xi') \end{bmatrix} d\xi_n)đ\xi',$$

$đ\xi' = (2\pi)^{-n+1}d\xi'$ and $C(x,\xi')$ is a simple closed curve in $\mathbf{C}_+$ enclosing the roots of $\tilde{p}_+(x,\xi) = 0$ in ξ_n. Then (MP)" with $\tilde{f}(x) = 0$ is reduced to the problem

$$\begin{bmatrix} \tilde{g}_1(x') \\ \vdots \\ \tilde{g}_\ell(x') \end{bmatrix} = (I-W) \begin{bmatrix} h_1(x') \\ \vdots \\ h_\ell(x') \end{bmatrix}$$

on the boundary, where I-W is a bounded linear operator on $\Pi_{j=1}^{\ell} H^{s-m_j}(\mathbb{R}^{n-1})$. We can prove that I-W has the inverse $(I-W)^{-1}$ if $h \geqq h_{a,T,s}$. Therefore, we can construct a solution v(x) of (MP)". For example, if $B_j(x',D) = D_n^{\ell-j}$ ($1 \leqq j \leqq \ell$; the Dirichlet conditions), we can consider the adjoint problem and construct solutions of the adjoint problem. This implies that (MP)" for the Dirichlet conditions has a unique solution in the Sobolev spaces. Since a solution v(x) of (MP)" with $\tilde{f}(x) = 0$ and $\tilde{g}_j(x') = 0$ ($1 \leqq j \leqq \ell$) is determined by its Dirichlet data, we have the equation on the boundary

$$L \begin{bmatrix} D_n^{\ell-1} v(x)|_{x_n=0} \\ \vdots \\ v(x)|_{x_n=0} \end{bmatrix} = \begin{bmatrix} 0 \\ \vdots \\ 0 \end{bmatrix} .$$

Similarly we can prove that L is invertible on $\Pi_{j=1}^{\ell} H^{s-\ell+j}(\mathbb{R}^{n-1})$ if $h \geqq h_{a,T,s}$, which shows uniqueness of solutions of (MP)". Then it is not hard to prove uniqueness of solutions of (MP)'. The above argument also gives another construction of solutions, using the results for the Dirichlet problem. Letting γ tend to ∞, we can obtain outer estimates of the supports of solutions of (MP)'. So we can prove that (MP)' has a unique solution. Regularities of solutions follows from partial hypoellipticity of $P^T(x,D)$ in x_n. By results for the Cauchy problem together we can prove Theorem 3.2 (see [21], [23]). In the forthcoming paper [12] we shall give a detailed proof of Theorem 3.2 and further results.

References

[1] M.F. Atiyah, R. Bott and L. Gårding, Lacunas for hyperbolic differential operators with constant coefficients, I, Acta Math. 124 (1970), 109-189.

[2] R. Beals, Hyperbolic equations and systems with multiple characteristics, Arch. Rat. Mech. Anal. 48 (1972), 123-152.

[3] M.D. Bronshtein, Smoothness of polynomials depending on parameters, Sib. Mat. Zh. 20 (1979), 493-501.

[4] M.D. Bronshtein, The Cauchy problem for hyperbolic operators with variable multiple characteristics, Trudy Moskov. Mat. Obšč. 41 (1980), 83-99.

[5] R. Hersh, Mixed problem in several variables, J. Math. Mech. 12 (1963), 317-334.

[6] L. Hörmander, Uniqueness theorems and wave front sets for solutions of linear differential equations with analytic coefficients, Comm. Pure Appl. Math. 26 (1971), 671-704.

[7] L. Hörmander, The Analysis of Linear Partial Differential Operators I, Springer, Berlin-Heidelberg-New York-Tokyo, 1983.

[8] K. Kajitani, A necessary condition for the well posed hyperbolic mixed problem with variable coefficients, J. Math. Kyoto Univ. 14 (1974), 231-242.

[9] K. Kajitani, The Cauchy problem for uniformly diagonalizable hyperbolic systems, Proc. Hyperbolic Equations and Related Topics, Taniguchi Symposium 1984, Kinokuniya, Tokyo.

[10] K. Kajitani, Well posedness in Gevrey class of the Cauchy problem for hyperbolic operators, Bull. Sci. Math. 111 (1987), 425-438.

[11] K. Kajitani and S. Wakabayashi, Microhyperbolic operators in Gevrey classes, to appear.

[12] K. Kajitani and S. Wakabayashi, The hyperbolic mixed problem in Gevrey classes, to appear.

[13] M. Kashiwara and T. Kawai, Micro-hyperbolic pseudo-differential operators I, J. Math. Soc. Japan 27 (1975), 359-404.

[14] H. Komatsu, Ultradistributions, I, Structure theorems and a characterization, J. Fac. Sci. Univ. Tokyo Sect. IA Math. 20 (1973), 25-105.

[15] P.D. Lax, Asymptotic solutions of oscillatory initial value problems, Duke Math. J. 24 (1957), 624-646.

[16] J. Leray, Hyperbolic Differential Equations, Princeton Univ. Press, Princeton, 1952.

[17] S. Mizohata, Some remarks on the Cauchy problem, J. Math. Kyoto Univ. 1 (1961), 109-127.
[18] R. Sakamoto, $\mathcal{E}$-well posedness for hyperbolic mixed problems with constant coefficients, J. Math. Kyoto Univ. 14 (1974), 93-118.
[19] Y. Shibata, A characterization of the hyperbolic mixed problems in a quarter space for differential operators with constant coefficients, Publ. RIMS, Kyoto Univ. 15 (1979), 357-399.
[20] S. Wakabayashi, A necessary condition for the mixed problem to be C^∞ well-posed, Comm. in Partial Differential Equations, 5 (1980), 1031-1064.
[21] S. Wakabayashi, Singularities of solutions of the Cauchy problem for hyperbolic systems in Gevrey classes, Japan. J. Math. 11 (1985), 157-201.
[22] S. Wakabayashi, Generalized Hamilton flows and singularities of solutions of the hyperbolic Cauchy problem, Proc. Hyperbolic Equations and Related Topics, Taniguchi Symposium 1984, Kinokuniya, Tokyo.
[23] S. Wakabayashi, Generalized flows and their applications, Proc. NATO Advanced Study Institutes on Advances in Microlocal Analysis, Series C, D. Reidel, 1986, 363-384.

Seiichiro Wakabayashi
Institute of Mathematics
University of Tsukuba
Ibaraki 305
Japan.

Kunihiko Kajitani
Institute of Mathematics
University of Tsukuba
Ibaraki 305
Japan.

C. ZUILY

Existence locale de solutions C^∞ pour l'équation de Monge-Ampère

ABSTRACT: Local existence of C^∞ solutions for the Monge-Ampère equation

We consider in this paper the n-dimensional Monge-Ampère equation

$$\det (u_{ij}) = f(x,u,\nabla u)$$

in the case when f is a non negative function in a neighbourhood of a point x in $\mathbb{R}^n$. When n = 2, C.S. Lin proved a local existence theorem for this type of equation in the scale of Sobolev spaces H^s. However, his method does not give the existence of a local C^∞ solution since the neighbourhood of existence decreases in an essential way when s increases. In this paper we first extend Lin's result to the arbitrary dimension n, with a slightly simpler proof. Then, using Bony's theory of paradifferential operators and some earlier results of C.J. Xu and C. Zuily, we prove the existence of a local C^∞ solution under some additional assumptions on f. In particular, our results apply to the problem of existence of C^∞ isometric embedding of 2-dimensional Riemannian manifolds in $\mathbb{R}^3$.

0. INTRODUCTION

De nombreux travaux ont été récemment consacrés à l'existence de solutions pour l'équation de Monge-Ampère réelle dans un ouvert Ω de $\mathbb{R}^n$,

$$\det(u_{ij}) = f(x,u,\nabla u) \tag{0.1}$$

ou pour des problèmes aux limites relatifs à cette équation (voir [2] et sa bibliographie). La plupart de ces travaux concernaient le cas elliptique où f est strictement positive. Le cas "dégénéré" où f est positive ou nulle n'a fait l'objet que de quelques publications. (Voir [3], [4], [5], [9]). Dans l'une d'entre elles [4] C.S. Lin démontrait en dimension 2 un théorème d'existence locale H^s, pour s assez grand, pour des seconds membres de même nature. Malheureusement l'ouvert d'existence dépend de s et sa taille tend vers zéro au fur et à mesure que s augmente. Le but de cet exposé est de montrer qu'avec des hypothèses additionnelles sur f on peut, en toute dimension, obtenir des solutions locales C^∞.

§1: NOTATIONS ET ENONCES DES RESULTATS

Dans tout ce qui suit f(y,u,p) sera *une fonction* C^∞, au voisinage d'un point $Z^0 = (y^0,u^0,p^0) \in \mathbb{R}^n \times \mathbb{R} \times \mathbb{R}^n$, *positive ou nulle.*

Le premier résultat est une extension à $\mathbb{R}^n$, $n \geq 2$, d'un résultat de C.S. Lin [4] obtenu en dimension n = 2.

THÉORÈME 1.1: Pour tout $s \in \mathbb{N}$, $s > [\frac{n}{2}] + 3$ il existe un voisinage de y^0 dans lequel l'équation (0,1) admet une solution convexe $u \in H^s$.

THÉORÈME 1. 2: Supposons en outre que $D_y^\alpha D_u^\ell D_p^\beta f(y^0,u^0,p^0) = 0$ pour tous $|\alpha| + \ell + |\beta| \leq k-1$ et qu'il existe $\alpha^* \in \mathbb{N}^n, |\alpha^*| = k$ tel que $D_y^{\alpha^*} f(y^0,u^0,p^0) \neq 0$. Alors l'équation (0.1) admet une solution convexe C^∞ dans un voisinage de y^0.

THÉORÈME 1. 3: On considère le cas où $f(y,u,p) = K(y)g(y,u,p)$ où $K(y^0) = 0$, $K \geq 0$ et $g(y^0,u^0,p^0) > 0$. Supposons qu'il existe un nombre fini d'hypersurfaces C^1, S_ℓ, $\ell = 0,\ldots,\ell_0$ et un voisinage V de y^0 tels que

$$V \cap K^{-1}(0) \subset \bigcup_{\ell=0}^{\ell_o} S_\ell$$

Alors (0.1) admet une solution locale convexe et C^∞ près de y^o.

§II. PREUVE DES RESULTATS

On peut bien entendu supposer que $Z^o = (0,0,0)$. Comme dans Lin [4] on fait un changement de fonction et un changement de variables; plus précisément on pose

$$u = \sum_{i=1}^{n-1} \frac{\sigma_i}{2} y_i^2 + \varepsilon^5 w(\varepsilon^{-2}y)$$
$$y_i = \varepsilon^2 x_i \quad 1 \leq i \leq n-1 \tag{2.1}$$

où ε est positif et petit.

L'équation (0.1) est transformée en l'équation

$$\det(\Phi_{ij}) = \det((1-\delta_i^n)\delta_i^j\sigma_i + \varepsilon w_{ij}) = \tilde{f} \tag{2.2}$$

où δ_i^j est le symbole de Kronecker. Les constantes σ_i sont choisies telles que

$$\sigma_1 > \sigma_2 > \dots > \sigma_{n-1} = 1$$

On pose $G(w) = \frac{1}{\varepsilon_n} \det(\Phi_{ij}) - \frac{1}{\varepsilon} \tilde{f}\chi(x')$ dans l'ensemble $\Omega = \{(x',x_n) \in \mathbb{R}^n : |x'| \leq \pi, |x_n| \leq x_o\}$ où χ est une troncature nulle au voisinage de $\mp \pi$, égale à 1 près de zéro et x_o est à choisir.

Le linéarisé de G en w est l'opérateur

$$L_G(w) = \sum_{i,j=1}^{n} \Phi^{ij}\partial_i\partial_j + \Sigma\, a_i\partial_i + a \tag{2.3}$$

où (Φ^{ij}) est la matrice des cofacteurs de la matrice (Φ_{ij}).

<u>LEMME 2. 1</u>: Supposons que w soit C^2 et que $\|w\|_{C^2} \leq 1$. Il existe une matrice orthogonale $T(x,\varepsilon)$ telle que:

a) $T(x,\varepsilon)\, \Phi_{ij}\, {}^tT(x,\varepsilon) = \mathrm{diag}(\lambda_1,\dots,\lambda_n)$

b) $T(x,\varepsilon)$ est régulière en (x,ε) dans $\bar{\Omega} \times [0,\varepsilon_0[$ où $\varepsilon_0 > 0$.

c) Il existe une constante C indépendante de w et de ε telle que

$$|T_{nn}(x,\varepsilon)-1| + \sum_{i,j=1}^{n} |\nabla_x T_{ij}(x,\varepsilon)| + \sum_{\ell=1}^{n-1} |\lambda_\ell (x,\varepsilon)-\sigma_\ell| + |\lambda_n(x,\varepsilon)| +$$
$$+ \sum_{\ell=1}^{n-1} |T_{\ell n}(x,\varepsilon)| \leq C\varepsilon. \tag{2.4}$$

A l'aide de ce lemme il est facile de démontrer le résultat suivant:

LEMME 2.2: Soit $w \in C^2$ telle que $\|w\|_{C^2} \leq 1$ et $\theta = \sup_{\Omega} |G(w)| \geq 0$. Alors l'opérateur

$$L = - L_G(w) - \theta\Delta \text{ où } \Delta = \sum_{i=1}^{n} \frac{\partial^2}{\partial x_i^2}$$

est pour ε assez petit un opérateur à symbole positif ou nul.

Ce résultat va nous permettre de poser un problème de Dirichlet pour l'opérateur L. Pour cela on travaillera dans des espaces de Sobolev formés de fonctions qui sont périodiques en les variables x_i', $i = 1,\ldots,n-1$ de période 2π. On notera H_s ces espaces et $\overset{\circ}{H}_s$ l'espace des fonctions s'annulant en $x_n = \mp x_0$; Ω désignera l'ouvert $\{|x_i'| < \pi$, $i = 1,\ldots,n-1$, $|x_n| < x_0\}$, $x_0 > 0$. Le résultat principal de ce paragraphe est le suivant.

THEOREME 2.3: Soit w une fonction régulière, périodique en x' telle que $\|w\|_{C^{[\frac{n}{2}]+3}} \leq 1$. Pour tout $s_0 \in \mathbb{N}$ il existe $\varepsilon(s_0)$ tel que le problème

$$L_G(w)\rho + \theta\Delta\rho = g \tag{2.5}$$

admet une solution unique $u \in \overset{\circ}{H}_s$ pour $g \in H_s$, $0 \leq s \leq s_0$, $0 < \varepsilon \leq \varepsilon(s_0)$. On a, de plus, l'inégalité

$$\|\rho\|_s \leq C_s(\|g\|_s + \|(w)\|_{s+4}\|\rho\|_{L^\infty}) \tag{2.6}$$

où C_s est une constante indépendante de w et de ε. Ici $\|(w)\|_{s+4}$ est égale à zéro si $s \leq [\frac{n}{2}] + 1$ et à $\|w\|_{s+4}$ si $s > [\frac{n}{2}] + 1$.

Décrivons les principales étapes de la preuve.

On commence par un changement de fonctions qui aura pour effet de donner à l'opérateur L(w) un terme constant qui soit grand; pour cela on conjugue L(w) par une exponentielle $e^{\lambda x_n^2}$. On procède ensuite à une régularisation elliptique i.e. on ajoute $\nu\Delta$(où $\nu > 0$ et Δ est le Laplacien) à l'opérateur obtenu; soit $L_\nu = \tilde{L}(w) + \nu\Delta$. On montre que le problème le Dirichlet homogène pour L_ν admet une solution unique $u \in C^\infty(\bar{\Omega})$. Pour cela on cherche une borne inférieure à la quantité $-(L_\nu\rho,\rho)$. C'est là que les lemmes 2.1 et 2.2 sont utiles. Après quelques calculs on obtient l'inégalité

$$-(L_\nu\rho,\rho) \geqq \int \{\nu|\nabla\rho|^2 + \frac{1}{2}\prod_{i=1}^{n-1} \lambda_i \, |\tilde{\rho}_n|^2 + [(\sigma+\theta)\lambda - C(1+\lambda^2 x_n^2) + O(\varepsilon)]\rho^2\}dx$$

ce qui pour λ assez grand fournit l'existence et l'unicité de la solution. De cette inégalité découle la suivante

$$\int |\rho_n|^2 \, dx + \|\rho\|_{L^2} \leqq C_0 \|g\|_{L^2}^2 \tag{2.7}$$

Elle correspond à l'étape s = o de l'inégalité (2.5). On obtient une inégalité H_s par récurrence sur s, partant de (2.6), encommençant par obtenir de la régularité tangentielle puis tirant la régularité normale de l'équation. Les détails sont donnés dans [10].

On utilise ensuite, pour montrer l'existence d'une solution du problème (2.1), le procédé de Nash-Moser (voir [6]).

On notera S_k l'opérateur de lissage qui consiste essentiellement à convoluer par la fonction $\mu_k^n\phi(\mu_k x)$ où $\phi \in \mathcal{S}$ est convenablement choisie et $\mu_k = \sigma^{\tau^k}$, $\sigma > 1$, $\tau > 1$ sont à choisir.

Le schéma d'approximation utilisé sera le suivant:

$$\begin{cases} w_0 = 0 \qquad w_{k+1} = w_k + S_k \, \rho_k \\ L(w_k)_k = g_k, \; \rho_k \in \mathring{H}_s(\Omega) \\ g_k = - G(w_k) \\ \theta_k = \sup_{\bar{\Omega}} |G(w_k)| \end{cases} \tag{2.8}$$

LEMME 2.4: Supposons $\|w_k\|_{C^{[\frac{n}{2}]+3}} \leq 1$ pour $k = 0,\dots,n$. Alors pour tout $0 \leq k \leq n$

$$\|g_k\|_s \leq C_s \{\|g_0\|_s + \|w_k\|_{s+2}\} \tag{2.9}$$

$$\|w_{k+1}\|_{s+4} \leq C_s^{k+1} \mu_{k+1}^{\beta} \|g_0\|_s \quad \text{où } \beta > \frac{4}{\tau-1} \tag{2.10}$$

$$\|g_{k+1}\|_{L^2} \leq \mu_{k+1}^{-\chi} \|g_0\|_{s^*} \quad \text{où } \chi > 0 \text{ et } s^* > 0. \tag{2.11}$$

Sous l'hypothèse du lemme 2.4 si w_k converge ce sera, grâce à (2.11), vers un w tel que $G(w) = 0$ c'est à dire une solution de notre problème.
La suite de la preuve va consister à prouver que w_k converge et satisfait à l'inégalité du Lemme 2.4. Plus précisément on montre qu'il existe une constante $C > 0$ telle que

$$\|w_k\|_{2[\frac{n}{2}]+4} \leq C \tag{2.12}$$

cela se fait par récurrence sur k et nous renvoyons à [10] pour les détails. Ceci permet de démontrer le théorème 1.1.
Pour le théorème 1.2 on utilise le théorème de régularité suivant dû à C.J. Xu [8]. Soit $F(x,u,\nabla u,D^2u) = 0$ où F est réelle et C^∞, une équation aux dérivées partielles non linéaire. A chaque solution réelle u on associe les champs de vecteurs réels $X_j = \sum_{k=1}^{n} \frac{\partial F}{\partial u_{jk}} \partial_k$, $1 \leq j \leq n$.

THEOREME 2.5: ([8]). Supposons que $u \in C^{\rho}_{loc}(\Omega)$ où $\rho > \text{Max}(4,r+2)$ où r est un entier ≥ 0 tel que les crochets des X_j d'ordre inférieur ou égal à r engendrent l'espace tangent en tout point de Ω. Alors $u \in C^\infty(\Omega)$.

Dans notre cas particulier on a:

$$\begin{cases} X_j = \varepsilon w_{nn} A_j \partial_j + \varepsilon \sum_{\ell=1}^{n} \sum_{j\neq i} w_{ij}\, B_{ij}\, (D^2w)\partial_\ell & 1 \leq j \leq n-1 \\ X_n = A_n \partial_n + \varepsilon \sum_{j,\ell\neq n} w_{nj}\, B_{nj\ell}(D^2w)\partial_\ell \end{cases}$$

où les $B_{ij\ell}$ sont des polynômes en D^2w, $A_n = \det(\delta_i^j \sigma_i + \varepsilon w_{ij}$, $1 \leq i, j \leq n-1)$ et A_ℓ est le cofacteur de $\sigma_\ell + \varepsilon w_{\ell\ell}$ dans A_n. Tous les A_i sont strictement positifs.

LEMME 2.6: Supposons $\partial_y^\alpha \partial_u^\ell \partial_p^\beta f(0,0,0) = 0$ pour $|\alpha| + \ell + |\beta| \leq k-1$. Supposons que $w \in C^{k+2}$ où $s > [\frac{n}{2}] + k + 3$ alors

$$|\partial^\alpha w| \leq C_\alpha \ \varepsilon^{2k-1}, \qquad |\alpha| \leq s - [\tfrac{n}{2}] - 1 \tag{2.13}$$

Ce lemme résulte de la formule de Taylor appliquée à $g_o = - G(w_o)$ et de l'inégalité $\|w_s\| \leq c_{s,s^*} \|g_o\|_{s^*}$ prouvée précédemment.

LEMME 2.7: Supposons de plus que $\partial_n^k f(0,0,0) > 0$. Alors si ε est assez petit on a

$$\partial_n^{k+2} w \geq C\varepsilon^{2k-1} \tag{2.14}$$

où C est une constante indépendante de ε.

On écrit pour cela l'équation $\det(\Phi_{ij}) = f$ sous la forme

$$w_{nn} + \varepsilon \sum_{i,j,k,m} w_{ij}\, w_{km}\, A_{ijkm}(D^2 w) = \frac{\tilde{f}}{\varepsilon}$$

on développe f autour de l'origine, on différentie les deux membres k fois, et on utilise (2.13).
A l'aide de ces deux lemmes et d'un calcul précis de $(adX_i)^k\ (X_i)$, $1 \leq i \leq n-1$, on montre facilement que les champs de vecteurs $(adX_n)^k(X_i)$, $i = 1,\ldots,n$, et X_n engendrent l'espace tangent ce qui, utilisant le théorème 2.5 démontre le théorème 1.2.
Donnons maintenant une idee de la preuve du théorème 1.3.
On peut tout de suite supposer que $y_o = 0$ et que les normales à l'origine aux surfaces S_ℓ, $\nu^\ell = (\nu_1^\ell,\ldots,\nu_n^\ell)$ satisfont la condition $\nu_n^\ell > 0$ dans Ω.
Au lieu de l'équation (2.2) nous considérons l'équation

$$\tilde{G}(w) = \frac{1}{\varepsilon} \det(\sigma_i \delta_i^j (1-\delta_i^n) + \varepsilon w_{ij}) = \frac{K}{\varepsilon} g\chi_1 + (1-\chi_1)\varepsilon^p g = K_1 g$$

où $\chi_1 \in C_0^\infty(\Omega)$, $\chi_1 = 1$ dans un voisinage V de zéro. D'après la régularité des solutions des équations elliptiques il est facile de voir que la solution construite au théorème 1.1. est C^∞ près de $\partial\Omega$ et que son support singulier est contenu dans $M = \bigcup_{\ell=0}^{\ell_0} S_\ell \cap V$. Il suffit donc de prouver que w est C^∞ près de M. On montre tout d'abord une inégalite de Poincaré.

<u>LEMME 2.8</u>: Soit $V = \sum_{j=1}^{n} v_j(x)\partial_j$ un champ de vecteur réel et C^1 dans $\bar{\Omega}$, non dégénéré.
Supposons que M est un ensemble compact et que pour chaque ℓ, $\sum_{j=1}^{n} \nu_j^\ell v_j(x) > 0$ sur M. Alors pour tout $\eta > 0$ on peut trouver un voisinage N_η de M tel que

$$\|u\|_0 \leqq \eta\|Vu\|_0$$

pour tout $u \in H^1$ telle que supp $u \subset N_\eta$.
Montrons comment le théorème 1.3 se déduit de ce lemme.

Supposons que la solution obtenue au théorème 1.1. soit dans $H_s(\Omega)$ pour $s > [\frac{n}{2}] + 6$. En utilisant la théorie para-différentielle de J.M. Bony [1] on obtient

$$\tilde{G}(w) - \sum_{|\alpha|\leqq 2} T_{\frac{\partial\tilde{G}}{\partial w_\alpha}} \partial^\alpha w \in H_{2s-4-[\frac{n}{2}]+\mu} \quad \forall\mu > 0 .$$

Le symbole de l'operateur linearise de $\tilde{G}$ en w est

$$\sigma(L_{\tilde{G}}(w)) = \sum_{i,j=1}^{n} \Phi^{ij}\xi_i\xi_j + \text{termes d'ordre 1}$$

Considérons le champ de vecteur $V = \sum_{j=1}^{n} \Phi^{jn}\, \partial_j$. La composante sur l'axe des x_n est égale à $\Phi^{nn} = \sigma + O(\varepsilon)$ où $\sigma = \prod_{j=1}^{n-1} \sigma_i$ tandis que les autres composantes sont $O(\varepsilon)$. On en déduit que V satisfait la condition exigée au lemme 2.8.
D'autre part par un résultat de O.A. Oleinik E.V. Radkevitch on a

$$\|Vu\|_0^2 \leqq C\{\mathrm{Re}(L_{\tilde{G}}(w)u,u) + \|u\|_0^2\} \quad u \in C_0^\infty(\Omega)$$

Par la théorie paradifférentielle on en déduit que

$$\|Vu\|_0^2 \leq C\{\mathrm{Re} \sum_\alpha (T_{a_\alpha} \partial^\alpha u,u) + \|u\|_0^2\}$$

où $a_\alpha = \frac{\partial \tilde{G}}{\partial w_\alpha}$. D'où, utilisant le lemme 2.2

$$\frac{1}{\eta} \|u\|_0^2 + \|Vu\|_0^2 \leq C \,\mathrm{Re} \,\Sigma(T_{a_\alpha} \partial^\alpha u,u) \tag{2.15}$$

Soit E_δ^s l'opérateur pseudo-différentiel de symbole $\phi(x)(1+|\xi|^2)^{\frac{s+1}{2}} \times (1 + \delta^2|\xi|^2)^{-2}$ où ϕ est une troncature égale à 1 près de M.
En utilisant l'inégalité (2.15) avec $E_\delta^s u$ au lieu de u et en utilisant la théorie paradifférentielle pour estimer les commutateurs on déduit que $u \in H_{s+1}$ près de M et en itérant ce procédé, que $u \in C^\infty$ près de M.

Bibliographie

[1] J.-M. Bony: Calcul symbolique et propagation des singularités pour les équations aux dérivées partielles non-linéaires, Ann. Scient. Ec. Norm. Sup., t. 14, 1981, 209-246.

[2] L. Caffarelli, L. Nirenberg, J. Spruck: The Dirichlet problem for non linear second order elliptic equations I: Monge-Ampere equations, Comm. on Pure and Applied Mathematics, Vol. XXXVII, 369-402, (1984).

[3] Hong Jiaxing: Surface in $\mathbb{R}^3$ with prescribed Gauss curvature, To appear in Chinese Ann. of Math.

[4] C.-S. Lin: The local isometric embedding in $\mathbb{R}^3$ of 2-dimensional Riemannian manifolds with non negative curvature, J. Diff. geometry 21 (1985), 213-230.

[5] C.-S. Lin: Isometric embedding in $\mathbb{R}^3$ of Riemannian manifolds with curvature vanishing clearly, Comm. pure and Appl. Math. Vol XXXIX, 867-887 (1986).

[6] J. Moser: A new technique for the construction of solutions of non linear partial differential equations, Proc. Nat. Acad. Sci. USA 47 (1961) 1824-1831.

[7] O.A. Oleinik - E.V. Radkevitch: Second order equations with non negative characteristic form, Plenum Press.

[8] C.J. Xu: Régularité des solutions des e.d.p. non linéaires, C.R. Acad. Sc. Paris, t. 300 (1985), p. 267-270 et article à paraître.

[9] C. Zuily: Sur la régularité des solutions non strictement convexes de l'équation de Monge-Ampère réelle, Prépublication d'Orsay 85 T 33 et Annali della Scuola Norm. di Pisa (to appear).

[10] J. Hong, C. Zuily: Existence of C^∞ local solutions for the Monge-Ampère equation. Inventiones math. 89, p. 645-661 (1987).

Claude Zuily
Département de Mathématiques
Université Paris XI
91405 Orsay Cedex
France.